OT 7:
Operator Theory: Advances and Applications
Vol. 7

Edited by

I. Gohberg

Editorial Office

Department of Mathematics
Tel-Aviv University
Ramat-Aviv (Israel)

Birkhäuser Verlag
Basel · Boston · Stuttgart

M. G. Krein

Topics in Differential and Integral Equations and Operator Theory

Edited by I. Gohberg

Translated from the Russian by A. Iacob

1983

Birkhäuser Verlag
Basel · Boston · Stuttgart

Volume Editorial Office

School of Mathematical Sciences
Tel-Aviv University
Ramat-Aviv (Israel)

Library of Congress Cataloging in Publication Data

Kreĭn, M. G. (Mark Grigor'evich), 1907 –
Topics in differential and integral equations and operator theory.
(Operator theory, advances and applications ; v. 7)
1. Differential equations, Linear. 2. Wiener-Hopf equations. 3. Selfadjoint operators. I. Gokhberg, I. TS. (Izrail' Tsudikovich), 1928 – . II. Title. III. Series.
QA 372. K 916 1983 515.3'54 83–8774
ISBN 3–7643–1517–2

CIP-Kurztitelaufnahme der Deutschen Bibliothek

Krejn, Mark G.:
Topics in differential and integral equations and operator theory / M. G. Krein. Ed. by I. Gohberg. Transl. from the Russian by A. Iacob. [Vol. ed. off. School of Math. Sciences, Tel-Aviv Univ., Ramat-Aviv (Israel)] – Basel ; Boston ; Stuttgart : Birkhäuser, 1983.
(Operator theory ; Vol. 7)
ISBN 3–7643–1517–2
NE: GT

Printed in Germany
ISBN 3–7643–1517–2

EDITORIAL INTRODUCTION

In this volume three important papers of M.G. Krein appear for the first time in English translation. Each of them is a short self-contained monograph, each a masterpiece of exposition. Although two of them were written more than twenty years ago, the passage of time has not decreased their value. They are as fresh and vital as if they had been written only yesterday. These papers contain a wealth of ideas, and will serve as a source of stimulation and inspiration for experts and beginners alike.

The first paper is dedicated to the theory of canonical linear differential equations, with periodic coefficients. It focuses on the study of linear Hamiltonian systems with bounded solutions which stay bounded under small perturbations of the system. The paper uses methods from operator theory in finite and infinite dimensional spaces and complex analysis. For an account of more recent literature which was generated by this paper see AMS Translations (2), Volume 93, 1970, pages 103-176 and Integral Equations and Operator Theory, Volume 5, Number 5, 1982, pages 718-757.

The second paper presents developments in perturbation theory for selfadjoint operators. It is comprised of six lectures which M.G. Krein presented to a mathematical summer school in 1963. These lectures are devoted to, (1) the theory of the spectral shift function which originated in the work of the physicist, I.M. Lifschits, and was developed by M.G. Krein, and (2) to the theory of wave and scattering operators. The theory of spectral shift functions appears here in full detail in English translation for the first time.

The third paper, the latest, appeared in Russian in 1976, and supplements the famous paper of M.G. Krein on Wiener-Hopf integral equations. It reduces a class of linear Wiener-Hopf equations to nonlinear integral equations. It turns out that the latter are preferable from many points of view. The origins of this theory can be traced back to the work of astrophysicists, V.A. Ambartsumyan and S. Chandrasekhar. This paper completes and extends the mathematical treatment.

This volume also includes three shorter papers written by M.G. Krein and his students. These also appear here for the first time in English translation.

As this volume went to press the editor and the publisher learned that M.G. Krein was awarded the Wolf Prize for 1983 in recognition of his many fundamental contributions to functional analysis and its applications. We would like to take this opportunity to extend to him our warmest congratulations.

I would like to express my thanks to André Iacob for his very dedicated work in preparing this translation.

THE EDITOR

Tel-Aviv
28 January, 1983

TABLE OF CONTENTS

THE BASIC PROPOSITIONS OF THE THEORY OF λ-ZONES OF STABILITY OF A CANONICAL SYSTEM OF LINEAR DIFFERENTIAL EQUATIONS WITH PERIODIC COEFFICIENTS*

M. G. Krein

Let S be a mechanical system with m degrees of freedom whose Hamiltonian H has the form

$$H = \frac{1}{2} \sum_{j,k=1}^{2m} h_{jk}(\omega t) x_j x_k + f(t) ,$$

where $x_1, \ldots, x_m$ are generalized coordinates, $x_{m+1}, \ldots, x_{2m}$ are the corresponding generalized momenta of S, $h_{jk}(t) = h_{kj}(t)$ $(j,k = 1,\ldots,m)$ are periodic functions of period T in the time variable t, and ω is a parameter.

It is well known that the differential equations of motion of the system S can be written as follows:

$$\frac{dx_j}{dt} = \frac{\partial H}{\partial x_{j+m}} , \quad \frac{dx_{j+m}}{dt} = -\frac{\partial H}{\partial x_j} \qquad (j = 1,\ldots,m).$$

With the help of the shortened notations from linear algebra, one may rewrite this system of differential equations as a single vector differential equation

$$\frac{dx}{dt} = JH(\omega t)x .$$

Here $x = (x_1,\ldots,x_{2m})$ is a 2m-dimensional vector-function, $H(t) = \|h_{jk}(t)\|_1^{2m}$ is a symmetric matrix-function, and

*Translation of: In Memory of A. A. Andronov, Izdat. Akad. Nauk. SSSR, Moscow (1955), 413-498.

$$J = \begin{pmatrix} 0 & I_m \\ -I_m & 0 \end{pmatrix},$$

where I_m denotes the m-th order identity matrix.

If in the last equation one changes the independent variable to $\tau = \omega t$ and then one replaces the letter τ by t, we are led to the vector differential equation

$$\frac{dx}{dt} = \lambda JH(t)x \quad (\lambda = \frac{1}{\omega}) \ . \tag{0.1}$$

Notice that any vector, second-order, differential equation

$$\frac{d^2y}{dt^2} + \mu P(t)y = 0 \ , \tag{0.2}$$

where $y = (y_1,\ldots,y_m)$ is an m-dimensional vector-function, and $P(t) = P(t + T)$ is an m-th order, symmetric matrix-function

$$P(t) = \|p_{jk}\|_1^m \ ,$$

can be reduced to form (0.1).

Indeed, assuming that $\mu \neq 0$ $(-\infty < \mu < \infty)$, setting

$$\lambda = \sqrt{|\mu|} > 0 \ , \quad \frac{dy}{dt} = \lambda z \ ,$$

and taking the vector $x = (x_1,\ldots,x_{2m})$ to be the direct sum of the vectors y and z, i.e.,

$$x_j = y_j \ , \quad x_{j+m} = z_j = \frac{1}{\lambda}\frac{dy_j}{dt} \quad (j = 1,\ldots,m),$$

equation (0.2) now becomes equivalent (in a certain sense) to equation (0.1), where

$$H(t) = \begin{pmatrix} \pm P(t) & 0 \\ 0 & I_m \end{pmatrix}.$$

A point $\lambda = \lambda_0$ $(-\infty < \lambda_0 < \infty)$ is called a *λ-point of stability of equation* (0.1) if all solutions to equation (0.1) are bounded (on the entire t-axis) when $\lambda = \lambda_0$.

If, in addition, for $\lambda = \lambda_0 \neq 0$ all the solutions to any equation

$$\frac{dx}{dt} = \lambda J H_1(t) x$$

with a periodic, symmetric matrix-function $H_1(t) = H_1(t + T)$ which is *sufficiently close* to the matrix-function $H(t)$ are also bounded, then we shall call $\lambda = \lambda_0$ a *λ-point of strong stability of equation* (0.1).

It goes without saying that the notion of proximity between two periodic matrix-functions $H(t)$ and $H_1(t)$ needs to be defined precisely. When the only condition imposed upon the elements of these matrices is that they be summable over the interval $(0,T)$, such a definition is given in §5 of this paper. [We remark that in §5 we give another definition of a λ-point of strong stability. The latter, according to Theorem 5.1, does not, at least, enlarge the set of λ-points of strong stability as we have defined it above. A more detailed analysis of these two definitions reveals their equivalence.] However, if we restrict ourselves to the case of *continuous* periodic matrix-functions $H(t)$, then, as it turns out, one obtains a definition of a point λ-point of strong stability equivalent to the general one, if by sufficiently close matrices $H(t)$ and $H_1(t)$ one understands matrices whose corresponding elements differ from each other on the entire axis by less than ε, where $\varepsilon > 0$ is small enough.

We also remark that we count the point of stability $\lambda = 0$ to be among the λ-points of strong stability of equation (0.1) if and only if one of its neighborhoods consists only of such points.

The set of λ-points of strong stability of equation (0.1) is open and so, when it is not empty, it decomposes into a finite or infinite collection of disjoint open intervals. The latter are called the *λ-zones of stability of equation* (0.1).

The μ-points of strong stability and the μ-zones of stability of equation (0.2) are similarly defined, the only difference being now that the point $\mu = 0$ can be never reckoned among not only the μ-points of strong stability, but even not the ordinary points of stability of equation (0.2). The reason is that when $\mu = 0$ equation (0.2) has unbounded solutions in the form of vector-functions which depend linearly on t.

One gets an equivalent definition of a μ-point of strong stability by saying that such a point is any $\mu_0 \neq 0$ $(-\infty < \mu_0 < \infty)$ for which the corresponding point $\lambda_0 = \sqrt{|\mu_0|}$ is a λ-point of strong stability for the equation (0.1) that is obtained from equation (0.2) as a result of the transformation indicated above. This is precisely the definition we shall work with in §9.

The problem of determining the zones of stability of equation (0.1), or of the more special equation (0.2), plays an important role in various problems of mechanics, and, in general, of physics (in parametric resonance problems, in the quantum-mechanical approach to the problem of the motion of an electron in a periodic field, and so on). At the present time, one has to deal with this problem in many technical situations too, when one sets out to investigate the dynamical stability of constructions.

For the scalar equation (0.2), the fundamental results concerning the existence of the μ-zones of stability and their location, the identification of the endpoints of successive μ-zones of stability as the characteristic numbers of the "periodic" and "antiperiodic" boundary value problem, which interlace according to a certain rule, and various estimates for the first μ-zone of stability, are all due to Lyapunov [18-22].

However, in spite of the importance of generalizing these remarkable studies of Lyapunov to vector differential equations, until recently nothing essential (apart from the information given by Lyapunov) has been done.

Further progress in the theory of equation (0.1) has been made possible only after, in this class of equations, the equations of *positive type* were singled out. The latter are characterized by the property that the form $(H(t)\xi,\xi)$ corresponding to their matrix-function $H(t)$ is nonnegative:

$$\sum_{j,k=1}^{2m} h_{jk}(t)\xi_j\xi_k \geqq 0 \quad (-\infty < t < \infty) , \tag{0.3}$$

and has a positive mean:

$$\sum_{j,k=1}^{2m} \left(\int_0^T h_{jk}(t)dt\right)\xi_j\xi_k > 0 \qquad \left(\sum_{j=1}^{2m} \xi_j^2 > 0\right) . \tag{0.4}$$

The isolation of this class was motivated by physical considerations. Namely, let an arbitrary conservative system S with m degrees of freedom perform small oscillations around a stable position of equilibrium where its potential energy attains a strict minimum (which can be discovered by examining the second-order terms in the expansion of the potential energy into a series of powers of the generalized coordinates). If one subjects S to a small, parametric, periodic excitation at the frequency ω, then in the linearized formulation of the problem the motion of system S will be described by an equation (0.1) of *positive type* (with $T = 2\pi$). [Moreover, in this case equality will be impossible already in (0.3) if not all $\xi_j = 0$, and so (0.4) will be a consequence of (0.3).]

At the beginning of 1950, the author discovered that the so-called multipliers of an equation (0.1) of positive type admit a special classification as multipliers of the first and second kind. Subsequently, with the use of new tools, a series of rules governing the motion of these multipliers as the parameter λ is continuously increased were revealed. This allowed one [9] to prove the existence of a *central* (i.e., containing the point $\lambda=0$) λ-zone of stability for an equation (0.1) of positive type, to give a rule of finding its endpoints, to find a sufficient criterion for a given λ to belong to this zone, and finally, to show that that there exists an infinite number of λ-zones of stability which extend to infinity on both sides of the λ-axis, under certain additional restriction imposed upon the matrix-function $H(t)$.

At the same time, one succeeded in going deeper into the theory of equations (0.2) of positive type. [We mean equations (0.2) with matrix-functions $P(t)$ such that the corresponding form $(P(t)\xi,\xi)$ is nonnegative and has a positive mean. It is obvious that in this case, and only in this one, equation (0.2) reduces to an equation (0.1) of positive type for $\mu > 0$.] In particular, the classical test of A. M. Lyapunov for μ to belong to the first zone of stability of the scalar equation (0.2) has been generalized to the case of a vector equation (0.2). Incidentally, let us comment that this result announced in [9], as well as the corresponding results concerning the central zone of stability of equation (0.1), have been significantly improved in the present paper (see §§ 7 and 9).

Let us remark, by the way, that the existence of the central zone of stability for an equation (0.1) of positive type reflects itself in the fact that a conservative system of the type discussed earlier preserves its stability under a small parametric excitation if the frequency ω of this excitation is large enough.

Soon after the author's report [9], M. G. Neigauz and V. B. Lidskii [16] pointed out several other important consequences of the theory of multipliers of the first and second kind. In particular, they showed that the sufficient criterion for a given μ to belong to some zone of stability, due to N. E. Zhukovskii [29], as well as other, stronger criteria obtained very recently by Soviet authors [13,26,27] for the scalar equation (0.2) as developments of the results of A. M. Lyapunov and N. E. Zhukovskii, can be extended to the vector equation (0.2).

In this paper we make the first complete analysis of the theory of multipliers of the first and second kind, and we present its main consequences to the theory of zones of stability for equations (0.1) and (0.2).

As the reader will discover by himself, new tools of contemporary matrix theory, analytic function theory and, in part, spectral theory of operators were required for the construction of this theory. The theory might have looked even more complete had we also enlisted the asymptotic methods which, combined with the theory of multipliers of the first and second kind, would allow us to prove the existence of an infinite number of zones of stability for a large class of equations (0.1) and (0.2) and to find asymptotic estimates for them.

Due to lack of space, the way that one applies the asymptotic methods is illustrated only on the example of equation (0.1) in two-dimensional space, where these methods can be discussed in some detail.

The same reason forced us to limit ourselves to brief remarks only (§ 9, no. 5) concerning the theory of multipliers of the first and second kind for equation (0.2), a theory that might have been constructed without the assumption that the form $(P(t)\xi,\xi)$ is nonnegative.

Nevertheless, we took the liberty of discussing the theory of the equations (0.1) and (0.2) under the assumption that $H(t)$ and $P(t)$ are some periodic, *Hermitian* matrix-functions (and not necessarily *real symmetric* matrix functions), the reason being that this more general assumption permitted us to clarify the mathematical ideas of the theory, without causing essential complications.

Had we been willing to encounter some difficulties in stating several propositions of the theory, we might have been able to generalize it also, with no special effort, to the case when the matrix J in equation (0.1) does not have the specific form indicated above, but is an arbitrary, nonsingular, Hermitian matrix with nondefinite form $(J\xi,\xi)$.

For readers who are specially interested in the theory of ε-parametric resonance, we comment that in order to understand § 10, which is devoted to this phenomenon, it is enough to be acquainted with the material in §§ 1 and 2 only.

§ 1. AUXILLIARY PROPOSITIONS FROM MATRIX THEORY

In this section a number of both well-known as well as new facts from matrix theory will be given. At the same time, we introduce the terminology and symbols to which we shall adhere. Concerning the known results of matrix theory that will be discussed below, we refer the reader to A. I. Mal'tsev course [24].

1. Let E_n denote the n-dimensional space of complex vectors $\xi = (\xi_1,\ldots,\xi_n)$ with the usual scalar product

$$(\xi,\eta) = \xi_1\bar{\eta}_1 + \ldots + \xi_n\bar{\eta}_n ,$$

where $\eta = (\eta_1,\ldots,\eta_n)$ and the bar denotes the complex conjugate.

The length (norm) of a vector $\xi \in E_n$ is denoted by $|\xi|$, and so

$$|\xi| = \sqrt{(\xi,\xi)} = [\,|\xi_1|^2 + \ldots + |\xi_n|^2\,]^{1/2} .$$

If $A = \|a_{jk}\|_1^n$ is a square matrix of order n and $\xi, \eta \in E_n$, then the equality

$$\eta = A\xi$$

expresses the linear transformation

$$\eta_j = a_{j1}\xi_1 + \dots + a_{jn}\xi_n \quad (j = 1,\dots,n).$$

By A^τ and A^* we shall denote the matrix transposed to the matrix A and the matrix transposed to the complex-conjugate matrix $\bar{A} = \|\bar{a}_{jk}\|_1^n$, respectively. Therefore,

$$(A\xi,\eta) = (\xi,A^*\eta)$$

for all $\xi,\eta \in E_n$.

We define the norm of the matrix A (denoted by $|A|$) by the equality

$$|A| = \max_{x \in E_n} \frac{|A\xi|}{|\xi|} .$$

If A is a Hermitian matrix $(A = A^*)$, then $|A|$ equals the largest of the absolute values of the eigenvalues of A. In general, $|A|$ is the square root of the largest eigenvalue of the matrix A^*A.

Along with the usual properties of the norm:

1) $|A| > 0$ if $A \neq 0$,

2) $|A + B| \leqq |A| + |B|$,

3) $|\lambda A| = |\lambda||A|$ (λ any scalar),

the norm thus introduced enjoys the following additional ones:

$$|A^*| = |A|, \quad \text{and} \quad |AB| \leqq |A||B|.$$

It is also obvious that

$$|a_{jk}| \leqq |A| \quad (j,k = 1,\dots,n).$$

The moment we define the norm in the space of matrices of order n, we have defined the notion of convergence of a sequence of matrices A_ν $(\nu = 1,2,\dots)$ to a matrix A. Incidentally, the latter means nothing more than element by element convergence.

Let

$$A(t) = \|a_{jk}(t)\|_1^n$$

be some matrix-function defined on the interval (α,β) and whose elements $a_{jk}(t)$ $(j,k = 1,\dots,n)$ are summable over this interval. Then the equality

$$C = \int_\alpha^\beta A(t)dt$$

means that $C = \|c_{jk}\|_1^n$, where

$$c_{jk} = \int_\alpha^\beta a_{jk}(t)dt \qquad (j,k = 1,\dots,n).$$

2. Let ρ be an eigenvalue of the matrix A, i.e.,

$$\det(A - \rho I_n) = 0 .$$

The set L_ρ of all vectors $\xi \in E_n$ satisfying

$$(A - \rho I_n)^n \xi = 0$$

is called the *root space* of the matrix A corresponding to the eigenvalue ρ.

If L_ρ consists exclusively of eigenvectors of A, then ρ is called an eigenvalue of *simple* type. In this case $\xi \in L_\rho$ if and only if $A\xi = \rho\xi$.

For ρ to be an eigenvalue of simple type it is necessary and sufficient that the multiplicity of ρ (i.e., its multiplicity k_ρ as a root of the equation $\det(A - \rho I_n) = 0$) be equal to the defect d_ρ (i.e., the difference between the order and the rank) of the matrix $A - \rho I_n$. If ρ is not of simple type, then $d_\rho < k_\rho$. If ρ is a simple eigenvalue (i.e., $k_\rho = 1$), then is of simple type and $k_\rho = d_\rho = 1$.

Let $\rho_1,\dots,\rho_s$ be all the distinct eigenvalue of the matrix A, and let $L_{\rho_1},\dots,L_{\rho_s}$ be the corresponding root spaces. Then E_n splits into a direct sum

$$E_n = L_{\rho_1} \dotplus \dots \dotplus L_{\rho_2} .$$

Therefore, each vector $\xi \in E_n$ can be uniquely represented as

$$\xi = \xi^{(1)} + \dots + \xi^{(s)} , \qquad (\xi^{(k)} \in L_{\rho_k};\ k = 1,\dots,s).$$

To this decomposition there corresponds a decomposition

of the identity matrix into a sum of projections:

$$I = P_1 + \dots + P_s ,$$

such that

$$P_k\xi = \xi^{(k)} \qquad (\xi \in E_n;\ k = 1,\dots,s) .$$

If Γ is a smooth, Jordan contour bounding a domain G in the complex plane, then

$$-\frac{1}{2\pi i}\oint_{\Gamma} (A - \lambda I_n)^{-1} d\lambda = \sum_{\rho_j \in G} P_j .$$

3. If H_1 and H_2 are two Hermitian matrices of order n, then we shall write

$$H_1 \leqq H_2 \quad (H_2 \geqq H_1)$$

whenever

$$(H_1\xi,\xi) \leqq (H_2\xi,\xi)$$

for all $\xi \in E_n$ $(\xi \neq 0)$. If we exclude equality in the last relation, then we write

$$H_1 < H_2 \quad (H_2 > H_1).$$

In what follows we shall be interested in the case $n = 2m$. In this case we set

$$J = J_n = \begin{pmatrix} 0 & I_m \\ -I_m & 0 \end{pmatrix} ,$$

where I_m is the m-th order identity matrix. Notice that

$$J = J^* = -J, \quad J^2 = -I_n .$$

Let us agree to call a vector $\xi \in E_n$ a *plus-vector*, a *minus-vector*, or a *null-vector* depending on whether the number $i(J\xi,\xi)$ is positive, negative, or zero.

A matrix U of order $n = 2m$ is called J-*expansive* if

$$iU^*JU > iJ ,$$

i.e.,

$$i(JU\xi,U\xi) > i(J\xi,\xi) \tag{1.1}$$

for all $\xi \in E_n$, $\xi \neq 0$.

Conversely, a matrix U will be called J-*contractive* if $iU^*JU < iJ$.

If e is an eigenvector of a J-expansive matrix U, i.e., $Ue = \rho e$, $(e \neq 0)$, then taking $\xi = e$ in (1.1.), we obtain

$$i(|\rho|^2 - 1)(Je,e) > 0 .$$

Therefore, either $|\rho| > 1$ and e is a plus-vector, or $|\rho| < 1$ and e is a minus vector.

We shall be interested in the case when U is a nonsingular matrix $(\det U \neq 0)$.

THEOREM 1.1. *Let U be a nonsingular J-expansive matrix. Then E_{2m} splits into a direct sum of two m-dimensional subspaces invariant under U,*

$$E_{2m} = E_+ \dotplus E_- ,$$

where E_+ and E_- consist of plus- and minus-vectors respectively (if one does not take into account the zero element). Moreover, all the eigenvalues of U in E_+ (E_-) have absolute values larger (smaller) than one.

This theorem is a particular case of a more general one that we proved in [10] (see also [12]) using Brower's fixed point theorem; however, Theorem 1.1 can be proven in a purely algebraic way with no difficulty. [In paper [10], a matrix U is called iJ-expansive if a weakened version of condition (1.1), where the sign $\geqq$ replaces the sign $>$, is satisfied.]

4. A matrix $U = \|u_{jk}\|_1^n$ is called J- *orthogonal* (or *symplectic*) if

$$U^\tau JU = J, \tag{1.2}$$

and J-*unitary* if

$$U^*JU = J. \tag{1.3}$$

For real matrices these two notions obviously coincide.

Both (1.2) and (1.3) imply that $\det U \neq 0$, and so the first of these relations can be rewritten as

$$U^\tau = JU^{-1}J^{-1},$$

while the second becomes

$$U^* = JU^{-1}J^{-1}.$$

This shows that the elementary divisors of a J-orthogonal (J-unitary) matrix which correspond to the eigenvalue $\rho \neq \pm 1$ (respectively, $|\rho| \neq 1$) appear in pairs $(\lambda - \rho)^k$, $(\lambda - \rho^{-1})^k$ (respectively, $(\lambda - \rho)^k$, $(\lambda - \bar{\rho}^{-1})^k$). We express this fact which, by the way, may be made more precise ([24], Ch. X), in the following way.

1°. *The spectrum of a J-orthogonal (J-unitary) matrix is skew-symmetric (respectively, symmetric) with respect to the unit circle.*

2°. *The spectrum of a real, J-orthogonal matrix is symmetric with respect to both the unit circle and the real line.*

The J-unitarity condition (1.3) can be restated in the form

$$(JU\xi, U\eta) = (J\xi, \eta) \ . \tag{1.4}$$

If

$$U\xi^{(1)} = \rho_1 \xi^{(1)} \ , \quad U\xi^{(2)} = \rho_2 \xi^{(2)}$$

and $\rho_1 \bar{\rho}_2 \neq 1$ then, upon taking $\xi = \xi^{(1)}$ and $\eta = \xi^{(2)}$ in (1.4), we see that

$$(J\xi^{(1)}, \xi^{(2)}) = 0 \ .$$

The following general fact can be easily proved too.

3°. *Let* L_{ρ_1} *and* L_{ρ_2} *be two root spaces of the J-unitary matrix* U. *If* $\rho_1 \bar{\rho}_2 \neq 1$, *then* L_{ρ_1} *and* L_{ρ_2} *are J-orthogonal, i.e.,*

$$(J\xi, \eta) = 0 \quad (\xi \in L_{\rho_1}, \eta \in L_{\rho_2}).$$

A subspace L is called *definite* if the form $(J\xi, \xi)$ has a constant sign on it and vanishes only for $\xi = 0$.

A subspace L is called *nondegenerate* if the form $(J\xi, \xi)$ is nondegenerate on it, i.e., if there is no nonzero vector in L which is J-orthogonal to the entire subspace L.

If L is nondegenerate (in particular, definite), then E_{2m} splits into a direct sum

$$E_{2m} = L \dotplus M \ , \tag{1.5}$$

where the complementary space M is J-orthogonal to L.

4°. *If the eigenspace* L *corresponding to the eigenvalue* ρ *of a J-unitary matrix* U *is nondegenerate, then* $|\rho| = 1$ *and* ρ *is an eigenvalue of simple type.*

Indeed, assuming that $\rho\bar{\rho} \neq 1$, 3° would imply that the subspace L is J-orthogonal to itself, which is impossible.

Let us briefly clarify the second statement in 4° too. To do this, we use (1.5). Since L is invariant under U, (1.4) implies that M is too. Let $\rho_2, \rho_3, \ldots, \rho_s$ be all the distinct eigenvalues of the matrix U in M. Then

$$M = L_{\rho_2} \dotplus L_{\rho_3} \dotplus \ldots \dotplus L_{\rho_s} .$$

Since all ρ_j $(j = 2,3,\ldots,s)$ differ from $\rho_1 = \rho$, (1.5) implies that $L = L_{\rho_1}$, as claimed.

5°. *Assume that the root space* L_ρ *corresponding to the eigenvalue* ρ $(|\rho| = 1)$ *of a J-unitary matrix* U *has the property that the form* $i(J\xi,\xi)$ *keeps a constant sign on* L_ρ. *Then* L_ρ *is a definite subspace, and so* ρ *is an eigenvalue of simple type.*

Indeed, suppose, for example, that

$$i(J\xi,\xi) \geqq 0 \quad \text{for all} \quad \xi \in L_\rho . \tag{1.6}$$

We have to show that equality is impossible when $\xi \neq 0$. Assume that for some $\xi^{(0)}$

$$(J\xi^{(0)},\xi^{(0)}) = 0 .$$

Considering inequality (1.6) for $\xi = \xi^{(0)} + \alpha\eta$, where $\eta \in L_\rho$ and α is a scalar, we easily conclude that $(J\xi^{(0)},\eta) = 0$ for all $\eta \in L_\rho$. In other words, the vector $\xi^{(0)}$ is J-orthogonal to L_ρ. Since $|\rho| = 1$, proposition 3° shows that $\xi^{(0)}$ is J-orthogonal to any other root space L_{ρ_j} $(\rho_j \neq \rho)$ of the matrix U. Therefore, $\xi^{(0)}$ is J-orthogonal - and the vector $J\xi^{(0)}$ is simply orthogonal - to all the vectors of E_n. Consequently, $J\xi^{(0)} = 0$ and $\xi^{(0)} = 0$.

Proposition 5° is proved.

5. We shall say that a J-unitary matrix U is of *stable type* if all its eigenspaces are of *definite type*. [A similar terminology has been proposed by I. M. Gel'fand and V. B.

Lidskii for real symplectic matrices.]

Therefore, if the matrix U is of stable type, then all its eigenvalues have absolute value one and are of simple type.

THEOREM 1.2. *If the J-unitary matrix U is of stable type, then all the J-unitary matrices V belonging to some δ-neighborhood*

$$|U - V| < \delta$$

of U are of stable type too.

PROOF. Let $\rho_1, \ldots, \rho_s$ be all the distinct eigenvalues of the given matrix U. Denote by $k_1, \ldots, k_s$ the multiplicities of $\rho_1, \ldots, \rho_s$, and let $L_1, \ldots, L_s$ be the corresponding eigenspaces. Consider s disjoint circles $\gamma_1, \ldots, \gamma_s$ centered at the points $\rho_1, \ldots, \rho_s$. Since in our case $L_j = L_{\rho_j}$ $(j = 1, \ldots, s)$, the projection matrices

$$P_j = \frac{1}{2\pi i} \oint_{\gamma_j} (\rho I_n - U)^{-1} d\rho \qquad (j = 1, \ldots, s)$$

will project E_n onto L_j $(j = 1, \ldots, s)$, respectively.

According to the hypothesis of the theorem, to each L_j there corresponds an $\ell_j > 0$ such that

$$|(J\xi, \xi)| \geq \ell_j(\xi, \xi) \quad \text{for all} \quad \xi \in L_j \quad (j = 1, \ldots, s). \tag{1.7}$$

Choose some constants $h_j < 1$ satisfying

$$2|P_j|h_j + h_j^2 < \tfrac{1}{2}\ell_j \; (j = 1, \ldots, s). \tag{1.8}$$

It is obvious that one can always find a small enough $\delta > 0$ such that the eigenvalues of any matrix V in the δ-neighborhood of U will lie inside the circles γ_j $(j = 1, \ldots, s)$ and, in addition, the following relations will be satisfied:

$$|P_j - Q_j| < h_j \quad (j = 1, \ldots, s), \tag{1.9}$$

where

$$Q_j = \frac{1}{2\pi i} \oint_{\gamma_j} (\rho I_n - V)^{-1} d\rho \quad (j = 1, \ldots, s) .$$

Let us show that the chosed number $\delta > 0$ satisfies the requirement of the theorem. Since $h_j < 1$, inequalities (1.9) imply the existence of the inverse matrices

$$(I_n - P_j + Q_j)^{-1} = \sum_{n=0}^{\infty} (P_j - Q_j)^n \quad (j = 1,\ldots,s).$$

If $\xi \in L_j$, then $\xi = P_j\xi$ and so

$$Q_j\xi = (I_n - P_j + Q_j)\xi \quad (\xi \in L_j).$$

Thus, as ξ runs over L_j, $Q_j\xi$ runs over the subspace $L_j' = Q_jL_j$ which has the same dimension k_j as does L_j $(j = 1,\ldots,s)$.

Since the projection matrices Q_j $(j = 1,\ldots,s)$ are mutually orthogonal and their sum is the identity matrix,

$$E_n = Q_1L_n \dot{+} \ldots \dot{+} Q_sL_n .$$

Now from the relations

$$k_1 + \ldots + k_s = n, \quad L_j' \subset Q_jE_n \quad (j = 1,\ldots,s),$$

we conclude that

$$L_j' = Q_jL_j = Q_jE_n$$

and, consequently, that L_j' is the direct sum of those root spaces of the matrix V which correspond to its eigenvalues which lie inside γ_j $(j = 1,\ldots,s)$.

Let us verify that the subspaces L_j' $(j = 1,\ldots,s)$ are definite. To this end, notice that

$$(JP_j\xi,P_j\xi) - (JQ_j\xi,Q_j\xi) = 2i\,\mathrm{Im}(JP_j\xi,(P_j - Q_j)\xi) -$$
$$- (J(P_j - Q_j)\xi,(P_j - Q_j)\xi)$$

whence, by (1.9),

$$|(JP_j\xi,P_j\xi) - (JQ_j\xi,Q_j\xi)| \leq 2h_j|P_j||\xi|^2 + h_j^2|\xi|^2 \leq$$
$$\leq \frac{1}{2}\,\ell_j(\xi,\xi) .$$

Taking into account (1.7), we see that

$$|(JQ_j\xi,Q_j\xi)| \leq \frac{1}{2}\,\ell_j(\xi,\xi)$$

for all $\xi \in L_j$, i.e., the subspace L_j' is definite $(j = 1,\ldots,s)$.

Now, given a matrix V in the δ-neighborhood of U

that we considered and which is J-unitary, it will be of stable type because all its root spaces are definite and hence so are its eigenspaces.

The theorem is proved.

6. An eigenvalue ρ of a J-unitary matrix of stable type is said to be of the *first (second) kind* if the form $i(J\xi,\xi)$ is positive (respectively, negative) on the eigenspace L_ρ.

Since the distinct eigenspace L_ρ of a J-unitary matrix U of stable type are mutually J-orthogonal and each of them is definite, one can state, in virtue of the inertia law for spaces with an indefinite metric (see [24], § 103), that the sum of the multiplicities of all eigenvalues of one and the same kind of a J-unitary matrix of stable type is precisely equal to m.

THEOREM 1.3. *Let* $\{U_\nu\}$ *be a sequence of J-unitary matrices of stable type which converges to the limit matrix* U:

$$U = \lim_{\nu\to\infty} U_\nu \quad .$$

For the matrix U *to be of stable type it is necessary and sufficient that the following condition be satisfied for some* $\varepsilon > 0$: *in the* ε*-neighborhood of any eigenvalue* ρ *of the matrix* U *one can find only eigenvalues of one and the same kind of all matrices* U_ν *with sufficiently large* ν.

PROOF. Since all the matrices U_ν $(\nu = 1,2,\dots)$ are J-unitary and their spectra lie on the unit circle, the limit matrix U has the same properties. Let ρ_0 be an eigenvalue of U, and let γ_ε be a circle of radius $\varepsilon > 0$ centered at ρ_0. Suppose that ε is so small that there are no other eigenvalues of U inside γ_ε. Then the projection matrix

$$P_0 = \frac{1}{2\pi i} \oint_{\gamma_\varepsilon} (\lambda I_n - U)^{-1} d\lambda$$

takes E_n onto the root space L_{ρ_0} of U.

Set

$$P_\nu = \frac{1}{2\pi i} \oint_{\gamma_\varepsilon} (\lambda I_n - U_\nu)^{-1} d\lambda \quad . \tag{1.10}$$

Then, oviously

$$P_0 = \lim_{\nu\to\infty} P_\nu .$$

The matrix P_ν projects E_n into the direct sum of mutually J-orthogonal eigenspaces of the matrix U_ν which correspond to its eigenvalues lying inside γ_ε. These eigenvalues converge to ρ_0 when $\nu \longrightarrow \infty$. Therefore, if the condition of the theorem is fulfilled, then for the chosen $\varepsilon > 0$ one can find N_ε such that the subspace $P_\nu E_n$ is definite for all $\nu > N_\varepsilon$. Then the form

$$i(J\xi,\xi) = i(JP_0\xi,P_0\xi) = \lim_{\nu\to\infty} i(JP_\nu\xi,P_\nu\xi) \qquad (\xi \in L_{\rho_0})$$

on L_{ρ_0} will have constant sign. Since $|\rho_0| = 1$, and according to 5°, the latter implies the definiteness of L_{ρ_0}.

Therefore, the sufficiency of the condition in the theorem is proved. Let us prove its necessity.

Suppose that the condition is not satisfied. Then the matrix U has an eigenvalue ρ_0 such that given any $\varepsilon > 0$ there is an infinite number of matrices U_ν $(\nu = 1,2,\ldots)$ having at least two distinct eigenvalues ρ_ν' and ρ_ν'' of different kinds in the ε-neighborhood of ρ_0. Again, we choose ε as was shown earlier, and consider the matrices P_ν $(\nu = 1,2,\ldots)$ defined by formula (1.10). As we just said, one can find an infinite set of values ν $(\nu = \nu_1,\nu_2,\ldots)$ such that to each one of them there corresponds a pair of vectors $\xi_\nu^{(1)}$ and $\xi_\nu^{(2)}$ of unit length satisfying

$$\begin{gathered} \xi_\nu^{(j)} = P_\nu\xi_\nu^{(j)} \qquad (j = 1,2) \\ i(J\xi_\nu^{(1)},\xi_\nu^{(1)}) > 0, \qquad i(J\xi_\nu^{(2)},\xi_\nu^{(2)}) < 0 . \end{gathered} \tag{1.11}$$

Since the unit sphere $(\xi,\xi) = 1$ is compact, one can assume, with no loss of generality, that there exist the limits

$$\xi^{(j)} = \lim_{k\to\infty} \xi_{\nu_k}^{(j)} \qquad (j = 1,2) .$$

Then, passing to the limit with respect to the sequence $\{\nu_k\}$ of values of ν in all relations (1.11), one obtains

$$(\xi^{(j)},\xi^{(j)}) = 1, \qquad \xi^{(j)} = P_0\xi^{(j)} \in L_{\rho_0} \qquad (j = 1,2)$$

and

$$i(J\xi^{(1)},\xi^{(1)}) \geqq 0 , \quad i(J\xi^{(2)},\xi^{(2)}) \leqq 0 .$$

Consequently, subspace L_{ρ_0} cannot be definite, which completes the proof of the theorem.

§ 2. THE CANONICAL SYSTEM OF DIFFERENTIAL EQUATIONS WITH PERIODIC COEFFICIENTS

In this section there are given, if one does not count slight additions, the familiar propositions from the theory of systems of linear differential equations with periodic coefficients.

1. Let

$$A(t) = \|a_{jk}(t)\|_1^n \qquad (0 \leqq t < \infty)$$

be some square, n-th order matrix function of the argument t, whose elements are summable on every finite interval $(0,\ell)$.

Consider the differential system

$$\frac{dx_j}{dt} = \sum_{k=1}^{n} a_{jk}(t)x_k \qquad (j = 1,\dots,n) ,$$

which we shall alternatively write as a single vector differential equation

$$\frac{dx}{dt} = A(t)x . \tag{2.1}$$

[A vector with constant coordinates ξ_j $(j = 1,\dots,n)$ will be denoted by ξ, while one with variable coordinates x_j $(j = 1,\dots,n)$ will be denoted by x.]

By a solution of this equation we shall understand any absolutely continuous (i.e., having absolutely continuous coordinates) vector-function $x = x(t)$ satisfying the integral equation

$$x(t) = x(t_0) + \int_0^t A(s)x(s)ds \qquad (0 \leqq t < \infty) , \tag{2.2}$$

where $t_0 \in [0,\infty)$ is arbitrary. Obviously, if $x(t)$ satisfies (2.2) for some $t_0 \in [0,\infty)$, then it satisfies (2.2) for any $t_0 \in [0,\infty)$. To each $\xi \in E_n$ there corresponds one and only one solution of equation (2.2), and so, of equation (2.1) too, satisfying $x(t_0) = \xi$.

The *matrizant* of equation (2.1) is the matrix-function

$$U(t) = \|u_{jk}\|_1^n \qquad (0 \leq t < \infty) ,$$

defined as the solution of the differential equation

$$\frac{dU}{dt} = A(t)U , \quad U(0) = I_n ,$$

where I_n is the identity matrix of order n.

Alternatively, $U(t)$ can be defined as the absolutely continuous solution of the integral equation

$$U(t) = I_n + \int_0^t A(s)U(s)ds \qquad (0 \leq t < \infty) .$$

This solution can be obtained in the form of a series

$$U(t) = I_n + \sum_{k=1}^{\infty} U_k(t) , \tag{2.3}$$

where

$$U_k(t) = \int_0^t A(s)U_{k-1}(s)ds \qquad (k = 1,2,\ldots; \ U_0 = I_n) .$$

Putting

$$a(t) = \int_0^t |A(s)|ds \qquad (0 \leq t < \infty) , \tag{2.4}$$

we get

$$|U_k(t)| \leq \int_0^t |U_{k-1}(s)|da(s) \qquad (k = 1,2,\ldots) .$$

By induction, this gives

$$|U_k(t)| \leq \frac{1}{k!} a^k(t) \qquad (k = 1,2,\ldots) , \tag{2.5}$$

showing that series (2.3) converges uniformly in any finite interval $[0,\ell]$ (i.e., the n^2 series formed by the the corresponding elements of the matrices $U_k(t)$ $(k = 1,2,\ldots)$ converge uniformly).

Using the matrizant, any solution $x = x(t)$ of equation (2.1) can be expressed in terms of its initial value $x_0 = x(0)$ via the formula

$$x(t) = U(t)x_0 .$$

It is known that (the Ostrogradskii-Liouville formula)

$$\det U(t) = \exp\left(\int_0^t \sum_{j=1}^{n} a_{jj}(s)ds\right) > 0 \qquad (0 \leq t < \infty) .$$

2. From now on we shall be interested in the case where $A(t)$ is a periodic matrix-function with period T:

$$A(t + T) = A(t) \qquad (0 \leq t < \infty).$$

In this case equation (2.1) has, along with each solution $x(t)$, the solution $x_1(t) = x(t + T)$.

A complex number ρ is called a *multiplier* of equation (2.1) if there exist a solution $x(t) \not\equiv 0$ of this equation with the property

$$x(t + T) = \rho x(t) \qquad (0 \leq t < \infty) . \tag{2.7}$$

Since $x(t)$ is uniquely determined by its value at any point, condition (2.7) will be satisfied for all $t \in [0,\infty)$ as soon as it is satisfied at a single arbitrary point - the point $t = 0$, for example, i.e.,

$$x(T) = \rho x(0) .$$

In virtue of (2.6), this equality means that

$$(U(T) - \rho I_n)x_0 = 0 . \tag{2.8}$$

The matrix $U(T)$ is called the *monodromy matrix* of equation (2.1). Therefore, a multiplier of equation (2.1) is nothing else but an eigenvalue of the monodromy matrix, i.e., a root of the algebraic equation

$$\Delta(\rho) = \det(U(T) - \rho I_n) = 0 . \tag{2.9}$$

From now on, by the *multiplicity* k_ρ of a multiplier ρ we shall understand its multiplicity as a root of equation (2.9).

Any solution $x = x(t)$ enjoying property (2.7) with a given ρ is said to *belong* to the given multiplier ρ.

It is evident that the maximal number of linearly independent solutions of equation (2.1) which belong to a given multiplier ρ is equal to the dimension of the linear manifold of all vectors $x_0 \in E_n$ satisfying (2.8), i.e., equals the defect

d_ρ (the difference between the order and the rank) of the matrix $U(T) - \rho I_n$.

Generally speaking, $d_\rho \leqq k_\rho$. If $d_\rho = k_\rho$, then, complying with the terminology of § 1, no. 2, ρ is called a multiplier of *simple type*.

3. We shall be interested in the case of a canonical equation (2.1). Then $n = 2m$ and (2.1) has the form

$$\frac{dx}{dt} = JH(t)x \, , \tag{2.10}$$

where

$$J = J_n = \begin{pmatrix} 0 & I_m \\ -I_m & 0 \end{pmatrix} , \quad \text{and} \quad H(t) = \| h_{jk}(t) \|_1^n \tag{2.11}$$

is a real, symmetric, matrix-function with period T.

As a generalization, we shall consider also the case where $H(t)$ is a Hermitian or complex symmetric matrix-function.

The matrizant $U(t)$ of equation (2.10) will be determined from the system

$$\frac{dU(t)}{dt} = JH(t)U(t) \, , \quad U(0) = I_n \, . \tag{2.12}$$

Therefore, given arbitrary vectors $\xi, \eta \in E_n$, we have, in case $H(t)$ is Hermitian

$$\frac{d}{dt} (JU\xi, U\eta) = - (HU\xi, U\eta) + (JU\xi, JHU\eta) = 0 \, ,$$

implying that

$$(JU\xi, U\eta)_{t=T} = (JU\xi, U\eta)_{t=0} = (J\xi, \eta) \, ,$$

i.e., that

$$U^*(T)JU(T) = J \, .$$

Now, if $H(t)$ is a complex, symmetric matrix, then passing to complex conjugates in (2.12), we have

$$\frac{d\bar{U}}{dt} = J\bar{H}(t)\bar{U} \, , \quad \bar{U}(0) = I_n \, ,$$

giving, for arbitrary $\xi, \eta \in E_n$,

$$\frac{d}{dt} (JU\xi, \bar{U}\eta) = - (HU\xi, \bar{U}\eta) + (JU\xi, J\bar{H}\bar{U}\eta) =$$

$$= -(HU\xi,\bar{U}\eta) + (U\xi,\bar{H}\bar{U}\eta) = 0 ,$$

as $\bar{H} = H^*$. Consequently, in this case

$$(JU\xi,\bar{U}\eta)_{t=T} = (JU\xi,\bar{U}\eta)_{t=0} = (J\xi,\eta) ,$$

which gives, using $(\bar{U})^* = U^\tau$,

$$U^\tau(T)JU(T) = J .$$

1°. *The monodromy matrix* $U(T)$ *of an equation* (2.10) *with a Hermitian (symmetric) matrix-function* $H(t)$ *is J-unitary (respectively, J-orthogonal).*

This implies (see propositions 1° and 2°, § 1)

THEOREM 2.1. *If* $H(t)$ *is a Hermitian (symmetric) matrix-function, then the spectrum of the monodromy matrix* $U(T)$ *of equation* (2.10) *is symmetric (skew-symmetric) with respect to the unit circle. If* $H(t)$ *is a real, symmetric, matrix-function, then this spectrum is symmetric with respect to both the unit circle and the real axis.*

This theorem (for the case of a symmetric matrix-function $H(t)$ has been proved independently by Lyapunov (see [18], p. 226) and H. Poincaré.

Suppose that some multiplier ρ of equation (2.1) is not of simple type. Then there always correspond to it (see [18] or [3]) two linearly independent solutions $x^{(0)}(t)$ and $x^{(1)}(t)$ such that

$$x^{(0)}(t + T) = \rho x^{(0)}(t) , \quad x^{(1)}(t + T) = \rho x^{(1)}(t) + x^{(0)}(t) , \qquad (0 \leqq t < \infty).$$

Combining this fact with Theorem 2.1 for the canonical system (2.10), we are led to the following conclusion.

2°. *All solutions of equation* (2.10) *are bounded on the entire interval* $[0,\infty)$ *if and only if all the multipliers of this equation have an absolute value equal to one and are of simple type.*

4. The problems arising in mechanics (see the Introduction) usually lead to an equation (2.10) which contains a parameter as a

factor in the right-hand side. This circumstance, as well as certain considerations of a purely mathematical nature, are reasons for considering, instead of equation (2.10), the equation

$$\frac{dx}{dt} = \lambda JH(t)x \ , \tag{0.1}$$

with a real, symmetric, matrix-function $H(t)$.

Generalizing this setting, we shall always assume from now on that $H(t)$ is a Hermitian matrix-function with period T.

The matrizant of equation (0.1) can be obtained as the sum of the series

$$U(t;\lambda) = I_n + \sum_{k=1}^{\infty} \lambda^k U_k(t) \ , \tag{2.13}$$

where the $U_k(t)$ $(k = 1,2,\dots)$ are computed recursively:

$$U_k(t) = J\int_0^t H(s)U_{k-1}(s)ds \qquad (k = 1,2,\dots;\ U_0 = I_n).$$

Since $(JH)^*JH = -H^2$,

$$|JH(t)| = |H(t)| = h_M(t) \qquad (0 \leq t < \infty),$$

where $h_M(t)$ is the maximum of the absolute values for the eigenvalues of $H(t)$ $(0 \leq t < \infty)$. In accordance with (2.4) and (2.5), the norm of each term from series (2.13) is no larger than the corresponding term of the series

$$1 + \sum_{k=1}^{\infty} \frac{1}{k!} |\lambda|^k \left(\int_0^t h_M(s)ds\right)^k .$$

Therefore series (2.13) converges uniformly in every cylinder $0 \leq t \leq L$, $|\lambda| \leq R$, where L,R are arbitrary, positive numbers.

In particular, the monodromy matrix $U(T;\lambda)$ of equation (0.1) will be an entire function:

$$U(T;\lambda) = I_n + S_1\lambda + S_2\lambda^2 + \dots \tag{2.14}$$

Notice that

$$S_1 = J\int_0^T H(t)dt \ .$$

The holomorphic matrix-function $\Gamma(t)$:

$$\Gamma(\lambda) = \ln U(T;\lambda) = \Gamma_1\lambda + \Gamma_2\lambda^2 + \ldots \quad (\Gamma_1 = S_1) \tag{2.15}$$

is well-defined in a small enough neighborhood of the point $\lambda = 0$. Moreover, if $\rho_j(\lambda)$ and $\gamma_j(\lambda)$ $(j = 1,\ldots,n)$ are the eigenvalues of the matrices $U(T;\lambda)$ and $\Gamma(\lambda)$, respectively, then by indexing these eigenvalues in a suitable way, we have

$$\rho_j(\lambda) = e^{\gamma_j(\lambda)} \qquad (j = 1,\ldots,n) \ . \tag{2.16}$$

Since

$$\lim_{\lambda\to 0} \frac{\Gamma(\lambda)}{\lambda} = S_1 \ ,$$

one can always find a sequence of real numbers λ_ν $(\nu = 1,2,\ldots)$ converging to zero, and such that

$$\lim_{\nu\to\infty} \frac{\gamma_j(\lambda_\nu)}{\lambda_\nu} = s_j \qquad (j = 1,\ldots,n) \tag{2.17}$$

where s_j $(j = 1,\ldots,n)$ are the eigenvalues of the matrix S_1.

According to 1°, for real λ all the eigenvalues of $U(T;\lambda)$ are disposed symmetrically with respect to the unit circle, and, consequently, all the eigenvalues of the matrix $\Gamma(\lambda)$ are disposed symmetrically with respect to the imaginary axis. Using (2.16), we deduce

3°. *The eigenvalues of the matrix* S_1 *are symmetrically disposed with respect to the imaginary axis.*

[Of course, this statement could be completed by the corresponding statement concerning the elementary divisors associated to the symmetrically disposed eigenvalues.]

Incidentally, 3° expresses a known property of any product of a nonsingular, real, skew-symmetric matrix with a Hermitian matrix - a form that the matrix S_1, in particular, has.

There is not much to add in order to get the following proposition.

THEOREM 2.2. *If all the eigenvalues* s_j $(j = 1,\ldots,n)$ *of the matrix* S_1 *lie on the imaginary axis and are distinct, then one can find a number* $\ell > 0$ *such that all the solutions of equation* (0.1) *are bounded when* $-\ell < \lambda < \ell$.

PROOF. From any sequence of real numbers λ one can

extract a subsequence $\{\lambda_\nu\}$ such that (2.16) is fulfilled. Therefore, assuming that all $s_j = \pm|s_j|i$ $(j = 1,\ldots,n)$ are distinct, one finds $\ell > 0$ such that for $-\ell < \lambda < \ell$ all the numbers $\gamma_j(\lambda)$ $(j = 1,\ldots,n)$ have distinct imaginary parts.

On the other hand, all the points $\gamma_j(\lambda)$ $(j = 1,\ldots,n)$ have to be symmetricaly disposed with respect to the imaginary axis. This shows that in the case under consideration they should lie on the imaginary axis itself. According to (2.16), the latter means that all the multipliers $\rho_j(\lambda)$ $(j = 1,\ldots,n)$ lie on the unit circle. Furthermore, for ℓ small enough, they will be all distinct.

The theorem is proved.

Let us remark that the first condition (that all eigenvalues s_j $(j = 1,\ldots,n)$ be purely imaginary) is satisfied whenever the Hermitian matrix

$$H_I = \int_0^T H(t)\,dt$$

is positive definite, i.e., $H_I > 0$.

Indeed, we have

$$\det(S_1 - i\sigma I_n) = \det(J(H_I + i\sigma J)) = (-1)^m \det(H_I + i\sigma J) \ .$$

On the other hand, since the matrix iJ is Hermitian, then for $H_I > 0$ all the roots of the equation $\det(H_I + i\sigma J) = 0$ will be real, as a well-known theorem from matrix theory shows (see [3], Ch. 10).

As we shall show in § 6 (Theorem 6.4), if $H_I > 0$ then the statement of Theorem 2.2 remains valid even when the matrix S_1 has multiple eigenvalues.

§ 3. SELFADJOINT BOUNDARY VALUE PROBLEMS FOR THE CANONICAL SYSTEM

1. Let

$$H(t) = \|h_{jk}(t)\|_1^n$$

be a Hermitian matrix-function whose elements are measurable and summable over the interval $[0,T]$.

We shall say that such a matrix-function is of class $P_n(T)$, and write $H(t) \in P_n(T)$, if

1) $H(t) \geqq 0 \quad (0 \leqq t \leqq T)$

and

2) $\int_0^T H(t)dt > 0$.

Alternatively, conditions 1) and 2) say that

$$(H(t)\xi,\xi) \geqq 0 \quad (0 \leqq t \leqq T) \quad \text{and} \quad \int_0^T (H(t)\xi,\xi)dt > 0 \tag{3.1}$$

for any $\xi \in E_n$, $\xi \neq 0$.

Now let $n = 2m$ and $J = J_n$.

LEMMA 3.1. *Let* $H(t) \in P_n(T)$. *Then for any solution*

$$x = x(t;\lambda) \neq 0$$

of equation (0.1)

$$\frac{dx}{dt} = \lambda JH(t)x ,$$

the equality

$$(Jx,x)_{t=T} - (Jx,x)_{t=0} = (\bar{\lambda} - \lambda)\int_0^T (H(t)x,x)dt \tag{3.2}$$

holds true, and

$$\int_0^T (H(t)x,x)dt > 0 . \tag{3.3}$$

PROOF. Indeed, equation (0.1) is equivalent to the following one

$$J\frac{dx}{dt} = - \lambda H(t)x . \tag{3.4}$$

On the other hand, given two arbitrary, absolutely continuous, vector-functions $x = x(t)$ and $y = y(t)$ $(0 \leqq t \leqq T)$, integration by parts yields

$$\int_0^T (J\frac{dx}{dt},y)dt - \int_0^T (x,J\frac{dy}{dt})dt = (Jx,y)_{t=T} - (Jx,y)_{t=0}.$$

Taking $y = x = x(t;\lambda)$ to be a solution of equation

(3.4), the last equality immediately implies (3.2). Moreover,

$$(H(t)x,x) \geqq 0 \qquad (0 \leqq t \leqq T) \tag{3.5}$$

in virtue of (3.1), and so

$$\int_0^T (H(t)x,x)dt \geqq 0 \ . \tag{3.6}$$

Suppose that one has equality in (3.6). Then one should have equality almost everywhere in (3.5). This would imply that $H(t)x = 0$ almost everywhere. [Indeed, if $A \geqq 0$ and $(A\xi,\xi) = 0$ for some ξ, then $A\xi = 0$.] Consequently, $\frac{dx}{dt} \equiv 0$, i.e., $x(t) \equiv \xi = \text{const} \neq 0$. On the other hand, if $x(t) \equiv \xi \neq 0$, then (3.1) excludes the equality in (3.6) - contradiction.

The lemma is proved.

A simple consequence of this lemma is

THEOREM 3.1. *Let* $H(t) \in P_n(T)$. *Then the monodromy matrix* $U(T;\lambda)$ *is J-unitary, J-expansive, or J-cotractive depending on whether* $\operatorname{Im}\lambda = 0$, > 0, or < 0.

[We keep calling $U(T;\lambda)$ the monodromy matrix, imagining that the matrix-function $H(t)$ is extended by periodicity: $H(t+T) = H(t)$.]

PROOF. Given any $\xi \in E_n$, $\xi \neq 0$, the vector-function

$$x(t;\lambda) = U(t;\lambda)\xi$$

is a solution of equation (0.1). If one applies identity (3.2) to it and then multiplies both sides by i, one obtains

$$i(JU(T;\lambda)\xi,U(T;\lambda)\xi) - i(J\xi,\xi) = 2 \operatorname{Im}\lambda \int_0^T (H(t)x,x)dt \ .$$

The left-hand side of this equality is 0, > 0, or < 0 depending on whether $\operatorname{Im}\lambda = 0$, > 0, or < 0, which completes the proof of the theorem.

2. In this section, E shall always denote an arbitrary J-unitary matrix, i.e.,

$$(JE\xi,E\xi) = (J\xi,\xi) \qquad (\xi \in E_n) \ .$$

THEOREM 3.2. *If* $H(t) \in P_n(T)$, *then all the characteristic numbers of the boundary value problem*

$$\frac{dx}{dt} = \lambda JH(t)x \ , \quad x(T) = Ex(0) \tag{3.7}$$

are real.

PROOF. Indeed, let $x = x(t;\lambda)$ be a fundamental solution of problem (3.7) (i.e., a nontrivial solution of system (3.7) for some value of λ). Then

$$(Jx,x)_{t=T} - (Jx,x)_{t=0} = (JEx(0),Ex(0)) - (Jx(0),x(0)) = 0.$$

On the other hand, this is possible only if $\lambda = \bar{\lambda}$, as (3.2) and (3.3) show.

The theorem is proved.

[It is evident that Theorem 3.2 remains valid should one replace the boundary condition $x(T) = Ex(0)$ by any other selfadjoint boundary condition of the form

$$Ax(0) + Bx(T) = 0 \ , \tag{*}$$

where (A,B) is a rectangular, $(n \times 2n)$-matrix of maximal rank n. More precisely, condition $(*)$ is said to be *selfadjoint* if for any absolutely continuous function $x(t)$ that satisfies $(*)$, the equality

$$(Jx,x)_{t=T} - (Jx,x)_{t=0} = 0$$

holds.

S. A. Orlov pointed out to the author that a necessary and sufficient condition for the selfadjointness of the general boundary condition $(*)$ is that the equality

$$A^*JA = B^*JB$$

hold.

The subsequent theorems 3.3 and 3.4 remain valid too when the boundary condition $x(T) = Ex(0)$ is replaced by the general selfadjoint condition $(*)$.]

3. If $U(t;\lambda)$ is the matrizant of equation (0.1), then obviously the characteristic numbers of the boundary value problem (3.7) are precisely the roots of the equation

$$\det(U(T;\lambda) - E) = 0 \tag{3.8}$$

Let

$$\lambda_1 \leqq \lambda_2 \leqq \ldots \tag{3.9}$$

and

$$\lambda_{-1} \geqq \lambda_{-2} \geqq \ldots \tag{3.10}$$

be, respectively, all the positive and negative characteristic numbers of the boundary value problem (3.7), where one assumes that each number λ_j appears in the sequence (3.9) or (3.10) a number of times equal to its multiplicity as a root of equation (3.8).

Notice that problem (3.7) has the additional characteristic number $\lambda_0 = 0$ if (and only if) $\det(E - I_n) = 0$.

We shall prove below (see no. 3) that the multiplicity of any characteristic number λ_j $(j = 0, \pm 1, \pm 2, \ldots)$ as a root of equation (3.8) is always equal to the number of linearly independent solutions of system (3.7) for $\lambda = \lambda_j$.

In order to emphasize the dependence of λ_j upon $H(t)$ $(0 \leqq t \leqq T)$, we shall write $\lambda_j(H)$.

THEOREM 3.3. *Let* $H_1(t)$ *and* $H_2(t)$ $(0 \leqq t \leqq T)$ *be two Hermitian matrix-functions of class* $P_n(T)$ *satisfying*

$$H_1(t) \leqq H_2(t) \quad (0 \leqq t \leqq T) .$$

Then

$$\lambda_j(H_1) \geqq \lambda_j(H_2), \quad \lambda_{-j}(H_1) \leqq \lambda_{-j}(H_2) \tag{3.11}$$

$$(j = 1,2,\ldots) .$$

PROOF. Consider the boundary value problem

$$\frac{dx}{dt} = \lambda J H_\varepsilon(t) x, \quad x(T) = E x(0) , \tag{3.12}$$

where

$$H_\varepsilon(t) = H_1(t) + \varepsilon(H_2(t) - H_1(t)) \quad (0 \leqq \varepsilon \leqq 1) .$$

The monodromy matrix $U_\varepsilon(T;\lambda)$ of equation (3.12) is obviously an analytic function of the parameter ε. Therefore, fixing any index j, the positive characteristic number $\lambda_\varepsilon = \lambda_j(H_\varepsilon)$ of problem (3.12) will be a piecewise-analytic function of ε.

Similarly, we claim that one can construct a piecewise-analytic solution ξ_ε to the equation

$$(U_\varepsilon(T;\lambda_\varepsilon) - E)\xi_\varepsilon = 0 .$$

But then the vector-function

$$x_\varepsilon(t) = U_\varepsilon(t;\lambda_\varepsilon)\xi_\varepsilon \quad (0 \leqq t \leqq T) ,$$

which is an eigen-vector-function of problem (3.12), i.e.,

$$\frac{dx_\varepsilon}{dt} = \lambda_\varepsilon J H_\varepsilon x_\varepsilon , \quad x_\varepsilon(T) = E x_\varepsilon(0) , \tag{3.13}$$

will be a piecewise-analytic with respect to ε, too.

Furthermore, we do not destroy the piecewise-analiticity of x_ε if we impose the normalizing condition

$$- \int_0^T (J\frac{dx_\varepsilon}{dt}, x_\varepsilon)dt = \lambda_\varepsilon \int_0^T (H_\varepsilon x_\varepsilon, x_\varepsilon)dt = 1 .$$

Now differentiate the terms of this equality with respect to ε. Differentiating the left-hand integral (at those points where the derivatives of λ_ε and $x_\varepsilon(t)$ with respect to ε exist), we get

$$\int_0^T (J\frac{d}{dt}\frac{\partial x_\varepsilon}{\partial \varepsilon}, x_\varepsilon)dt + \int_0^T (J\frac{dx_\varepsilon}{dt}, \frac{\partial x_\varepsilon}{\partial \varepsilon})dt = 0 \tag{3.14}$$

Differentiating the second integral with respect to ε gives

$$\frac{d\lambda_\varepsilon}{d\varepsilon}\int_0^T (H_\varepsilon x_\varepsilon, x_\varepsilon)dt + \lambda_\varepsilon \int_0^T ((H_2 - H_1)x_\varepsilon, x_\varepsilon)dt + \\ + \lambda_\varepsilon \int_0^T (H_\varepsilon \frac{\partial x_\varepsilon}{\partial \varepsilon}, x_\varepsilon)dt + \lambda_\varepsilon \int_0^T (H_\varepsilon x_\varepsilon, \frac{\partial x_\varepsilon}{\partial \varepsilon})dt = 0 \tag{3.15}$$

On the other hand, according to (3.13)

$$\lambda_\varepsilon \int_0^T (H_\varepsilon x_\varepsilon, \frac{\partial x_\varepsilon}{\partial \varepsilon})dt = - \int_0^T (J\frac{dx_\varepsilon}{dt}, \frac{\partial x_\varepsilon}{\partial \varepsilon})dt$$

and

$$\lambda_\varepsilon \int_0^T (H_\varepsilon \frac{\partial x_\varepsilon}{\partial \varepsilon}, x_\varepsilon)dt = \lambda_\varepsilon \int_0^T (\frac{\partial x_\varepsilon}{\partial \varepsilon}, H_\varepsilon x_\varepsilon)dt = \\ = - \int_0^T (\frac{\partial x_\varepsilon}{\partial \varepsilon}, J\frac{dx_\varepsilon}{dt})dt = \int_0^T (J\frac{\partial x_\varepsilon}{\partial \varepsilon}, \frac{dx_\varepsilon}{dt})dt = \\ = - \int_0^T (J\frac{d}{dt}\frac{\partial x_\varepsilon}{\partial \varepsilon}, x_\varepsilon)dt$$

To obtain the last equality written above, we applied integration by parts and the fact that not only x_ε, but also $\partial x_\varepsilon/\partial\varepsilon$ satisfies the given boundary condition $x(T) = Ex(0)$.

Therefore, in virtue of (3.14), the sum of the last two integrals in (3.15) is zero, whence

$$\frac{d\lambda_\varepsilon}{d\varepsilon} = -\lambda_\varepsilon^2 \int_0^T ((H_2 - H_1)x_\varepsilon, x_\varepsilon)dt \leqq 0 ,$$

i.e., $\lambda_\varepsilon = \lambda(\varepsilon)$ is a nonincreasing function of ε. Since $\lambda(0) = \lambda_j(H_1)$ and $\lambda(1) = \lambda_j(H_2)$, we have proved the first group of inequalities (3.11). The proof of the second groups is similar.

4. In paper [10], we proved Theorem 3.2 while employing other, more difficult, but at the same time more profound considerations. In particular, these considerations allow us to prove the following proposition too.

THEOREM 3.4. *The multiplicity* ν_j *(as a root of equation* (3.8)*) of any characteristic number* λ_j *of the boundary value problem* (3.7) *is equal to the number* d_j *of linearly independent, fundamental vector-functions corresponding to* λ_j.

PROOF. The number d_j is obviously equal to the defect of the matrix $U(T;\lambda_j) - E$. Set

$$V(\lambda) = U(T;\lambda) - E$$

and let $D(\lambda)$ denote the determinant of $V(\lambda)$. Consider the ring R_j of all power series

$$\sum_{k=0}^{\infty} c_k(\lambda - \lambda_j)^k$$

having a radius of convergence different from zero.

The rank r_j, and so the defect $d_j = n - r_j$ of matrix $V(\lambda)$ for $\lambda = \lambda_j$ do not change as a result of the following elementary operations that we might perform on this matrix:

a) permute two rows or two columns,

b) add the elements of some row (column), all multiplied by the same function $f(\lambda) \in R_j$, to the corresponding elements of another row (column).

It is clear that the ring R_j has the property that

$f \in R_j$ and $f(\lambda_j) \neq 0$ imply $f^{-1} \in R_j$.

Moreover, since $D(\lambda) \not\equiv 0$, one can use elementary operations in the usual way (see [3], Ch. XII) and reduce $V(\lambda)$ to the diagonal matrix

$$W(\lambda) = \|w_i(\lambda)\delta_{ik}\|_1^n , \tag{3.16}$$

where $w_i(\lambda) \in R_j$, $w_i(\lambda_j) = 0$ $(i = 1,\dots,d_j)$, and

$$w_{d_j+p}(\lambda) \equiv 1 \quad (p = 1,\dots,r_j) .$$

We shall have also

$$W(\lambda) = A(\lambda)V(\lambda)B(\lambda) \tag{3.17}$$

where $A(\lambda)$ and $B(\lambda)$ are some matrices with elements from R_j and having determinants equal to one:

$$\det A(\lambda) \equiv \det B(\lambda) \equiv 1 . \tag{3.18}$$

From (3.16), (3.17) and (3.18) we conclude that

$$D(\lambda) = w_1(\lambda) \dots w_{d_j}(\lambda) .$$

Therefore, our claim will be proved if we can show that the point λ_j is a zero of first order for each of the functions $w_i(\lambda)$. This is equivalent to the matrix-function

$$W^{-1}(\lambda) = \|w_i^{-1}(\lambda)\delta_{ik}\|_1^n = B^{-1}(\lambda)V^{-1}(\lambda)A^{-1}(\lambda)$$

having at the point $\lambda = \lambda_j$ a pole of order at most one. Finally, this will happen if we can show that the matrix $V^{-1}(\lambda)$ has at $\lambda = \lambda_j$ a pole of order one. Indeed, (3.18) guarantees that the elements of the matrices $A^{-1}(\lambda)$ and $B^{-1}(\lambda)$ are functions which belong to R_j and so are holomorphic in a neighborhood of the point $\lambda = \lambda_j$.

In other words, in order to complete the proof of the theorem, it suffices to establish the following result, which itself is useful in many situations (see § 6 and paper [11]).

THEOREM 3.4. *The matrix-function* $(U(T;\lambda) - E)^{-1}$ *in the variable* λ *can be expanded into an absolutely convergent series*

$$(U(T;\lambda) - E)^{-1} = -\tfrac{1}{2}E^{-1} - \tfrac{1}{2}E^{-1}J\{A_0 - A_1\lambda + \frac{B_0}{\lambda} + \lambda\sum_{j\neq 0}\frac{B_j}{\lambda_j(\lambda-\lambda_j)}\}, \tag{3.19}$$

where A_0, A_1 *and* B_j $(j = 0, \pm 1, \pm 2, \ldots)$ *are Hermitian matrices satisfying*

$$A_1 \geqq 0, \quad B_j \geqq 0 \quad (j = 0, 1, 2, \ldots) .$$

[Here "absolute convergence" means absolute convergence of the n^2 series formed by the corresponding elements of the matrix terms of series (3.19).]

PROOF. By Theorem 3.1,

$$i(JU_\lambda \xi, U_\lambda \xi) > i(J\xi, \xi) \tag{3.20}$$

for $\operatorname{Im} \lambda > 0$ and all $\xi \in E_n$, $\xi \neq 0$ $(U_\lambda = U(T;\lambda))$.

Let us introduce the matrix

$$Z_\lambda = -i(U_\lambda + E)(U_\lambda - E)^{-1} .$$

If $\eta = i(U_\lambda - E)\xi$, then $Z_\lambda \eta = (U_\lambda + E)\xi$, and so

$$\xi = \frac{1}{2} E^{-1}(Z_\lambda \eta + i\eta), \quad U_\lambda \xi = \frac{1}{2} (Z_\lambda \eta - i\eta) .$$

Inserting these expressions for ξ and $U_\lambda \xi$ in (3.20), we obtain

$$-(JZ_\lambda \eta, \eta) + (J\eta, Z_\lambda \eta) > 0 \quad (\operatorname{Im} \lambda > 0) . \tag{3.21}$$

When $\operatorname{Im} \lambda > 0$, $\det(U_\lambda - E) \neq 0$. Consequently, as ξ runs over the entire space E_n, the vector η runs over the entire space E_n too, and this shows that (3.21) holds for any vector $\eta \in E_n$.

Consider the meromorphic function

$$F_\eta(\lambda) = i(JZ_\lambda \eta, \eta) .$$

Since

$$(J\eta, Z_\lambda \eta) = -(\eta, JZ_\lambda \eta) = -\overline{(JZ_\lambda \eta, \eta)} ,$$

inequality (3.21) means that

$$\operatorname{Im} F_\eta(\lambda) < 0 \quad \text{for} \quad \operatorname{Im} \lambda > 0 .$$

Similarly, one may prove that

$$\operatorname{Im} F_\eta(\lambda) > 0 \quad \text{for} \quad \operatorname{Im} \lambda < 0 .$$

Now according to a well-known theorem of N. G. Chebotarev (see [2], p. 197) such a function always admits the absolutely convergent expansion

$$F_\eta(\lambda) = \alpha_0 - \alpha_1 \lambda + \frac{\beta_0}{\lambda} + \lambda \sum_{j \neq 0} \frac{\beta_j}{\lambda_j(\lambda - \lambda_j)} , \tag{3.22}$$

where α_0 is real,

$$\alpha_1 \geqq 0, \quad \beta_j \geqq 0 \quad (j = 0, \pm 1, \pm 2, \ldots) \ , \tag{3.23}$$

and the poles λ_j $(j = \pm 1, \pm 2, \ldots)$ are real too.

For the case in which we are interested, the poles can be only the roots of equation (3.8). Allowing β_j to be zero too, we shall assume that the sum in (3.22) is extended over *all the distinct* roots of equation (3.8).

Since $F_\eta(\lambda)$ is a bilinear form in the coordinates $\eta_1, \ldots, \eta_n$ of the vector η and the conjugate variables $\bar{\eta}_1, \ldots, \bar{\eta}_n$, it is straightforward that the quantities α_0, α_1, and β_j $(j = 0, 1, 2, \ldots)$ are such too. Therefore, one can find matrices A_0, A_1, and B_j $(j = 0, 1, 2, \ldots)$ of order n, such that

$$\alpha_0 = (A_0\eta, \eta), \quad \alpha_1 = (A_1\eta, \eta), \quad \alpha_j = (B_j\eta, \eta)$$
$$(j = 0, \pm 1, \pm 2, \ldots) \ .$$

Therefore, (3.22) leads to the following expansion for iJZ_λ:

$$iJZ_\lambda = A_0 - A_1\lambda + \frac{B_0}{\lambda} + \lambda \sum_{j \neq 0} \frac{B_j}{\lambda_j(\lambda - \lambda_j)} \ .$$

Since the numbers α_0, α_1, and β_j are real, we see that our matrices are Hermitian. Moreover, (3.23) implies that

$$A_1 \geqq 0, \quad B_j \geqq 0 \quad (j = 0, \pm 1, \pm 2, \ldots) \ .$$

Finally, to obtain (3.19) it remains to observe that

$$iJZ_\lambda = J(U_\lambda + E)(U_\lambda - E)^{-1} = J + 2JE(U_\lambda - E)^{-1} \ .$$

The theorem is proved.

In what follows, we shall apply the theorems of this section only in the case $E = \rho I_n$, where $|\rho| = 1$.

Our present study of the boundary value problem (3.7) could be broadened in many respects by bringing into play the methods of the theory of integral equations and of the general theory of compact Hermitian operators. We will have a glimpse of these methods in § 6 , no.3, for the important example $E = - I_n$.

§ 4. MULTIPLIERS OF THE FIRST AND SECOND KIND OF A CANONICAL SYSTEM OF POSITIVE TYPE

1. Returning to the study of the solutions of the differential equation

$$\frac{dx}{dt} = \lambda JH(t)x \qquad (0.1)$$

with periodic, Hermitian, matrix-function $H(t) = H(t + T)$, we shall always assume, unless we mention otherwise, that (0.1) is an equation of *positive type*, i.e., that the function $H(t)$ is of class $P_n(T)$ in the interval $[0,T]$ (see § 3, no. 1).

In what follows the next result plays an important role.

THEOREM 4.1. *Given any nonreal* λ, *among the multipliers of equation* (0.1) *there are precisely* m *having modulus greater than one, and precisely* m *having modulus less than one.*

PROOF. If $\operatorname{Im} \lambda \neq 0$, then the monodromy matrix $U(T;\lambda)$ is J-expansive or J-contractive, depending on whether $\operatorname{Im} \lambda > 0$ or $\operatorname{Im} \lambda < 0$. Applying Theorem 1.1 to $U(T;\lambda)$ we get Theorem 4.1 immediately.

Let us remark that when $H(t)$ is a real, symmetric matrix, the theorem can be proved without appealing to the special algebraic result 1.1, by using Theorem 2.1 (Lyapunov-Poincaré). Indeed, the latter shows that for any λ the multipliers of equation (0.1) are symmetrically situated with respect to the unit circle. It follows that in order to prove 4.1, one need only show that none of the multipliers lies on the unit circle when $\operatorname{Im} \lambda \neq 0$.

Assuming the contrary, i.e., that for some λ $(\operatorname{Im} \lambda \neq 0)$ one of the multipliers, $\rho_0(\lambda)$, lies on the unit circle, we would conclude that λ is a characteristic number of the boundary value problem (3.7) with $E = \rho_0(\lambda) I_n$, and this would contradict Theorem 3.2.

2. As we already know (§ 2, no. 3), the multipliers of equation (0.1) are the roots of the algebaric equation

$$\Delta(\rho;\lambda) = \det(U(T;\lambda) - \rho I_n) = 0 .$$

In expanded form, this equation looks like

$$\rho^{2m} + A_1(\lambda)\rho^{2m-1} + \ldots + A_{2m-1}(\lambda)\rho + 1 = 0 , \qquad (4.1)$$

where $A_k(\lambda)$ $(k = 1,\ldots,2m-1)$ are entire functions of λ satisfying (according to Theorem 2.1)

$$A_k(\lambda) = \bar{A}_{2m-k}(\lambda) \qquad (k = 1,\ldots,2m-1) .$$

If equation (4.1) is irreducible (i.e., the left-hand side cannot be represented as the product of two polynomials in λ, whose coefficients are entire functions of λ), then it defines a multivalued analytic function with a 2m-sheeted Riemann surface R. In the contrary case, equation (4.1) defines a finite $(< 2m)$ number of multivalued analytic functions with Riemann surfaces having a finite number of sheets, the total number of sheets being equal to 2m. In this situation, R will denote the disconnected union of these individual Riemann surfaces lying over the complex λ-plane.

Therefore, equation (4.1) defines a single-valued analytic function $\rho(\lambda)$ on R.

Notice that if a is an arbitrary point of the Riemann surface R lying over a point λ_0, then in any δ-neighborhood of a (which contains no ramification points aside from a), the function $\rho(\lambda)$ admits the expansion

$$\rho(\lambda) = \rho(a) + \sum_{k=1}^{\infty} c_k \zeta^k ,$$

where

$$\zeta = (\lambda - \lambda_0)^{1/\nu}$$

and $\nu \geqq 1$ is the ramification order of R at the point a (i.e., the number of sheets in a δ-neighborhood of a).

3. Now considering that λ takes values only in the upper half plane $\operatorname{Im} \lambda > 0$, we introduce the following notions.

We shall say that the multiplier $\rho_j(\lambda)$ $(\operatorname{Im} \lambda > 0)$ is of the *first* (respectively, *second*) *kind* if it lies inside (respectively, outside) the unit circle.

Suppose, for some real $\lambda = \lambda_0$, that the point ρ_0 $(|\rho_0| = 1)$ is a p-multiple multiplier (i.e., ρ_0 is a p-multiple root of equation (4.1) for $\lambda = \lambda_0$). Enclose ρ_0 with a circle

$\gamma: |\rho - \rho_0| = r$ of radius r sufficiently small to ensure that no other multipliers of equation (0.1) for $\lambda = \lambda_0$ fall inside γ. Then there exists $\delta > 0$ such that for $|\lambda - \lambda_0| < \delta$ one finds precisely p multipliers inside γ, if each is counted according to its multiplicity.

Consider the open half-heighborhood

$$|\lambda - \lambda_0| < \delta, \quad \text{Im } \lambda > 0$$

of the point λ_0. Suppose that, for some λ in this half-neighborhood, one finds among the multipliers falling inside γ, precisely p_1 ($p_2 = p - p_1$) which lie outside (respectively, inside) the unit circle. Then this will be true for all λ in this half-neighborhood: indeed, as long as λ does not cross the real axis, no multiplier can cross the unit circle, as Theorem 4.1 shows.

Accordingly, we shall say that for $\lambda = \lambda_0$ there are to be found p_1 multipliers of the first kind and p_2 multipliers of the second kind at the point ρ_0.

LEMMA 4.1. *Suppose that in some neighborhood* $|\lambda-\lambda_0| < \delta$ *of the real point* λ_0 *there exists a single-valued, regular branch* $\rho_0(\lambda)$ *of the function* $\rho = \rho(\lambda)$, *which gives a multiplier of the first kind of equation* (0.1) *for any* λ *from the upper half-neighborhood*

$$|\lambda - \lambda_0| < \delta, \quad \text{Im } \lambda > 0 \ . \tag{4.2}$$

If, in addition, $|\rho_0(\lambda_0)| = 1$, *then*

$$|\rho_0(\lambda)| = 1, \quad i\rho_0^{-1}(\lambda)\rho_0'(\lambda) > 0 \quad \text{for} \quad \lambda_0-\delta < \lambda < \lambda_0+\delta$$

and so, as λ *increases from* $\lambda_0 - \delta$ *to* $\lambda_0 + \delta$, *the multiplier* $\rho_0(\lambda)$ *moves counterclockwise on the unit circle.*

PROOF. Indeed, by hypothesis, one has the expansion

$$\rho_0(\lambda) = \rho_0(\lambda_0) + \sum_{k=1}^{\infty} c_k(\lambda - \lambda_0)^k \qquad (|\lambda - \lambda_0| < \delta) \ .$$

The coefficient $c_1 = \rho_0'(\lambda_0) \neq 0$: if not, the map $\lambda \longrightarrow \rho_0(\lambda)$ would be quasiconformal at the point λ_0 (i.e., any angle made by lines originating from the point λ_0 would be taken by the map $\lambda \longrightarrow \rho_0(\lambda)$ into an angle ℓ times bigger, where ℓ is an integer larger than one) and it could not take the half-neighborhood (4.2) into a domain that lies entirely outside the unit circle.

On the other hand, if $\rho_0'(\lambda_0) \neq 0$, and so the map $\lambda \longrightarrow \rho_0(\lambda)$ is conformal at the point $\lambda = \lambda_0$, then one can find points in the lower half-neighborhood

$$|\lambda - \lambda_0| < \delta, \quad \text{Im } \lambda < 0 \tag{4.3}$$

which are taken by this map into points that lie inside the unit circle. But then, as Theorem 3.1 shows, $\rho_0(\lambda)$ would lie inside the unit circle for all λ in the half-neighborhood (4.3). Therefore, on the joint boundary of the two half-neighborhoods (4.2) and (4.3), one has $|\rho_0(\lambda)| = 1$, i.e.,

$$|\rho_0(\lambda)| = 1 \quad \text{for} \quad \lambda_0 - \delta < \lambda < \lambda_0 + \delta.$$

The same considerations which proved that $\rho_0'(\lambda_0) \neq 0$ show that $\rho_0'(\lambda) \neq 0$ for $\lambda_0 - \delta < \lambda < \lambda_0 + \delta$. The fact that the map $\lambda \longrightarrow \rho_0(\lambda)$ is conformal and satisfies $|\rho_0(\lambda)| > 1$ for $\text{Im } \lambda > 0$ implies that, as λ increases, $\rho_0(\lambda)$ moves clockwise on the unit circle. Therefore, the direction of the vector $\rho_0'(\lambda)$ is obtained from that of the vector $\rho_0(\lambda)$ by a clockwise rotation through an angle $\pi/2$ of the latter, and so $i\rho_0^{-1}(\lambda)\rho_0'(\lambda) > 0$ for $\lambda_0 - \delta < \lambda < \lambda_0 + \delta$.

The lemma is proved.

THEOREM 4.2. *Suppose that at the point* ρ_0 *of the unit circle, and for real* $\lambda = \lambda_0$, *there are only multipliers of the first kind of equation* (0.1), *and their number is* p. *Then in some half-neighborhood*

$$|\lambda - \lambda_0| < \delta, \quad \text{Im } \lambda \geqq 0 \tag{4.4}$$

of the point λ_0, *one can distinguish* p *single-valued, analytic branches* $\rho_j(\lambda)$ $(j = 1,\dots,p)$ *of the function* $\rho(\lambda)$, *each giving, for any* λ $(\text{Im } \lambda > 0)$ *in this half-neighborhood, a multiplier of the first kind, and each satisfying*

$$|\rho_j(\lambda)| = 1, \quad 0 < i\rho_j^{-1}(\lambda)\rho_j'(\lambda) < \infty \tag{4.5}$$

$$(\lambda_0 - \delta < \lambda < \lambda_0 + \delta, \quad j = 1,\dots,p) .$$

[If equation (4.1) is not irreducible, then it may happen that some of these branches are identical. However, given any in the half-neighborhood (4.4), the polynomial $\Delta(\rho;\lambda)$ in ρ will be divisible by the polynomial $(\rho - \rho_1(\lambda))\ldots(\rho - \rho_p(\lambda))$. This is precisely the sense in which we use the word "distinguish".]

PROOF. Denote by $a_1,\ldots,a_q$ the various points of the "Riemann" surface R lying over λ_0 and such that $\rho(\lambda)$ takes the value ρ_0. Let $\nu_1,\ldots,\nu_q$ be the ramification orders of the surface R at the points $a_1,\ldots,a_q$. Then

$$\nu_1 + \ldots + \nu_q = p \tag{4.6}$$

Now let a and ν be respectively one of the points a_j $(j = 1,\ldots,q)$ and the ramification order of R at a. Then in some ν-sheeted δ-neighborhood of a one has the expansion

$$\rho(\lambda) - \rho_0 = \sum_{k=\ell}^{\infty} c_k(\lambda - \lambda_0)^{k/\nu} \quad (c_\ell \neq 0) \;, \tag{4.7}$$

where the right-hand series starts with the first power $(\lambda-\lambda_0)^{\ell/\nu}$ whose coefficient c_ℓ is different from zero.

Using (4.7) one can define in the half-neighborhood (4.2) exactly ν distinct, sigle-valued, analytic branches $\rho_j(\lambda)$ $(j = 1,\ldots,\nu)$ of the function $\rho(\lambda)$ corresponding to the ν distinct values of $\zeta = (\lambda - \lambda_0)^{1/\nu}$. At any point $\lambda \neq \lambda_0$ of the interval $(\lambda_0 - \delta,\lambda_0 + \delta)$ each function $\rho_j(\lambda)$ $(j = 1,\ldots,\nu)$ satisfies the conditions of Theorem 4.2. Consequently, condition (4.5) is satisfied by each of the functions $\rho_j(\lambda)$ $(j = 1,\ldots,\nu)$ for all $\lambda \neq \lambda_0$ in the above indicated interval. If $\ell = \nu$ in the expansion (4.7), then (4.5) is satisfied at $\lambda = \lambda_0$ too.

Let us show that this is indeed the case, i.e., $\ell = \nu$. In fact, suppose $\ell > \nu$. Then the map $\lambda \longrightarrow \rho(\lambda)$ enlarges $\ell/\nu > 1$ times each angle made by lines originating from λ_0. We see that, for any $\delta > 0$, each of the functions $\rho_j(\lambda)$ $(j = 1,\ldots,\nu)$ takes values lying inside the unit circle at the points λ $(\operatorname{Im}\lambda > 0)$ of the half-neighborhood (4.2). This is impossible, because we assumed that for $\lambda = \lambda_0$ there are to be found only multipliers of the first kind at the point ρ_0.

Therefore, $\ell \leq \nu$. Assuming that $\ell < \nu$, each function $\rho_j(\lambda)$ $(j = 1,\ldots,\nu)$ would map the interval $\lambda_0 - \delta < \lambda < \lambda_0 + \delta$ into an arc having at ρ_0 an angular point with the inner angle measuring $\ell\pi/\nu < \pi$. It follows that for arbitrarily small δ

one would find a point λ^* in the interval $(\lambda_0 - \delta, \lambda_0 + \delta)$ such that $|\rho_j(\lambda^*)| \neq 1$.

By Theorem 2.1, for $\lambda = \lambda^*$, there exist multipliers $\rho_j^*(\lambda^*)$ which are symmetric to $\rho_j(\lambda^*)$ with respect to the unit circle. But then, when $\varepsilon > 0$ is small enough, there will exist, for $\lambda = \lambda^* + i\varepsilon$ and in a neighborhood of the point ρ_0, multipliers of both the first and second kind. Since λ^* can be taken arbitrarily close to λ_0, we would again contradict the hypothesis of the theorem.

We conclude that to each point a_j $(j = 1,\dots,q)$ there correspond ν_j $(j = 1,\dots,q)$ single-valued, analytic branches of the function $\rho(\lambda)$ in the half-neighborhood (4.4), satisfying all the requirements of the theorem.

The theorem is proved.

One can reformulate this theorem in a less precise, but a more picturesque way, as follows.

Given some real λ_0, suppose that at the point ρ_0 of the unit circle there are only multipliers of the first kind, a total number $p \geqq 1$. Then one can always find $\delta > 0$ such that, as λ starts from the value λ_0 and varies continuously in the interval $(\lambda_0 - \delta, \lambda_0 + \delta)$, exactly p multipliers emerge from the point ρ_0 and move clockwise or counterclockwise on the unit circle, depending on whether λ increases or decreases.

One can state a similar proposition in the case where, at the point ρ_0 $(|\rho_0| = 1)$, one finds only multipliers of the second kind for λ_0 real. The difference is that this time the multipliers of second kind will move counterclockwise (clockwise) as the real parameter λ increases (decreases).

At the same time, we reach the following conclusion.

Suppose that for some real λ_0 several multipliers meet at the point ρ_0 $(|\rho_0| = 1)$, and, as λ starts from λ_0 and varies continuously in one of the intervals $(\lambda_0, \lambda_0 + \delta)$ or $(\lambda_0 - \delta, \lambda_0)$, some of the multipliers emerging from the point ρ_0 leave the unit circle. Then among those multipliers that meet at ρ_0 there are multipliers of distinct kinds.

4. Now we show that if there are to be found only

multipliers of one kind at the point ρ_0 $(|\rho_0| = 1)$, then ρ_0 is a multiplier of simple type, i.e., ρ_0 is an eigenvalue of simple type of the matrix $U(T;\lambda_0)$.

For the sake of definiteness, assume that, when $\lambda = \lambda_0$, there are p multipliers of the first kind at the point ρ_0 (p is the multiplicity of ρ_0). Consider the half-neighborhood (4.4) and the functions $\rho_j(\lambda)$ $(j = 1,\dots,p)$ defined in it, whose existence is guaranteed by Theorem 4.2. Form the polynomial in ρ

$$P(\rho;\lambda) = (\rho - \rho_1(\lambda)) \dots (\rho - \rho_p(\lambda)) =$$

$$= \rho^p + B_1(\lambda)\ {}^{p-1} + \dots + B_p(\lambda) .$$

By analytically continuing the functions $\rho_j(\lambda)$ $(j = 1, \dots,p)$ from the upper half of the neighborhood $|\lambda - \lambda_0| < \delta$ into its lower half, the functions $\rho_j(\lambda)$ $(j = 1,\dots,p)$ are transformed into each other and so $B_j(\lambda)$ $(j = 1,\dots,p)$ are holomorphic functions in the neighborhood $|\lambda - \lambda_0| < \delta$. The polynomial $\Delta(\rho;\lambda)$ is divisible by $P(\rho;\lambda)$ for all λ in this neighborhood, whence

$$\Delta(\rho;\lambda) = P(\rho;\lambda)Q(\rho;\lambda) ,$$

where

$$Q(\rho;\lambda) = \rho^{n-p} + C_1(\lambda)\rho^{n-p-1} + \dots + C_{n-p}(\lambda) ,$$

and $C_k(\lambda)$ $(k = 1,\dots,n-p)$ are some functions, holomorphic in the same neighborhood $|\lambda - \lambda_0| < \delta$.

Since $P(\rho;\lambda_0) = (\rho - \rho_0)^p$, $Q(\rho;\lambda_0)$ does not vanish at $\rho = \rho_0$, i.e.,

$$Q(\rho_0;\lambda_0) \neq 0 . \qquad (4.8)$$

Consider the function of λ

$$\Delta(\rho_0;\lambda) = P(\rho_0;\lambda)Q(\rho_0;\lambda) ,$$

holomorphic for $|\lambda - \lambda_0| < \delta$. The number λ_0 is a zero of $\Delta(\rho_0;\lambda)$. By virtue of (4.8), its multiplicity equals the multiplicity of λ_0 as a zero of the function

$$P(\rho_0;\lambda) = \prod_{j=1}^{p} (\rho_0 - \rho_j(\lambda)) .$$

On the other hand,

$$\lim_{\lambda\to\lambda_0} \frac{\rho_j(\lambda) - \rho_0}{\lambda - \lambda_0} = \rho_j'(\lambda_0) \neq 0 \quad (j = 1,\ldots,p) ,$$

as Theorem 4.2 shows. Therefore, λ_0 is a p-multiple zero of the entire function $\Delta(\rho_0;\lambda)$.

The equation

$$\Delta(\rho_0;\lambda) = \det(U(T;\lambda) - \rho_0 I_n) = 0$$

is the characteristic equation for the boundary value problem

$$\frac{dx}{dt} = \lambda JH(t)x, \quad x(T) = \rho_0 x(0) , \tag{4.9}$$

and by Theorem 3.4, the multiplicity of its root λ_0 as a characteristic number of problem (4.9) is precisely the defect of the matrix $U(T;\lambda_0) - \rho_0 I_n$.

Our assertion is proved. It is only part of the following important proposition.

THEOREM 4.3. *In order that at the point* ρ_0 $(|\rho_0| = 1)$ *there will be only multipliers of the first kind (only multipliers of the second kind) for* $\lambda = \lambda_0$, *it is necessary and sufficient that* ρ_0 *be an eigenvalue of the first (respectively, second) kind of the monodromy matrix* $U(T;\lambda_0)$, *i.e., that all the eigenvectors of* $U(T;\lambda_0)$ *corresponding to* ρ_0 *be plus- (respectively, minus-) vectors.*

[This will imply that ρ_0 is an eigenvalue of simple type of the matrix $U(T;\lambda_0)$ (according to proposition 4°, § 1).]

PROOF. Split E_n into the direct sum of the root spaces of the matrix $U(T;\lambda_0)$

$$E_n = L_{\rho^{(1)}} \dotplus \ldots \dotplus L_{\rho^{(k)}} ,$$

and let

$$I_n = P_1 + \ldots + P_k$$

be the corresponding splitting of the identity matrix into a direct sum of projection matrices (see § 1 , no. 2), where one takes $\rho^{(1)} = \rho_0$.

Consider a circle γ centered at the point ρ_0 with small enough radius r so that none of the multipliers $\rho^{(j)}$ $(j = 2,3,\ldots,k)$ falls inside γ. Then

$$P_1 = \frac{1}{2\pi i}\oint_\gamma (\rho I_n - U(T;\lambda_0))^{-1}d\rho \ .$$

If p is the multiplicity of the multiplier ρ_0, then for small enough $h > 0$ one finds inside γ precisely p eigenvalues (counted with their multiplicities) of the J-expansive matrix $U(T;\lambda_0 + ih)$. Denote by P_h the sum of the projections corresponding to those root spaces of the matrix $U(T;\lambda_0 + ih)$ that arise from eigenvalues lying inside γ. Then

$$P_h = \frac{1}{2\pi i}\oint_\gamma (\rho I_n - U(T;\lambda_0 + ih))^{-1}d\rho \ .$$

Obviously,

$$P_1 = \lim_{h\to 0} P_h \ . \tag{4.10}$$

Now let $\xi \neq 0$ be some eigenvector of the matrix $U(T;\lambda_0)$ corresponding to the eigenvalue ρ_0: $U(T;\lambda_0)\xi = \rho_0\xi$. Then $P_1\xi = \xi$, and so

$$\xi = \lim_{h\to 0} P_h\xi \ . \tag{4.11}$$

Suppose that at ρ_0 there are to be found, for example, only multipliers of the first kind. Then, as we proved above, L_{ρ_0} coincides with the eigenspace of the matrix $U(T;\lambda_0)$ corresponding to the eigenvalue ρ_0. Moreover, recalling the definition of a multiplier of the first kind, one can assert that all the eigenvalues of the matrix $U(T;\lambda_0 + ih)$ lying inside γ have modulus larger than one for small enough h (when their number is exactly p).

According to Theorem 1.1, the direct sum L_h of the root spaces of the matrix $U(T;\lambda_0 + ih)$ corresponding to these eigenvalues consists exclusively of plus-vectors.

Therefore, if $\eta_h = P_h\xi \neq 0$, then

$$i(J\eta_h,\eta_h) > 0 \ .$$

Passing to the limit $h \longrightarrow 0$, we obtain

$$(J\xi,\xi) \geqq 0 \quad \text{for all} \quad \xi \in L_{\rho_0} \ . \tag{4.12}$$

Let us show that equality is never attained for $\xi \neq 0$. To see this, notice that if $(J\xi^{(0)}, \xi^{(0)}) = 0$ for some $\xi^{(0)} \in L_{\rho_0}$, then, by virtue of (4.12),

$$(J\xi^{(0)}, \eta) = 0 \quad \text{for all} \quad \eta \in L_{\rho_0} . \tag{4.13}$$

[To get (4.13), consider (4.12) for $\xi = \xi^{(0)} + \alpha\eta$, where $\eta \in L_{\rho_0}$ and α is a scalar.]

On the other hand, according to proposition 3°, § 1, $\xi^{(0)}$ is J-orthogonal to each of the subspaces $L_{\rho^{(k)}}$ $(k = 2,3, \ldots, p)$, i.e.,

$$(J\xi^{(0)}, \eta) = 0 \quad \text{for all} \quad \eta \in L_{\rho^{(k)}} \quad (k = 2,3,\ldots,p) .$$

Combined with (4.13), this shows that the vector $\xi^{(0)}$ is J-orthogonal - and the vector $J\xi^{(0)}$ is simply orthogonal - to any vector of E_n, whence $J\xi^{(0)} = 0$ and $\xi^{(0)} = 0$.

This completes the proof of the first part of the theorem. Now we prove the sufficiency part.

Let ρ_0 be a multiplier of simple type of the matrix $U(T;\lambda_0)$, and assume that all the eigenvectors corresponding to ρ_0 are, for example, plus-vectors. Then one can find $h > 0$ such that $(J\xi,\xi) > h$ for all $\xi \in L_{\rho_0}$ satisfying $(\xi,\xi) = 1$. By (4.10), one can find $h_0 > 0$ such that

$$\begin{gathered} i(JP_h\xi, P_h\xi) > h \quad \text{for all} \quad \xi \in L_{\rho_0}, \\ (\xi,\xi) = 1, \quad 0 < h < h_0. \end{gathered} \tag{4.14}$$

As ξ runs over E_n, $P_h\xi$ runs over the subspace $L_h = P_hE_n$, which is the direct sum of the root spaces of the matrix $U(T;\lambda_0 + ih)$ corresponding to the eigenvalues that lie inside γ. If h is small enough, then $\dim L_h = p = \dim L_{\rho_0}$.

On the other hand, if $h > 0$ is small enough, then $\det(I_n - P_1 + P_h)$ will be arbitrarily close to one, and so the matrix $C_h = I_n - P_1 + P_h$ will be nonsingular. Therefore, if ξ runs over L_{ρ_0}, then

$$C_h\xi = \xi - P_1\xi + P_h\xi = P_h\xi \in L_h$$

runs over some p-dimensional subspace which, as such, is identical to L_h. According to (4.14), for small enough $h > 0$ the subspace L_h consist only of plus-vectors and the zero vector. Consequently, the same property is enjoyed by the root spaces of the matrix $U(T;\lambda_0 + ih)$ which are contained in L_h. Since $U(T;\lambda_0 + ih)$ is J-expansive, Theorem 1.1 shows that these root spaces correspond to eigenvalues having modulus larger than one.

In other words, we proved that for small enough $h > 0$ all the eigenvalues of the matrix $U(T;\lambda_0 + ih)$ that lie inside γ have modulus larger than one. But this means that for $\lambda = \lambda_0$ there are only multipliers of the first kind at the point ρ_0.

The theorem is proved.

§ 5. ZONES OF STABILITY OF A CANONICAL SYSTEM WITH A PARAMETER

1. Consider the canonical differential equation

$$\frac{dx}{dt} = JH(t)x \tag{5.1}$$

with a summable, Hermitian matrix-function $H(t) = H(t + T)$.

We shall say that equation (5.1) is of *stable type* if its monodromy matrix is of stable type (for the definition of this subclass of J-unitary matrices, see § 1, no. 6).

Since all the eigenvalues of a J-unitary matrix of stable type are of simple type and have absolute value one, all the solutions of a differential equation (5.1) of stable type are bounded on the entire axis.

THEOREM 5.1. *If equation* (5.1) *is of stable type, then there is a* $\delta > 0$ *associated to it such that any other differential equation*

$$\frac{dx}{dt} = JH_1(t)x \tag{5.2}$$

with a summable, Hermitian, matrix-function $H_1(t) = H_1(t+T)$ *will be of stable type provided that*

$$\int_0^T |H(t) - H_1(t)|dt < \delta .$$

PROOF. By virtue of Theorem 1.2, it suffices to establish the following lemma in order to prove Theorem 5.1.

LEMMA 5.1. *Let* U_0 *be the monodromy matrix of the differential equation in* E_n

$$\frac{dx}{dt} = A(t)x \tag{5.1'}$$

having a summable, periodic, matrix-function $A(t) = A(t + T)$ *of order* n. *Then for any* $\varepsilon > 0$ *one can find* $\delta = \delta_\varepsilon > 0$ *such that given any summable, periodic, matrix-function* $B(t) = B(t + T)$ *satisfying*

$$\int_0^T |A(t) - B(t)|dt < \delta , \tag{5.3}$$

the monodromy matrix V_0 *of the equation*

$$\frac{dx}{dt} = B(t)x \tag{5.2'}$$

will belong to the ε*-neighborhood of the matrix* U_0:

$$|U_0 - V_0| < \varepsilon.$$

PROOF OF THE LEMMA. Denote by $U(t)$ and $V(t)$ the matrizants of the equations (5.1') and (5.2') respectively, so that

$$U_0 = U(T), \quad V_0 = V(T) .$$

We shall use the fact that given the system

$$\frac{dX}{dt} = A(t)X + F(t), \quad X(0) = I_n , \tag{5.4}$$

where the n-th order matrix-function $F(t)$ $(0 \leqq t < \infty)$ is summable on each finite interval $(0,\ell)$, and $X(t)$ is the unknown, n-th order, matrix-function, its solution is given by the formula

$$X(t) = U(t)[I_n + \int_0^t U^{-1}(s)F(s)ds]$$

(which can be obtained simply by employing the substitution $X(t) = U(t)Y(t)$ in (5.4)).

Since $V(t)$ satisfies (5.4) with

$$F(t) = (B(t) - A(t))V(t) ,$$

we have

$$V(t) = U(t) + U(t)\int_0^t U^{-1}(s)(B(s) - A(s))V(s)ds .$$

It follows that

$$|V(T) - U(T)| \leq |U(T)| \int_0^T |U^{-1}(s)||B(s) - A(s)||V(s)|ds \ .$$

The estimate (2.5) gives

$$|U(t)| \leq e^{a(t)}, \quad \text{with} \quad a(t) = \int_0^t |A(s)|ds \ ,$$

whence, applying (2.5) one again, it results from (5.3) that

$$|V(t)| \leq e^{a(t)+\delta t} \leq e^{a(T)+\delta T} \qquad (0 \leq t \leq T) \ .$$

Moreover, since the matrix-function $U_1(t) = [U^{-1}(t)]^{\tau}$ satisfies the system

$$\frac{dU_1}{dt} = -A(t)U_1 \ , \quad U_1(0) = I_n \ ,$$

we have also

$$|U^{-1}(t)| = |U_1(t)| \leq e^{a(t)} \leq e^{a(T)} \qquad (0 \leq t \leq T) \ .$$

Finally, (5.5) gives, provided that (5.3) holds,

$$|V(T) - U(T)| \leq e^{3a(T)+\delta T} \int_0^T |A(t) - B(t)|dt$$

$$< \delta e^{3a(T)+\delta T} \ .$$

This proves Lemma 5.1, and, together with it, Theorem 5.1.

2. A real number $\lambda = \lambda_0 \neq 0$ will be called a point of *strong stability* of the equation

$$\frac{dx}{dt} = \lambda JH(t)x \tag{0.1}$$

having a summable, Hermitian, matrix-function $H(t) = H(t + T)$ if for $\lambda = \lambda_0$ this equation is of stable type, i.e., if for $\lambda = \lambda_0$ the monodromy matrix $U(T;\lambda)$ is of stable type.

The point $\lambda = 0$ is said to be a point of strong stability of equation (0.1) if all the point $\lambda \neq 0$ belonging to one of its neighborhoods are points of strong stability.

The first part of the following theorem is a consequence of Theorem 5.1.

THEOREM 5.2. *The points* λ *of strong stability of equation* (0.1) *form an open set. This set is nonempty whenever equation* (0.1) *is of positive type.*

In fact, by stating now the second part of Theorem 5.2, we have jumped ahead: this part is a straightforward consequence of Theorem 6.1, according to which $\lambda = 0$ is a point of strong stability for any equation (0.1) of positive type.

By Theorem 5.2, the set of points of strong stability of equation (0.1) consists, as soon as it is not empty, of a finite or infinite number of open intervals.

We shall call these intervals the *zones of stability* of equation (0.1).

If (0.1) is of positive type, then among the zones of stability there is one containing the point $\lambda = 0$. We call it the *central zone of stability*.

We remark that, according to Theorem 4.3, when equation (0.1) is of positive type, a point $\lambda = \lambda_0 \neq 0$ is a point of strong stability if for $\lambda = \lambda_0$ all the multipliers of equation (0.1) have absolute value one and there are no equal multipliers of distinct kinds among them.

3. We shall point out below a fairly general kind of equation (0.1) of positive type for which one is able to give an upper bound for the length of a zone of stability and this bound does not depend upon the position of the zone to the right or to the left of the central zone.

To this end, we need

LEMMA 5.2. *Suppose that for a real* $\lambda = \lambda_0$ *one finds only multipliers of the first kind of equation* (0.1) *at the point* ρ_0 $(|\rho_0| = 1)$, *a number of* p. *Moreover, in a half-neighborhood*

$$|\lambda - \lambda_0| < \delta, \quad \operatorname{Im} \lambda > 0 \tag{5.6}$$

of the point λ_0, *assume that the* p *single-valued analytic branches* $\rho_j(\lambda)$ $(j = 1,\dots,p)$ *that were considered in Theorem* 4.2, *have been chosen. Then to each branch* $\rho_j(\lambda)$ $(j = 1,\dots,p)$ *one can associate a plus-vector* $x_0^{(j)}$ $(j = 1,\dots,p)$ *such that*

$$-\rho_j^{-1}(\lambda_0)\rho_j'(\lambda_0)(Jx_0^{(j)},x_0^{(j)}) = \int_0^T (H(t)x^{(j)},x^{(j)})\,dt, \tag{5.7}$$

where $x^{(j)}(t) = U(t;\lambda_0)x_0^{(j)}$ $(j = 1,\dots,p)$.

PROOF. If $0 < h < \delta$, then to the point $\lambda = \lambda_0 + ih$ in the half-neighborhood (5.6) one can associate a vector $x^{(j)}(\lambda)$ $(j = 1,\dots,p)$ such that

$$U(T;\lambda)x^{(j)}(\lambda) = \rho_j(\lambda)x^{(j)}(\lambda), \quad (x^{(j)}(\lambda),x^{(j)}(\lambda))=1. \tag{5.8}$$

Now the solution

$$x^{(j)}(t;\lambda) = U(t;\lambda)x^{(j)}(\lambda)$$

of equation (0.1) has the property that

$$x^{(j)}(T;\lambda) = \rho_j(\lambda)x^{(j)}(\lambda) .$$

Applying relation (3.2) to this solution, one obtains

$$(1 - |\rho_j(\lambda)|^2)(Jx^{(j)}(\lambda),x^{(j)}(\lambda)) = \\ = 2ih \int_0^T (H(t)x^{(j)}(t;\lambda),x^{(j)}(t;\lambda))dt \quad (\lambda = \lambda_0 + ih) . \tag{5.9}$$

Using the compactness of the unit sphere $(\xi,\xi) = 1$, pick a sequence $\lambda_k = \lambda_0 + ih_k$, $h_k \longrightarrow 0$, $h_k > 0$ $(k = 1,2,\dots)$ such that the limit

$$\lim_{k\to\infty} x^{(j)}(\lambda_k) = x_0^{(j)}$$

exists. Then, by (5.8),

$$U(T;\lambda_0)x_0^{(j)} = \rho_0 x_0^{(j)} \quad \text{and} \quad (x_0^{(j)},x_0^{(j)}) = 1 .$$

Now notice that

$$1 - |\rho_j(\lambda_0 + ih)|^2 = 1 - \frac{\rho_j(\lambda_0 + ih)}{\rho_j(\lambda_0 - ih)} ,$$

whence

$$\lim_{h\to 0} \frac{1 - |\rho_j(\lambda_0 + ih)|^2}{2ih} = - \rho_j^{-1}(\lambda_0)\rho_j'(\lambda_0) .$$

Therefore, upon setting $\lambda = \lambda_k$ and $h = h_k$ in both sides of equality (5.9), and subsequently dividing them by $2ih_k$ and passing to the limit $k \longrightarrow \infty$, we obtain the required equality (5.7).

The fact that $x_0^{(j)}$ $(j = 1,\dots,p)$ are plus-vectors is a consequence of Theorem 4.3, although it does also result from the relation (5.7) itself if one takes into account that

$$- i\rho_j^{-1}(\lambda_0)\rho_j'(\lambda_0) < 0 ,$$

by Theorem 4.2.

4. Let us remark that differentiation on the real axis gives

$$- i\rho_j^{-1}(\lambda)\rho_j'(\lambda) = \frac{d}{d\lambda} \arg \rho_j(\lambda) \quad (\lambda_0 - \delta < \lambda < \lambda_0 + \delta).$$

Since

$$\frac{d}{dt}(Jx,x) = (\bar{\lambda} - \lambda)(H(t)x,x)$$

for any solution $x = x(t;\lambda)$ of equation (0.1), we have $(Jx,x) = \text{const}$ for real $\lambda = \lambda_0$. Consequently,

$$|(Jx_0,x_0)| = |(Jx_\mu,x_\mu)| \leqq (x_\mu,x_\mu) \quad (x_0=x(0;\lambda_0)) , \tag{5.10}$$

where x_μ is that value of the vector-function $x(t) = x(t;\lambda_0)$ having the smallest norm:

$$(x_\mu,x_\mu) = \min_{0\leqq t\leqq T} (x(t),x(t)) .$$

On the other hand,

$$(H(t)x,x) \geqq h_\mu(t)(x,x) \geqq h_\mu(t)(x_\mu,x_\mu) , \tag{5.11}$$

where $h_\mu(t)$ is the smallest eigenvalue of the matrix $H(t)$.

Applying the estimates (5.10) and (5.11) to $x = x^{(j)}(t;\lambda)$, we obtain from (5.7)

$$- \frac{d}{d\lambda_0} \arg \rho_j(\lambda_0) \geqq \int_0^T h_\mu(t)dt . \tag{5.12}$$

This inequality allows us to prove the following proposition.

THEOREM 5.3. *If the set of all points* t *for which the matrix* $H(t)$ *is nonsingular has measure zero, then the length of any noncentral zone of stability of equation* (0.1) *is no larger than*

$$\pi\left(\int_0^T h_\mu(t)dt\right)^{-1} , \tag{5.13}$$

where $h_\mu(t)$ *is the smallest eigenvalue of* $H(t)$.

The same number (5.13) *is an upper bound for the length of each of the two parts into which the central zone of stability is divided by the point* $\lambda = 0$.

PROOF. Let (α,β) be either a noncentral zone of stability or one of the parts into which $\lambda = 0$ divides the central zone of stability. Then for each λ inside (α,β), all the multipliers of equation (0.1) lie on the unit circle, and among them are precisely m multipliers of the first kind and precisely m multipliers of the second kind (see § 1, no. 6, p. 16). Moreover, multipliers of distinct kinds cannot be equal. Given any $\lambda \in (\alpha,\beta)$, let us index the multipliers of the first kind in the order in which they appear on the unit circle as one moves clockwise, which ensures that their variation with λ is continuous. In this way we obtain m multiplier functions $\rho_1(\lambda), \ldots, \rho_m(\lambda)$. Each of these functions will satisfy (5.12) at any interior point $\lambda_0 \in (\alpha,\beta)$.

Therefore, as λ runs over the interval (α,β), each of the multipliers describes an arc of length at most $\chi(\alpha - \beta)$ in the clockwise direction, where χ denotes the right-hand integral in (5.2).

A similar discussion can be made for the multipliers of the second kind, the only difference being that now the motion is in the opposite direction. Since multipliers of distinct kinds can meet only when $\lambda = \alpha$ and $\lambda = \beta$, we conclude that $2\chi(\beta - \alpha) \leqq 2\pi$, i.e., $\beta - \alpha \leqq \pi/\chi$, as claimed in the theorem.

5. A number of the previous results can be generalized to the differential equation

$$\frac{dx}{dt} = JH(t;\mu)x \,, \tag{5.14}$$

where $H(t;\mu) = H(t + T;\mu)$ $(0 \leqq t < \infty,\ \alpha < \mu < \beta)$ is a matrix-function summable with respect to t over the interval $[0,T]$, and depending continuously in the mean on the parameter μ, i.e.,

$$\lim_{\mu' \to \mu} \int_0^T |H(t;\mu') - H(t;\mu)|\,dt = 0 \qquad (\alpha < \mu < \beta).$$

According to Lemma 5.1, the last condition ensures that the monodromy matrix $U(T;\mu)$ of equation (5.14) depends continuously upon μ.

Generalizing the definition given in no. 2 for equation (0.1), we shall say that $\mu = \mu_0 \in (\alpha,\beta)$ is a *point of strong stability* of equation (5.14) if the monodromy matrix $U(T;\mu)$ of this equation is of stable type for $\mu = \mu_0$.

In virtue of Theorem 1.2, we can assert, as we did earlier, that *the points* μ *of strong stability of equation* (5.14) *form an open set.* This set decomposes into open intervals (if it is not empty), which will be called the *zones of stability* of equation (5.14).

A multiplier $\rho(\mu)$ of equation (5.14) is said to be of the *first (second) kind* if it is an eigenvalue of the first (respectively, second) kind of the J-unitary matrix $U(T;\mu)$ (see § 1, no. 6).

We shall be interested in matrix-functions $H(t;\mu)$ satisfying

1) $H(t;\mu) \in P_n(T)$,

2) $H(t;\mu') \leqq H(t;\mu'')$ for $\mu' < \mu''$.

In this case we shall say that equation (5.14) is of *positive type.*

An equation (0.1) of positive type is obviously a particular case of equation (5.14) of such type.

THEOREM 5.3. *Let* (μ_1,μ_2) *be an arbitrary zone of stability of an equation* (5.14) *of positive type. Let us index the multipliers of the first kind* $\rho_1(\mu),\ldots,\rho_m(\mu)$ $(\mu_1 < \mu < \mu_2)$ *of this equation in such a way that, under an appropriate definition of their arguments, the latter will depend continuously upon* μ *and will satisfy*

$$\arg \rho_1(\mu) \geqq \ldots \geqq \arg \rho_m(\mu) \qquad (\mu_1 < \mu < \mu_2) .$$

Then $\arg \rho_j(\mu)$ $(j = 1,\ldots,m)$ *is a nonincreasing function of* μ *in the interval* (μ_1,μ_2).

A similar statement (where the word "nonincreasing" is replaced by "nondecreasing") holds for the multipliers of the second kind of equation (5.14).

PROOF. Let μ_0 be an arbitrary number in the zone of stability (μ_1,μ_2). Consider a multiplier $\rho_j(\mu)$. Set $\rho_j(\mu_0) = \mu_0$ and, in order to clarify the basic idea of the proof, first assume that ρ_0 is a simple multiplier, i.e., a simple eigenvalue of the matrix $U(T;\mu_0)$.

For a fixed μ, consider the boundary value problem

$$\frac{dx}{dt} = \lambda JH(t;\mu)x, \quad x(T) = \rho_0 x(0) . \tag{5.15}$$

Now let

$$\lambda_1(\mu) \leqq \lambda_2(\mu) \leqq \ldots$$

be the sequence of positive characteristic number for this problem.

According to Theorem 3.3,

$$\lambda_k(\mu') \geqq \lambda_k(\mu'') \quad \text{for} \quad \mu' < \mu'' \quad (k = 1,2,\ldots) . \tag{5.16}$$

For $\mu = \mu_0$, boundary value problem (5.15) has a simple characteristic number equal to 1. Suppose

$$\lambda_{q-1}(\mu_0) < \lambda_q(\mu_0) = 1 < \lambda_{q+1}(\mu_0).$$

Now let $\mu_1 > \mu_0$ be close enough to μ_0. We show that in this case

$$\arg \rho_j(\mu_1) \leqq \arg \rho_j(\mu_0) = \arg \rho_0. \tag{5.17}$$

In virtue of (5.16), $\lambda_q(\mu_1) \leqq \lambda_q(\mu_0) = 1$. If $\lambda_q(\mu_1) = 1$, then, in general,

$$\lambda_q(\mu) = 1 \quad \text{for} \quad \mu_0 \leqq \mu \leqq \mu_1 ,$$

whence

$$\rho_j(\mu) = \rho_0 \quad \text{for} \quad \mu_0 \leqq \mu \leqq \mu_1 .$$

If, however, $\lambda_q(\mu_1) < 1$, then consider the equation

$$\frac{dx}{dt} = \lambda JH(t;\mu_1)x . \tag{5.18}$$

When $\lambda = \lambda_q(\mu_1)$, ρ_0 is a multiplier of the first kind for (5.18). As Theorem 4.2 shows, when λ varies continuously from $\lambda_q(\mu_1)$ to 1, this multiplier moves clockwise on the unit circle from the point ρ_0 and reaches the point $\rho_j(\mu_1)$. Therefore, in this case (5.17) holds with the sign $<$.

Now let ρ_0 be a multiplier of multiplicity ν. As ρ_0 is of the first kind, and, in any case, of simple type, the boundary value problem (5.15) has $\lambda = 1$ as characteristic number of multiplicity exactly ν for $\mu = \mu_0$.

Let

$$\lambda_{q-1}(\mu_0) < \lambda_q(\mu_0) = \lambda_{q+1}(\mu_0) = \ldots = \lambda_{q+\nu-1}(\mu_0) = 1 < \lambda_{q+\nu}(\mu_0) .$$

Then we shall have, when $\mu_1 > \mu_0$ and μ_1 is close enough to μ_0, that

$$\lambda_{q-1}(\mu_0) < \lambda_q(\mu_1) \leqq \lambda_{q+1}(\mu_1) \leqq \ldots \leqq \lambda_{q+\nu-1}(\mu_1) \leqq 1 < \lambda_{q+\nu}(\mu_0) .$$

Considering equation (5.18) again, we now discover, on the basis of Theorem 4.3, that as λ varies continuously from $\lambda_q(\mu_1) - \varepsilon$ to 1 (where $\varepsilon > 0$ is small enough), exactly ν multipliers pass, moving clockwise, through the point ρ_0 (provided that $\lambda_q(\mu_1) < 1$).

Once again, this implies that (5.17) holds for those values of j for which $\rho_j(\mu_0) = \rho_0$.

Since the choice of $\mu_0 \in (\mu_1, \mu_2)$ was arbitrary, this completes the proof of the theorem.

REMARK. If one of the endpoints μ_1 or μ_2 of the zone of stability lies in the interior of the interval (α, β), then, as μ tends to this endpoint from the interior of the zone, the members of at least one pair of multipliers of distinct kinds will tend to coincide. Indeed, assuming the contrary, Theorem 1.3 ensures that at this endpoint the monodromy matrix would be of stable type, i.e., this endpoint would be a point of strong stability, which is impossible.

§ 6. THE CENTRAL ZONE OF STABILITY OF A CANONICAL SYSTEM (0.1) OF POSITIVE TYPE

1. Assuming that the matrix-function $H(t)$ is of class $P_n(T)$, consider the boundary value problem

$$\frac{dx}{dt} = \lambda JH(t)x, \quad x(0) + x(T) = 0 . \tag{6.1}$$

By Theorem 3.2, all the characteristic numbers of this problem are real.

Denote by Λ_+ and Λ_- the smallest positive and, respectively, the largest negative characteristic numbers of problem (6.1). One has

THEOREM 6.1. *The open interval* (Λ_-, Λ_+) *is contained in the central zone of stability of equation* (0.1). *If the matrix-function* $H(t)$ *is real, then this interval is precisely the central zone of stability of equation* (0.1).

PROOF. If one performs the change $t = -\tau$, $\lambda = -\mu$ in (6.1), then the equation preserves the same type. Therefore, it suffices to consider only nonnegative values of λ.

First assume that all the eigenvalues of the matrix

$$S_1 = J\int_0^T H(t)dt = JH_I$$

are distinct. Then one can write them in the form $i\sigma_{\pm j}$ $(j = 1, \dots, m)$, where

$$\sigma_{-m} < \dots < \sigma_{-1} < 0 < \sigma_1 < \dots < \sigma_m .$$

When $H(t)$ is real, we shall have also

$$\sigma_{-j} = -\sigma_j \quad (j = 1,\dots,m).$$

The arguments given in the proof of Theorem 2.1 show that here, when $\lambda > 0$ is small enough, all the multipliers lie on the unit circle, while when complex λ has small enough absolute value these multipliers $\rho_j(\lambda)$, suitably reindexed with indices $j = \pm 1,\dots,\pm m$, will admit the representation

$$\rho_j(\lambda) = e^{i\sigma_j\lambda + o(\lambda)} .$$

Putting here $\lambda = \delta e^{i\phi}$, where $\delta > 0$ is small enough, and letting ϕ vary continuously from 0 to $\pi/2$, we can see ourselves that for small enough $\lambda > 0$ the multipliers $\rho_j(\lambda)$ $(j = 1,\dots,m)$ are of the second kind and lie on the open upper semicircle, while the multipliers $\rho_j(\lambda)$ $(j = -1,\dots,-m)$ are of the first kind and lie on the open lower semicircle. Moreover, when $H(t)$ is real, we have

$$\rho_{-j}(\lambda) = \overline{\rho_j}(\lambda) \quad (j = 1,\dots,m) .$$

Now suppose that λ increases continuously, starting from a sufficiently small positive value. Then, by Theorem 4.2, the multipliers $\rho_j(\lambda)$ $(j = 1,\dots,m)$ will move counterclockwise, while the multipliers $\rho_j(\lambda)$ $(j = -1,\dots,-m)$ will move clockwise. In doing so, some of the multipliers may overtake some of the others, but they will not jump off the unit circle as long as two multipliers of distinct kinds do not meet. Such an encounter could take place for the first time when either both multipliers $\rho_m(\lambda)$ and $\rho_{-m}(\lambda)$ simultaneously reach the point -1, or after one of these multipliers passes through the point -1. Since any value of λ for which one of the multipliers becomes equal to -1 is a characteristic number of the boundary value problem (6.1), an encounter of multipliers of distinct kinds cannot occur for $0 < \lambda < \Lambda_+$. One should add that when $H(t)$ is a real matrix-function, then for $\lambda = \Lambda_+$ two multipliers of distinct kinds must indeed meet at the point -1, namely the multipliers $\rho_m(\lambda)$ and $\rho_{-m}(\lambda) = \overline{\rho_m}(\lambda)$. This completes the proof of the theorem for the case under consideration, when all the eigenvalues of S_1 are distinct.

We emphasize that we have proved at the same time that, in this case, when $0 < \lambda < \Lambda_+$, precisely m multipliers of the second kind of equation (0.1) lie on the open upper semicircle $\rho = e^{i\phi}$, $(0 < \phi < \pi)$, and precisely m multipliers of the first kind lie on the open lower semicircle $\rho = e^{i\phi}$, $(-\pi < \phi < 0)$. This implies that the first characteristic number of the boundary value problem

$$\frac{dx}{dt} = \lambda JH(t)x, \quad x(0) - x(T) = 0 \tag{6.2}$$

is larger than the first positive characteristic number of the boundary value problem (6.1).

Now suppose that the matrix S_1 has multiple eigenvalues. In this case we first choose a Hermitian matrix $D > 0$ such that for all $\varepsilon > 0$ the eigenvalues of the matrix

$$S_1(\varepsilon) = J(H_I + \varepsilon D) ,$$

or, what amounts to the same, the roots σ of the equation

$$\det(H_I + \varepsilon D + i\sigma J) = 0$$

are all distinct.

[This can be done as follows. Since $H_I > 0$ and J is skew-symmetric, one can find a system of vectors $v_k \in E_{2m}$ $(k = \pm 1, \ldots, \pm m)$, such that [24]

$$H_I v_k = -i\sigma_k J v_k \quad (k = \pm 1, \ldots, \pm m) ,$$

$$\sigma_{-m} \leqq \dots \leqq \sigma_{-1} < 0 < \sigma_1 \leqq \dots \leqq \sigma_m ,$$

and

$$(H_I v_j, v_k) = \delta_{jk} \quad (j,k = \pm 1,\dots,\pm m) .$$

Let d_k $(k = \pm 1,\dots,\pm m)$ be arbitrary numbers satisfying the single condition

$$d_{-m} < \dots < d_{-1} < 0 < d_1 < \dots < d_m .$$

Then the required matrix D can be defined by the following relations

$$Dv_k = -id_k Jv_k = \frac{d_k}{\sigma_k} H_I v_k \quad (k = \pm 1,\dots,\pm m) .$$

Indeed, given any vector $\xi = \sum c_k v_k$, one has, for D so defined,

$$(D\xi,\xi) = \sum |c_k|^2 \frac{d_k}{\sigma_k} ,$$

whence $D > 0$. Moreover, since

$$(H_I + \varepsilon D)v_k = -i(\sigma_k + \varepsilon d_k)Jv_k \quad (k = \pm 1,\dots,\pm m) ,$$

the eigenvalues of the matrix $J(H_I + \varepsilon D)$ are precisely the distinct numbers

$$i(\sigma_k + \varepsilon d_k) \quad (k = \pm 1,\dots,\pm m) .]$$

Now form the differential equation

$$\frac{dx}{dt} = \lambda J H_\varepsilon(t)x \quad (H_\varepsilon(t) = H(t) + \frac{\varepsilon}{T} D). \tag{6.3}$$

Since

$$J \int_0^T H_\varepsilon(t)dt = J(H_I + \varepsilon D) ,$$

the conclusions obtained earlier are valid for system (6.3).

Therefore, if $\Lambda_+(\varepsilon)$ is the first positive value of λ for which equation (6.3) has a nontrivial antiperiodic solution $x_c(t)$ $(x_c(T) = - x_c(0))$, then when

$$0 < \lambda < \Lambda_+(\varepsilon) \tag{6.4}$$

the monodromy matrix $U_\varepsilon(T;\lambda)$ of equation (6.3) will be of stable type, and there are m multipliers (eigenvalues of $U_\varepsilon(T;\lambda)$) of the second kind lying on the open upper semicircle, while m multipliers of the first kind lie on the open lower semicircle. Since

$$\Lambda_+ = \lim_{\varepsilon\to 0} \Lambda_+(\varepsilon) ,$$

given the inequality $0 < \lambda < \Lambda_+$ and $\varepsilon > 0$ small enough, inequality (6.4) will be also valid; hence the matrix $U_\varepsilon(T;\lambda)$ will be of stable type. Then, according to Theorem 2.3, the matrix

$$U(T;\lambda) = \lim_{\varepsilon\to 0} U_\varepsilon(T;\lambda)$$

will be of stable type too, and it will have m multipliers of the second (first) kind lying on the open upper (lower) semicircle. Notice that Theorem 1.3 can be applied provided that all the multipliers of $U(T;\lambda)$ differ from -1 and 1. That they differ from -1 for $0 < \lambda < \Lambda_+$ is obvious. Assume that for some $\lambda = \mu_0$ $(0 < \mu_0 < \Lambda_+)$ one of these multipliers equals 1. This means that μ_0 is a characteristic number of the boundary value problem (6.1). But then, replacing $H(t)$ by $H_\varepsilon(t)$ in (6.2), we would obtain that for $\varepsilon > 0$ small enough, the first positive characteristic number of the system

$$\frac{dx}{dt} = \lambda H_\varepsilon(t)x\ , \qquad x(0) - x(T) = 0$$

is smaller than $\Lambda_+(\varepsilon)$, which is impossible.

To complete the proof of the theorem, it remains to say again that for a real matrix-function $H(t)$ the multipliers of the first and second kind are symmetrically disposed with respect to the real axis and, as λ increases to Λ_+, two conjugate (and so, of distinct kinds) multipliers meet at the point -1.

At the same time, we see that the claim concerning the first positive characteristic numbers of the boundary value problems (6.1) and (6.2) is valid in the most general case.

2. In Theorem 6.1 we have assumed tacitly that the numbers Λ_+ and Λ_- exist. It turns out that this assumption is always correct when $H(t) \in P_n(T)$.

THEOREM 6.2. *The boundary value problem* (6.1) *has at least one positive and one negative characteristic number.*

In order to prove this theorem, we make use of some of the delicate tools from the theory of entire functions.

PROOF. The assertion of the theorem is that the equation

$$\det(U(T;\lambda) + I_n) = 0 \qquad (n = 2m) \tag{6.5}$$

has at least one positive and one negative root.

In virtue of Theorem 3.2, one is able to say that the roots λ_j of equation (6.5) are real, whatever their number.

Moreover, by Theorem 3.5, one has the expansion

$$2(U(T;\lambda) + I_n)^{-1} = I_n + J\{-A_1\lambda + \lambda \sum_j \frac{B_j}{\lambda_j(\lambda - \lambda_j)}\} , \qquad (6.6)$$

where

$$A_1 \geqq 0, \quad B_j \geqq 0 \quad (j = \pm 1, \pm 2, \ldots) .$$

Assuming that equation (6.5) has no roots at all, (6.6) implies that

$$(U(T;\lambda) + I_n)^{-1} = \frac{1}{2} (I_n - JA_1\lambda) ,$$

whence

$$U(T;\lambda) = -I_n + 2(I_n - JA_1\lambda)^{-1} .$$

It follows that for sufficiently small λ

$$U(T;\lambda) = I_n + 2JA_1\lambda + 2(JA_1)^2\lambda^2 + \ldots .$$

On the other hand, recalling (2.14) we have

$$U(T;\lambda) = I_n + JH_I\lambda + \ldots$$

and so

$$A_1 = \frac{1}{2} H_I , \quad JA_1 = \frac{1}{2} S_1 .$$

Now let $v \neq 0$ be one of the eigenvectors of S_1; then $S_1 v = i\sigma v$ $(\sigma \neq 0)$, whence

$$(U(T;\lambda) + I_n)v = \frac{4}{2 - i\sigma\lambda} v ,$$

which contradicts the fact that $U(T;\lambda)$ is an entire matrix-function.

The existence of at least one characteristic number for the boundary value problem (6.1) is thus proved.

We now show that problem (6.1) actually has characteristic numbers of both signs.

Assume that this is not so, and suppose that problem (6.1) has no positive characteristic numbers. Then the arguments used in the proof of Theorem 6.1 will imply that, for any $\lambda > 0$ all the multipliers $\rho_j(\lambda)$ of equation (0.1) lie on the unit circle. Since

$$\Delta(\rho;\lambda) = \prod_j (\rho - \rho_j(\lambda)) = \det(U(T;\lambda) - \rho I_n) ,$$

we shall have

$$|\Delta(-1;\lambda)| \leqq 2^{2m} \quad \text{for} \quad 0 < \lambda < \infty \,. \tag{6.7}$$

On the other hand, according to (6.6)

$$2(U(T;-\lambda^2) + I_n)^{-1} = I_n + JA_1\lambda^2 + \frac{\lambda^2}{2} \sum_{k\in K} \frac{1}{\zeta_k^3} \left(\frac{1}{\lambda+\zeta_k} - \frac{1}{\lambda-\zeta_k}\right) JB_k ,$$

where

$$\zeta_k = \sqrt{-\lambda_k} > 0 \qquad (k \in K)$$

and K is either a finite or an infinite set of indices.

In paper [11] we studied entire matrix-functions $W(\lambda) = \|w_{jk}(\lambda)\|_1^n$ satisfying the condition

$$W^{-1}(\lambda) = A_0 + A_1\lambda + \dots + A_p\lambda^p + \lambda^p \sum_{k\in K} \frac{C_k}{\lambda - \zeta_k} ,$$

where the numbers ζ_k are real, the matrices $A_0, A_1, \dots, A_p$, and C_k $(k \in K)$ have order n, and

$$\sum_{k \in K} \frac{|C_k|}{|\zeta_k|} < \infty \,.$$

In particular, we showed that the determinant of such an entire matrix-function is at most of exponential type, i.e.,

$$|\det W(\lambda)| \leqq \alpha e^{\beta|\lambda|},$$

where α, β are some positive constants.

Applying this result to the matrix-function

$$W(\lambda) = U(T;-\lambda^2) + I_n ,$$

we find that

$$|\Delta(-1;\lambda)| = |\det(U(T;\lambda) + I_n)| \leqq \alpha e^{\beta\sqrt{|\lambda|}} \,.$$

According to well-known results concerning entire functions of order less than one, we can assert that

$$\Delta(-1;\lambda) = 2^{2m} \prod_{k\in K} \left(1 - \frac{\lambda}{\lambda_k}\right) .$$

This leads to a contradiction because all the numbers λ_k $(k \in K)$ are negative and so the right-hand side tends to $+\infty$ as $\lambda \longrightarrow +\infty$, which is impossible by (6.7).

The theorem is proved.

3. It seems rather surprising that we had to use such strong tools from function theory in order to prove the last theorem. It might be possible to prove this theorem in a more simple way by using methods from the theory of integral equations, and, in general, from operator theory.

By these methods one does indeed reach the goal quickly when there are certain special assumptions on the Hermitian matrix-function $H(t)$. Let us explain briefly what we mean by this, especially because this allows us to complete in a certain respect Theorem 6.1.

Set

$$g(t) = \begin{cases} \frac{1}{2} & (0 \leqq t \leqq T) \\ \frac{1}{2} & (-T \leqq t < 0) . \end{cases}$$

This function has the following property: if

$$\frac{d\phi}{dt} = \psi , \quad \phi(0) + \phi(T) = 0 \qquad (0 \leqq t \leqq T) ,$$

then

$$\phi(t) = \int_0^T g(t - s)\psi(s)ds \qquad (0 \leqq t \leqq T) ,$$

and conversely, the last relation implies the first two.

It follows that system (6.1) is equivalent to the following integral equation

$$x(t) = \lambda\int_0^T g(t - s)JH(s)x(s)ds ,$$

which we write also as

$$x(t) = \lambda\int_0^T G(t - s)H(s)x(s)ds , \tag{6.8}$$

setting $G(t) = g(t)J$.

Due to the fact that $G(-t) = G^*(t)$, $G(t - s)$ is a Hermitian (symmetric) matrix kernel, and one can look upon (6.8) as a weighted integral equation, to which one can apply the usual results from the theory of positive, scalar, weighted, integral equations with Hermitian kernels, suitably generalized (see [14] and [6]). However, everything becomes much more transparent if one uses the general theory of compact operators.

Denote by L the set of all continuous vector-functions $x(t)$ $(0 \leqq t \leqq T)$ taking values in E_{2m}. For $x, y \in L$, set

$$\{x,y\} = \int_0^T (H(t)x,y)dt \ .$$

A function $x \in L$ will be called *degenerate* if $\{x,x\} = 0$ or, equivalently, if $H(t)x(t) = 0$ (almost everywhere).

Identifying every two functions $x_1, x_2 \in L$ whose difference is degenerate, we transform L into a pre-Hilbert space $\hat{L}$ with the scalar product $\{x,y\}$.

To each $x \in L$ one can associate the new function

$$y(t) = Ax(t) = \int_0^T G(t-s)H(s)x(s)ds \ . \tag{6.9}$$

Since the integral (6.9) vanishes on a degenerate function, one can look upon $x \longrightarrow Ax$ as an operator in $\hat{L}$. As such, A is Hermitian and compact. Moreover, A is also compact in the uniform norm

$$||x(\cdot)|| = \max_{0 \leqq t \leqq T} \sqrt{(x(t),x(t))},$$

relative to which $\hat{L}$ is complete.

Since, in addition,

$$\{x,x\} \leqq \int_0^T h_M(t)(x(t),x(t))dt \leqq ||x||^2 \int_0^T \operatorname{Sp} H(t)dt \ ,$$

where $h_M(t)$ is the largest eigenvalue of the matrix $H(t)$, one can apply here the theory of compact operators in a space with two norms (see [15]), which enables us, in particular, to avoid the completion of $\hat{L}$ to a Hilbert space.

According to this theory, equation (6.8) will have as many characteristic numbers as the dimension of $A\hat{L}$. Here each characteristic number λ must be counted a number of times equal to the dimension of the set of eigenvectors (solutions $x(t)$ of equation (6.8)) in $\hat{L}$ corresponding to this number λ.

Everything becomes simpler if there are no intervals (a,b) on which the matrix $H(t)$ is degenerate almost everywhere. In this case

$$\{x,x\} = \int_0^T (H(t)x(t),x(t))dt > 0$$

for any $x(t) \not\equiv 0$, and $\hat{L} = L$. Then the operator A is not degenerate, i.e., $\{x,x\} > 0$ implies $\{Ax,Ax\} > 0$. Indeed, $x(t) \not\equiv 0$ implies $y(t) = Ax(t) \not\equiv 0$, because

$$\frac{dy}{dt} = JH(t)x, \quad -\int_0^T (J\frac{dy}{dt},x)\,dt = \{x,x\} > 0 \ .$$

It follows that for equation (6.8), and hence for system (6.1) too, there exist a complete orthonormal system of eigenvectors $x^{(k)}(t)$ $(k \in K)$:

$$\frac{dx^{(k)}}{dt} = \lambda_k JH(t)x^{(k)}, \quad \{x^{(k)},x^{(\ell)}\} = \delta_{k\ell} \quad (k,\ell \in K) \ .$$

Thus, there exists an infinity of distinct characteristic numbers. On the basis of some general considerations, one may be persuaded that among these numbers there is an infinite number of both positive and negative ones.

If one brings into play more subtle analytic tools, then one can prove the existence of an infinity of characteristic numbers of both signs under significantly fewer requirements on $H(t)$.

Let $i\sigma_j(t)$ $(j = \pm 1,\ldots,\pm m)$ be all the eigenvalues of the matrix $JH(t)$, counted with their multiplicities:

$$\sigma_{-m}(t) \leqq \ldots \leqq \sigma_{-1}(t) \leqq 0 \leqq \sigma_1(t) \leqq \ldots \leqq \sigma_m(t) \ .$$

Denote by $n_+(r)$ and $n_-(r)$ $(r > 0)$ the numbers of characteristic numbers of the boundary value problem (6.1) that lie inside the intervals $(0,r)$ and $(-r,0)$, respectively. Then, as it turns out, the following relations hold:

$$\lim_{r\to\infty} \frac{n_+(r)}{r} = \lim_{r\to\infty} \frac{n_-(r)}{r} = \frac{1}{2\pi}\int_0^T \sum_{j=-m}^{m} |\sigma_j(t)|\,dt \ .$$

However, even this result, whose proof requires in the general case rather difficult tools, does not offer the possibility of obtaining Theorem 6.1 for arbitrary matrix-functions $H(t) \in$ $\in P_n(T)$. Nevertheless, several useful consequences can be extracted from it. In particular, it shows that if the boundary value problem (6.1) has a finite number of characteristic numbers of a given sign, or if there is an infinity of such numbers which form however a sparse sequence, then all $\sigma_j(t) \equiv 0$ almost everywhere, and so the matrix $JH(t)$ is nilpotent almost everywhere (i.e., its n-th power is equal to zero). As one can show, in this case the rank of the matrix $H(t)$ is not larger than m $(n = 2m)$.

4. To emphasize the dependence of the numbers $\Lambda_\pm$ upon $H(t)$, we shall write

$$\Lambda_\pm = \Lambda_\pm(H) \ .$$

Now let $H_1(t) = H_1(t+T)$ and $H_2(t) = H_2(t+T)$ $(0 \leqq t < \infty)$ be two Hermitian matrix functions of class $P_n(T)$.

In virtue of Theorem 3.3,

$$\Lambda_+(H_1) \geqq \Lambda_+(H_2), \quad \Lambda_-(H_1) \leqq \Lambda_-(H_2) \tag{6.10}$$

if $H_1(t) \leqq H_2(t)$ $(0 \leqq t \leqq T)$, whence one obtains the following proposition very easily.

THEOREM 6.3. *The following estimates are valid for the numbers* $\Lambda_\pm$:

$$\pi\left(\int_0^T h_M(t)dt\right)^{-1} \leqq |\Lambda_\pm| \leqq \pi\left(\int_0^T h_\mu(t)dt\right)^{-1} , \tag{6.11}$$

where $h_\mu(t)$ *and* $h_M(t)$ *are respectively the smallest and the largest eigenvalues of the matrix* $H(t)$.

PROOF. Indeed, if $H_0(t)$ has the form

$$H_0(t) = h(t)I_n ,$$

then, as one can easily see,

$$\Lambda_+(H_0) = -\Lambda_-(H_0) = \pi\left(\int_0^T h(t)dt\right)^{-1} .$$

On the other hand,

$$h_\mu(t)I_n \leqq H(t) \leqq h_M(t)I_n$$

and so the estimates (6.11) are corollaries of the general rule (6.10).

We remark that the upper bound in (6.11) is also a simple consequence of formula (5.12).

As for the lower bound for $|\Lambda_\pm|$ given in (6.11), it can be replaced, as we show in the next section, by other bounds which are both more simple to calculate and more precise, for many important cases.

5. We conclude this section by a theorem in which one drops the first condition (see § 3, no. 1) ensuring that $H(t)$ belongs to the class $P_n(T)$ (i.e., one admits that the form $(H(t)\xi,\xi)$ may be nonpositive or even nondefinite on a set of values of t having positive measure), but one still retains the

condition of *positivity in the mean*, i.e.,

$$\int_0^T H(t)dt > 0 \quad . \tag{6.12}$$

In a certain respect, this theorem completes both Theorem 2.3 and Theorem 6.1.

THEOREM 6.4. *If the matrix function*

$$H(t) = H(t+T)$$

of equation (0.1) *satisfies condition* (6.12), *then this equation has a zone of stability containing the point* $\lambda = 0$.

[That is, all the points λ in some neighborhood $(-\ell,\ell)$ of $\lambda = 0$ are points of strong stability of equation (0.1) (see § 5 , no. 1).]

PROOF. By (6.12), one can find $L > 0$ such that

$$\int_0^T (H(t)\xi,\xi)dt \geqq L(\xi,\xi) \qquad (\xi \in E_n).$$

It is obvious that there exists $\varepsilon > 0$ such that for any continuous vector-function $x = x(t)$ $(0 \leqq t \leqq T)$ satisfying the conditions

$$|x(0)| = 1, \quad |x(t) - x(0)| < \varepsilon \quad (0 \leqq t \leqq T)$$

the following inequality holds

$$\int_0^T (H(t)x,x)dt > 0 \quad . \tag{6.13}$$

On the other hand, given such an $\varepsilon > 0$, one can always find $\delta_\varepsilon > 0$ such that for $|\lambda| < \delta_\varepsilon$ any solution $x = x(t;\lambda)$ of equation (0.1) with initial condition $x_0 = x(0;\lambda)$, where $(x_0,x_0) = 1$, will satisfy the inequality

$$|x_0 - x(t;\lambda)| < \varepsilon \quad (0 \leqq t \leqq T) \quad .$$

For δ_ε so chosen and $|\lambda| < \delta_\varepsilon$, condition (6.13) will be fulfilled for any solution $x = x(t)$ of (0.1). It is easy to trace the appearance of this condition back and see that it is precisely the one which formed the basis of the whole discussion in §§ 3 and 4. Therefore, all the basic results of these sections remain true in the case under consideration too, with the only constraint being that now λ may take values only in the disk $|\lambda| < \delta_\varepsilon$. Using the arguments given during the proof of Theorem 6.1, we show the existence of an interval $(-\ell,\ell)$, all of whose points are points of strong stability of equation (0.1), as claimed in Theorem 6.4.

§ 7. TESTS FOR λ TO BELONG TO THE CENTRAL ZONE OF STABILITY

1. To formulate these tests, we must first introduce some notation.

Let $A = \|a_{jk}\|_1^n \neq 0$ be a square matrix with nonnegative elements. By Perron's theorem [4], among the eigenvalues of A having the largest modulus, there will be at least one that is positive. We denote it by $M(A)$. We shall need the following lemma.

LEMMA 7.1. *Suppose that for the matrix* $A = \|a_{jk}\|_1^n \neq 0$ *having nonnegative elements one can find a vector* $\xi = (\xi_1,\ldots,\xi_n)$ $\neq 0$ *with nonnegative coordinates satisfying*

$$\xi \leqq A\xi \; . \tag{7.1}$$

Then $M(A) \geqq 1$.

Inequality (7.1) is understood to mean that each component of the vector ξ is not larger than the corresponding component of the vector $\eta = A\xi$.

PROOF. Inequality (7.1) obviously remains valid if one applies some power A^p $(p = 1,2,\ldots)$ of the matrix A to both sides. If one does so succesively for $p = 1,2,\ldots$, one finds that

$$\xi \leqq A^p\xi \quad (p = 1,2,\ldots) \; .$$

Now assuming that $M(A) < 1$, all the eigenvalues of A will have modulus smaller than one, and so we shall have $A^p \longrightarrow 0$ as $p \longrightarrow \infty$. Since $\xi \neq 0$ by hypothesis, this is impossible, and the lemma is proved.

2. If $B = \|b_{jk}\|_1^n$ is any matrix, then we shall denote

by B_a the matrix obtained by replacing all elements of B by their absolute values. Therefore, if $A = B_a$, then

$$a_{jk} = |b_{jk}| \quad (j,k = 1,\dots,n).$$

In particular,

$$J_a = \begin{pmatrix} 0 & I_m \\ I_m & 0 \end{pmatrix} .$$

TEST I_n. *A real* λ *belongs to the central zone of stability of an equation* (0.1) *of positive type whenever*

$$|\lambda| \leqq 2M^{-1}(C) , \tag{7.2}$$

where

$$C = J_a \int_0^T H_a(t)dt .$$

PROOF. Let $x^+(t) \not\equiv 0$ be a solution of equation (0.1) for $\lambda = \Lambda_+$, satisfying (see § 6 , no. 1)

$$x^+(t + T) = - x^+(t) . \tag{7.3}$$

Denote by $x_j^+(t)$ $(j = 1,\dots,n)$ the coordinates of the vector-function $x^+(t)$, and set

$$\xi_j = \max_{0 \leqq t \leqq T} |x_j^+(t)| = x_j^+(\tau_j) \quad (j = 1,\dots,n) . \tag{7.4}$$

Putting $JH(t) = \|a_{jk}(t)\|_1^n$, we have

$$\frac{dx_j^+}{dt} = \Lambda_+ \sum_{k=1}^n a_{jk}(t)x_k^+ \quad (j = 1,\dots,n) .$$

Integrating the j-th of these equalities from τ_j to $\tau_j + T$, we find

$$- 2x_j^+(\tau_j) = \Lambda_+ \sum_{k=1}^n \int_{\tau_j}^{\tau_j+T} a_{jk}(t)x_k^+(t)dt \quad (j = 1,\dots,n) .$$

Taking absolute values of both sides and using (7.4), we obtain

$$2\xi_j \leqq \Lambda_+ \sum_{k=1}^n \xi_k \int_{\tau_j}^{\tau_j+T} |a_{jk}(t)|dt = \Lambda_+ \sum_{k=1}^n c_{jk}\xi_k$$

$(j = 1,\dots,n)$, where the matrix C is given by

$$C = \|c_{jk}\|_1^n = \|\int_0^T |a_{jk}(t)|dt\|_1^n = J_a\int_0^T H_a(t)dt .$$

Therefore, for $\xi = (\xi_1,\ldots,\xi_n)$ one has that

$$\xi \leqq \frac{1}{2}\Lambda_+ C ,$$

and Lemma 7.1 is applicable to the matrix $A = \frac{1}{2}\Lambda_+ C$. We obtain

$$\frac{1}{2}\Lambda_+ M(C) \geqq 1, \quad \text{or} \quad \Lambda_+ \geqq \frac{2}{M(C)} .$$

The inequality

$$-\Lambda_- \geqq \frac{2}{M(C)}$$

can be established similarly. We conclude that if λ satisfies inequality (7.2), then

$$-\Lambda_- < \lambda < \Lambda_+ ,$$

and hence λ belongs to the central zone of stability.

3. Consider the particular case where $n = 2$ $(m = 1)$ and $H(t) \in \mathcal{P}_n(T)$ is a real matrix function. Setting

$$H(t) = \begin{pmatrix} a(t) & b(t) \\ b(t) & c(t) \end{pmatrix} ,$$

system (0.1) can be written in the form

$$\begin{aligned} \frac{dx_1}{dt} &= \lambda(b(t)x_1 + c(t)x_2), \\ \frac{dx_2}{dt} &= -\lambda(a(t)x_1 + b(t)x_2). \end{aligned} \tag{7.5}$$

Introduce the notations

$$I_a = \int_0^T a(t)dt, \quad I_b = \int_0^T |b(t)|dt, \quad I_c = \int_0^T c(t)dt .$$

Then in our case $M(C)$ is the largest root of the equation

$$\begin{vmatrix} I_b - \mu & I_c \\ I_a & I_b - \mu \end{vmatrix} = 0 .$$

Therefore, test I_n reduces to the following statement in the case $n = 2$ under consideration.

TEST I_2. *All the solutions of system* (7.5) *are bounded whenever the real* λ *satisfies the inequality*

$$|\lambda| < 2(I_b + \sqrt{I_a I_c})^{-1} . \tag{7.6}$$

Here one assumes, of course, that the periodic, second order, matrix-function $H(t)$ is of class $P_2(T)$ (this condition is written out in detail at the beginning of § 8).

The test for the boundedness of the solutions of system (7.5) stated above includes, as a particular case, the known test of A. M. Lyapunov for the boundedness of the solutions of the differential equation

$$\frac{d^2y}{dt^2} + \lambda^2 p(t)y = 0 \qquad (p(t) = p(t + T) \geq 0, \int_0^T p(t)dt > 0).$$

Indeed, this equation is equivalent for $\lambda \neq 0$ to the following differential system $(x_1 = y)$:

$$\frac{dx_1}{dt} = \lambda x_2, \quad \frac{dx_2}{dt} = - \lambda p x_1 ,$$

which is a particular case of system (7.5), with $a(t) \equiv 1$, $b(t) \equiv 0$, and $c(t) \equiv p(t)$. In this case inequality (7.6) reduces for $\lambda \neq 0$ to the Lyapunov inequality

$$\lambda^2 < \frac{4}{T} (\int_0^T p(t)dt)^{-1} . \tag{7.7}$$

We remark that when $b(t) \equiv 0$ test I_2 was already known (see, for example, [23]). Notice however that in this particular case test I_2 is not essentially new in comparison with the Lyapunov test.

In fact, if $b \equiv 0$ and $1/c(t)$ is integrable, then upon setting

$$\tau = \int_0^t c(s)ds$$

we easily see that system (7.5) is equivalent to a single equation

$$\frac{d^2x_1}{d\tau^2} + \lambda^2 p(\tau)x_1 = 0$$

where

$$p(\tau) = \left.\frac{a(t)}{c(t)}\right|_{t=t(\tau)} , \quad p(\tau) = p(\tau + \Omega), \quad \Omega = \int_0^T c(t)dt .$$

For equation (7.8), Lyapunov's inequality (7.7) means that

$$\lambda^2 < 4\, I_a^{-1} I_c^{-1} .$$

If $1/c(t)$ is not integrable, we obtain the desired result if we replace $c(t)$ by $c(t) + \varepsilon$, where ε is an arbitrarily small positive constant, and then use continuity arguments.

4. Returning to the general case of equation (0.1), we comment that the method used to derive test I_n allows one to establish a number of other tests for determining whether λ belongs to the central zone of stability. In fact, this method actually leads to as many tests as there are integer divisors q of the number $n = 2m$.

Let us explain this in the case $q = 2$.

Write the matrix $H(t)$ in the form

$$H(t) = \begin{pmatrix} A(t) & B(t) \\ B^*(t) & C(t) \end{pmatrix}$$

where $A(t)$, $B(t)$, $C(t)$ are m-th order square matrices, $A(t)$ and $C(t)$ are Hermitian matrices to which correspond nonnegative Hermitian forms, i.e., $A(t) \geqq 0$, $C(t) \geqq 0$, and $B^*(t)$ denotes the conjugate transpose of the matrix $B(t)$.

Splitting the n-dimensional vector x into the direct sum of two m-dimensional vectors y and z:

$$x = y \dotplus z ,$$

we can recast the differential equation (0.1) into the system

$$\frac{dy}{dt} = \lambda(B^*y + Cz), \quad \frac{dz}{dt} = -\lambda(Ay + Bz) .$$

As before, let $x^+(t) = y^+(t) \dotplus z^+(t)$ be a solution to equation (0.1) for $\lambda = \Lambda_+$ satisfying (7.3). Set

$$\eta = \max_{0\leqq t\leqq T} |y^+(t)| = |y^+(\tau)|, \quad \zeta = \max_{0\leqq t\leqq T} |z^+(t)| = |z^+(\sigma)|.$$

Let $\alpha(t)$, $\beta(t)$ and $\gamma(t)$ denote the norms (see § 1, no. 1) of the matrices $A(t)$, $B(t)$ and $C(t)$, so that

$$\alpha(t) = |A(t)| = \max_{u\in E_m} \{|A(t)u|/|u|\} ,$$

$$\beta(t) = |B(t)| = |B^*(t)|, \quad \gamma(t) = |C(t)| .$$

Integrating the following equation term by term

$$\frac{dy^+}{dt} = \Lambda_+(B^*y^+ + Cz^+)$$

from τ to $\tau + T$ and then taking the norms of both sides, we obtain

$$\begin{aligned} 2\eta &= \Lambda_+\left|\int_\tau^{\tau+T} B^*(t)y^+(t)dt + \int_\tau^{\tau+T} C(t)z^+(t)dt\right| \leqq \\ &\leqq \Lambda_+\left(\int_\tau^{\tau+T} |B^*(t)y^+(t)|dt + \int_\tau^{\tau+T} |C(t)z^+(t)|dt\right). \end{aligned} \tag{7.9}$$

Since

$$|B^*(t)y^+(t)| \leqq \beta(t)|y^+(t)| \leqq \beta(t)\eta ,$$

$$|C(t)z^+(t)| \leqq \gamma(t)|z^+(t)| \leqq \gamma(t)\zeta ,$$

we conclude from (7.9) that

$$2\eta \leqq \Lambda_+\left(\eta\int_0^T \beta(t)dt + \zeta\int_0^T \gamma(t)dt\right) . \tag{7.10}$$

Similarly, starting with the equation

$$\frac{dz^+}{dt} = -\Lambda_+(Ay^+ + Bz^+) ,$$

we get

$$2\zeta \leqq \Lambda_+\left(\eta\int_0^T \alpha(t)dt + \zeta\int_0^T \beta(t)dt\right) . \tag{7.11}$$

Using Lemma 7.1, (7.10) and (7.11) yield

$$2 \leqq \Lambda_+\left\{\int_0^T \beta(t)dt + \left(\int_0^T \alpha(t)dt\cdot\int_0^T \gamma(t)dt\right)^{1/2}\right\} .$$

The resulting bound for Λ_+ enables us to formulate the following result.

TEST II. *A real λ belongs to the central zone of stability of equation (0.1) whenever*

$$|\lambda| < 2\left\{\int_0^T \beta(t)dt + \left(\int_0^T \alpha(t)dt\cdot\int_0^T \gamma(t)dt\right)^{1/2}\right\} ,$$

where $\alpha(t)$, $\beta(t)$, *and* $\gamma(t)$ *denote the norms of the matrices* $A(t)$, $B(t)$, *and* $C(t)$, *respectively, which figure in the representation*

$$H(t) = \begin{pmatrix} A(t) & B(t) \\ B^*(t) & C(t) \end{pmatrix} .$$

Test II includes test I_2 as a particular case, as did test I_n.

§ 8. THE CASE OF A SECOND-ORDER CANONICAL SYSTEM

1. As we already mentioned in § 7, when $n = 2$ the real canonical equation (0.1) has the form

$$\frac{dx_1}{dt} = \lambda(bx_1 + cx_2), \quad \frac{dx_2}{dt} = -\lambda(ax_1 + bx_2) . \tag{8.1}$$

For a real, periodic matrix-function

$$H(t) = \begin{pmatrix} a(t) & b(t) \\ b(t) & c(t) \end{pmatrix}, \quad H(t + T) = H(t) ,$$

the condition that $H(t) \in P_2(T)$ is easily seen to be equivalent to the following set of conditions:

$$a(t) \geqq 0, \quad c(t) \geqq 0, \quad a(t)c(t) - b^2(t) \geqq 0 ,$$

and

$$\int_0^T a(t)dt \cdot \int_0^T c(t)dt - \left(\int_0^T b(t)dt\right)^2 > 0 .$$

Let

$$U(t;\lambda) = \begin{pmatrix} u_{11}(t;\lambda) & u_{12}(t;\lambda) \\ u_{21}(t;\lambda) & u_{22}(t;\lambda) \end{pmatrix}$$

be the matrizant of system (8.1). Since $\det U(t;\lambda) \equiv 1$, the multipliers ρ_1 and ρ_2 of system (8.1) will be the roots of the quadratic equation

$$\rho^2 - 2A(\lambda)\rho + 1 = 0 ,$$

where

$$2A(\lambda) = u_{11}(T;\lambda) + u_{22}(T;\lambda) .$$

It follows that $\rho_1(\lambda)\rho_2(\lambda) \equiv 1$ and

$$\rho_{1,2}(\lambda) = A(\lambda) \pm (A^2(\lambda) - 1)^{1/2} . \tag{8.2}$$

Assuming that $\operatorname{Im} \lambda \geqq 0$, denote by $\rho_1(\lambda)$ $(\rho_2(\lambda))$ that multiplier of the first (second) kind of system (8.1). Thus

$$|\rho_1(\lambda)| < 1, \quad |\rho_2(\lambda)| > 1 \quad \text{for} \quad \operatorname{Im} \lambda > 0 . \tag{8.3}$$

Since

$$U(t;\lambda) = \begin{pmatrix} 1 & 0 \\ 0 & 1 \end{pmatrix} + \lambda \int_0^t \begin{pmatrix} b & c \\ -a & -b \end{pmatrix} dt + \lambda^2 C_2 + \lambda^3 C_3 + \ldots ,$$

where C_k $(k = 2,3,\ldots)$ are certain second-order matrices,

$$A(\lambda) = 1 + \frac{1}{2} a_2 \lambda^2 + \ldots \quad (a_2 = \operatorname{Sp} C_2).$$

A straightforward computation of C_2 leads us to the conclusion that

$$a_2 < 0 . \tag{8.4}$$

However, one can reach the same conclusion from the general arguments that follow. Had a_2 been 0, the function $\rho(\lambda)$ would have the following expansion in the vicinity of the point $\lambda = 0$:

$$\rho(\lambda) = 1 + \sum_{k=3}^{\infty} \gamma_k \lambda^{k/2} ,$$

according to which each branch of $\rho(\lambda)$ would take values both inside and outside the unit disk in any half-neighborhood $|\lambda| < r$, $\operatorname{Im} \lambda > 0$.

Therefore, $a_2 \neq 0$. On the other hand, applying Theorem 6.1 for λ real and $|\lambda|$ sufficiently small, we see that the multipliers $\rho_{1,2}(\lambda)$ must lie on the unit circle, i.e., $|A(\lambda)| < 1$, whence (8.4).

At the zeros of the real entire function

$$F(\lambda) = A^2(\lambda) - 1 ,$$

the multipliers $\rho_{1,2}(\lambda)$ are ± 1, and this implies, by virtue of (8.3), that the zeros of $F(\lambda)$ are all real.

Following a line of argument similar to that taken earlier for the case $\lambda_0 = 0$, it is not difficult to show that the multiplicity of any zero λ_0 of the function $F(\lambda)$ is at most two.

If λ_0 is a simple zero of $F(\lambda)$, then λ_0 will be a second-order ramification point for $\rho(\lambda)$, and $\rho(\lambda)$ will have, in the vicinity of this point, the expansion

$$\rho(\lambda) = \rho(\lambda_0) + \sum_{k=1}^{\infty} c_k(\lambda - \lambda_0)^{k/2} \quad (\rho(\lambda_0) = \pm 1, \quad c_1 \neq 0).$$

If, however, λ_0 is a zero of multiplicity two for $F(\lambda)$, then, in the vicinity of the point $\lambda = \lambda_0$, one can distinguish tho holomorphic branches of the function $\rho(\lambda)$:

$$\rho_j(\lambda) = \rho(\lambda_0) + \sum_{k=1}^{\infty} c_k^{(j)}(\lambda - \lambda_0)^k$$

$$(\rho(\lambda_0) = \pm 1, \quad c_1^{(j)} \neq 0, \ j = 1,2) .$$

2. In our paper [11] it was shown that each of the entire functions $u_{jk}(T;\lambda)$ $(j,k = 1,2)$ belongs to the class (N) of entire functions $f(\lambda)$ which is characterized by the following two properties:

$$1) \quad \overline{\lim_{|\zeta|\to\infty}} \frac{\ln|f(\zeta)|}{|\zeta|} < \infty \qquad (\zeta \text{ any complex number}),$$

$$2) \quad \int_{-\infty}^{\infty} \frac{\ln^+|f(\lambda)|}{1 + \lambda^2} d\lambda < \infty .$$

On the other hand, B. Ya. Levin [17] established a general result according to which any function $f(\lambda)$ of class (N) can be always be expanded into a product which converges absolutely in any bounded subset of the complex plane

$$f(\lambda) = c\lambda^p \lim_{R\to\infty} \prod_{|\alpha_j|<R} (1 - \frac{\lambda}{\alpha_j}) \qquad (p \geqq 0),$$

where $\{\alpha_j\}$ is the set of all zeros of the entire function $f(\lambda)$.

Obviously, if $u_{jk}(T;\lambda)$ $(j,k = 1,2)$ are functions of class (N), then $A(\lambda)$ and $F(\lambda)$ will be also.

THEOREM 8.1. *All the roots of the entire function* $F(\lambda) = A^2(\lambda) - 1$ *are real, and among them there are at least one positive and one negative root. The nonzero roots of the entire function* $F(\lambda)$ *can be indexed, taking into account their multiplicities, in such a way that*

$$\ldots < \lambda_{-4} \leqq \lambda_{-3} \leqq \lambda_{-2} \leqq \lambda_{-1} < 0 < \lambda_1 \leqq \lambda_2 \leqq \lambda_3 \leqq \lambda_4 \leqq \ldots \quad (8.5)$$

Moreover, one has the product expansion that converges uniformly in each bounded subset of the complex plane

$$A^2(\lambda) = a_2\lambda^2 \lim_{R\to\infty} \prod_{|\lambda_j|<R} \left(1 - \frac{\lambda}{\lambda_j}\right) \qquad (a_2 < 0) . \tag{8.6}$$

In any interval $(\lambda_{2k}, \lambda_{2k+1})$ $(k = 0,1,2,\ldots;\ \lambda_0 = 0)$, *the function* $(-1)^{k-1}A(\lambda)$ *is increasing, and*

$$-1 < A(\lambda) < 1 \quad \textit{for} \quad \lambda_{2k} < \lambda < \lambda_{2k+1} . \tag{8.7}$$

If for some $k = 1,2,\ldots,$ *one has* $\lambda_{2k-1} < \lambda_{2k}$, *then*

$$(-1)^k A(\lambda) > 1 \quad \textit{for} \quad \lambda_{2k-1} < \lambda < \lambda_{2k} , \tag{8.8}$$

and inside the interval $(\lambda_{2k-1}, \lambda_{2k})$ *there lies only one maximum point for the function* $(-1)^k A(\lambda)$. *If, however,* $\lambda_{2k-1} = \lambda_{2k}$, *then the function* $(-1)^k A(\lambda)$ *has a maximum equal to* 1 *at the point* λ_{2k-1}.

The function $A(\lambda)$ *behaves similarly on the negative intervals* $(\lambda_{2k-1}, \lambda_{2k})$ *and* $(\lambda_{2k-2}, \lambda_{2k-1})$ $(k = 0,-1,-2,\ldots)$.

If one constructs the graph of the function $\mu = A(\lambda)$ in the real plane (λ, μ), the claims of Theorem 8.1 concerning the behavior of $A(\lambda)$ translate themselves into the fact that this graph has precisely the shape shown in Fig. 1 (where the case $\lambda_{2k-1} = \lambda_{2k}$ is not drawn.

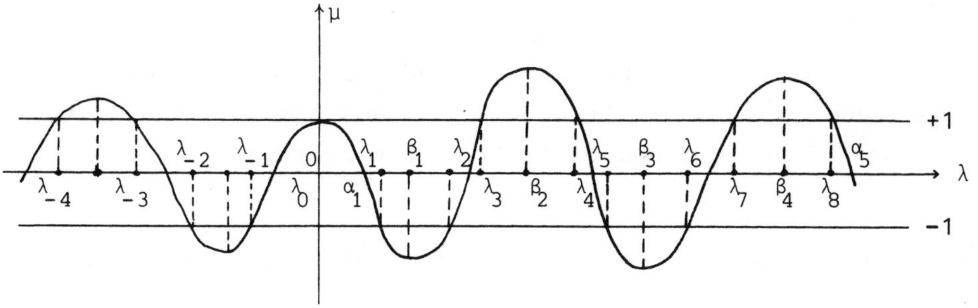

Fig. 1

PROOF. The realness of the zeros of the function $F(\lambda)$ and the possibility of representing the latter in the form (8.6) have been proved in the discussion preceding the theorem.

The existence of the zeros $\lambda_{\pm 1}$ is a consequence of the general Theorem 6.2. Consider the polynomial

$$P_R(\lambda) = a_2\lambda^2 \prod_{|\lambda_j|<R} (1 - \frac{\lambda}{\lambda_j}) \qquad (0 < R < \infty) .$$

Since all its zeros λ_j are real, the same holds for the zeros of the derivative $P_R'(\lambda)$. It follows that all the zeros of the function $F'(\lambda)$ are real, because $P_R'(\lambda)$ converges uniformly to $F'(\lambda) = 2A(\lambda)A'(\lambda)$.

For real λ

$$\frac{d}{d\lambda} \frac{F'(\lambda)}{F(\lambda)} = - \frac{2}{\lambda^2} - \sum_j \frac{1}{(\lambda - \lambda_j)^2} , \tag{8.9}$$

and therefore, each zero μ_0 of the function $F'(\lambda)$ which is not a zero of $F(\lambda)$ is simple $(F''(\mu_0)/F(\mu_0) < 0)$. If, however, $F'(\mu_0) = F(\mu_0) = 0$, then μ_0, being a zero of $F(\lambda)$ of multiplicity two, is a zero of multiplicity one for $F'(\lambda)$. Thus $F'(\lambda)$ has only simple zeros.

From (8.9) it results also that if one discards the common zeros of $F(\lambda)$ and $F'(\lambda)$, then the remaining zeros of these functions interlace.

In fact, in virtue of (8.9), one can assert, for any pair $\lambda_k < \lambda_{k+1}$, that as λ monotonically increases from λ_k to λ_{k+1}, the function $F'(\lambda)/F(\lambda)$ decreases from $-\infty$ to ∞, and so $F'(\lambda)$ passes through zero once and only once.

At the same time, we showed that the zeros of the functions $A(\lambda)$ and $A'(\lambda)$ are real and simple. The fact that they interlace is a consequaence of the relation

$$\frac{d}{d\lambda} \frac{A'(\lambda)}{A(\lambda)} < 0 \qquad (-\infty < \lambda < \infty) ,$$

whose derivation for $A(\lambda) \in (N)$ is similar to that given for $F(\lambda)$.

As one can easily see, all the other claims of the theorem follow from the realness, the simplicity, and the interlacing properties of the zeros of the functions $A(\lambda)$, $A'(\lambda)$, as well as those of the functions $F(\lambda)$ and $A(\lambda)A'(\lambda)$.

The theorem is proved.

Denote by

$\alpha_1 < \alpha_2 < \ldots$ and $\beta_1 < \beta_2 < \ldots$

the successive positive zeros of the functions $A(\lambda)$ and $A'(\lambda)$, respectively. Then what the theorem asserts is that

$$\lambda_{2k-2} < \alpha_k < \lambda_{2k-1}, \quad \lambda_{2k-1} \leqq \beta_k \leqq \lambda_{2k} \quad (k = 1,2,\ldots),$$

where equality is excluded as soon as $\lambda_{2k-1} < \lambda_{2k}$.

3. The theorem permits one to draw the motion of the multipliers as λ increases continuously from $-\infty$ to ∞.

Here we describe this picture only for the case where λ varies from 0 to ∞, because the picture for λ varying from 0 to $-\infty$ is completely analogous.

THEOREM 8.2. *As* λ *increases continuously from* λ_{2k} *to* λ_{2k+1} $(k = 0,1,2,\ldots;\ \lambda_0 = 0)$, *the multipliers of the first and second kind,* $\rho_1(\lambda)$ *and* $\rho_2(\lambda)$, *move clockwise and counterclockwise, respectively, on the semicircles of the unit circle, beginning from the point* $(-1)^k$ *and until the point* $(-1)^{k+1}$.

When $\lambda_{2k+1} < \lambda_{2k+2}$, *the multipliers* $\rho_1(\lambda)$ *and* $\rho_2(\lambda)$, *having met at the point* $(-1)^{k+1}$ *for* $\lambda = \lambda_{2k+1}$, *continue their motion, and as* λ *further increases to reach a certain value* β_{k+1} $(\lambda_{2k+1} < \beta_{k+1} < \lambda_{2k+2})$, *they move along the real axis away from the point* $(-1)^{k+1}$ *in opposite directions. Then, as* λ *increases from* β_{k+1} *to* λ_{2k+2}, *they reverse direction and meet again at the point* $(-1)^{k+1}$.

If, however, $\lambda_{2k+1} = \lambda_{2k+2}$, *then as* λ *passes through* λ_{2k+1}, $\rho_1(\lambda)$ *and* $\rho_2(\lambda)$ *meet at the point* $(-1)^{k+1}$ *and afterwards continue their motion on the unit circle, each in its own direction.*

This theorem is a straightforward corollary of Theorem 8.1 and formula (8.2).

We remark that the first statement of Theorem 8.2 also results directly from the fact that $|\rho_1(\lambda)| \leqq 1$ when $\operatorname{Im} \lambda \geqq 0$, and its generalization is Theorem 4.2.

This part of the theorem has been given by us already in our note [9].

We made the third statement of the theorem in order to complete the picture, without saying anything new, in addition to the first statement.

If, for some value $\lambda = \lambda'$, one of the multipliers is $\rho(\lambda') = 1$ $(\rho(\lambda') = -1)$,
then in this case, and only in this case, system (8.1) has a non-trivial solution $x = (x_1, x_2) \neq 0$ satisfying

$$x(T) = x(0) \quad (\text{respectively, } x(T) = -x(0)) .$$

It follows that the numbers

$$\lambda_{\pm(4k+1)}, \ \lambda_{\pm(4k+2)} \quad (k = 0,1,2,\ldots) \tag{8.10}$$

form the spectrum (the set of characteristic numbers) of the boundary value problem

$$\begin{aligned} \frac{dx_1}{dt} &= \lambda(bx_1 + cx_2), \quad x_1(0) + x_1(T) = 0, \\ \frac{dx_2}{dt} &= -\lambda(ax_1 + bx_2), \quad x_2(0) + x_2(T) = 0. \end{aligned} \tag{8.11}$$

Similarly, the numbers

$$0,0,\lambda_{\pm(4k+3)}, \ \lambda_{\pm(4k+4)} \quad (k = 0,1,2,\ldots) \tag{8.12}$$

form the spectrum of the boundary value problem

$$\begin{aligned} \frac{dx_1}{dt} &= \lambda(bx_1 + cx_2), \quad x_1(0) - x_1(T) = 0, \\ \frac{dx_2}{dt} &= -\lambda(ax_1 + bx_2), \quad x_2(0) - x_2(T) = 0. \end{aligned} \tag{8.13}$$

Here each characteristic number appears in the spectrum (8.10) or (8.12) a number of times equal to the number of linearly independent solutions to the problem (8.11) or (8.13), respectively, that correspond to it.

The last claim results from the general theorem 3.4, because the sets (8.10) and (8.12) are the sets of roots of the respective equations

$$A(\lambda) + 1 = 0, \quad A(\lambda) - 1 = 0,$$

and each root is counted according to its multiplicity (as a root of the corresponding equation).

It is easy to show that if some λ' is a simple characteristic number of problem (8.11) or (8.13), then system (8.1) has for $\lambda = \lambda'$ two linearly independent solutions $x^{(0)}(t)$

and $x^{(1)}(t)$ such that

$$x^{(0)}(t+T) \equiv \pm x^{(0)}(t), \quad x^{(1)}(t+T) \equiv \pm x^{(1)}(t) + x^{(0)}(t).$$

If λ' is of multiplicity two, then, obviously, each solution $x(t)$ of system (8.1) will have the property $x(t+T) \equiv \equiv \pm x(t)$ (and hence it will be bounded).

According to the general definition (see § 5), the zones of stability for system (8.1) are precisely the open intervals

$$(\lambda_{2k}, \lambda_{2k+1}) \quad (k = 0,1,2,\ldots)$$

on the half line $\lambda > 0$, and the corresponding open intervals

$$(\lambda_{-2k-1}, \lambda_{-2k}) \quad (k = 0,1,2,\ldots)$$

on the half line $\lambda < 0$, with the addition of the open interval $(\lambda_{-1}, \lambda_1)$.

Therefore, the characteristic numbers of the boundary value problems (8.11) and (8.13) *interlace in pairs.*

The open interval $(\lambda_k, \lambda_{k+1})$ $(k = 0, \pm 1, \pm 2, \ldots)$ is a zone of stability of system (8.1) if and only if its endpoints are characteristic numbers of distinct kinds (of distinct boundary value problems).

As we know (see § 7, no. 3), the differential equation

$$\frac{d^2y}{dt^2} + \lambda^2 p(t) y = 0 , \tag{8.14}$$

where $p(t) = p(t+T) \geqq 0$ and $p(t) \not\equiv 0$, is equivalent, in a certain sense, to a differential system of the form (8.1).

If one reformulates the assertions concerning the boundary value problems (8.11) and (8.13) to fit the corresponding boundary value problems posed for equation (8.14), then one recovers results that were proved for the first time by A. M. Lyapunov [20]. Subsequently, Lyapunov generalized these results to include the case when the function $p(t)$ takes values of both signs [21].

Using the method based on the investigation of the behavior of the multipliers for equation (8.14) in the upper half plane, one can quite simply obtain, in this more general case too, the theorems of Lyapunov, which he announced without proofs in the note [21] (see § 9, and also [8]).

4. Below we shall establish the asymptotic properties of the numbers λ_k $(k = \pm 1, \pm 2, \ldots)$. To do this, we need a number of propositions due to V. A. Yakubovich [26,27] and concerning the geometric character of the behavior of the solutions to system (8.1).

1°. Consider a real λ, real x_1, x_2 $(x_1^2 + x_2^2 < 1)$, and set

$$x_1 = r \cos \theta , \quad x_2 = r \sin \theta .$$

Therefore, r and θ are the polar coordinates of the point (x_1, x_2) in the phase plane.

It results from (8.1) that

$$x_1 \frac{dx_2}{dt} - x_2 \frac{dx_1}{dt} = -\lambda[a(t)x_1^2 + 2b(t)x_1x_2 + c(t)x_2^2] .$$

Substituting the expressions for x_1 and x_2 in terms of r and θ into this, we obtain

$$\frac{d\theta}{dt} = -\lambda[a(t)\cos^2\theta + 2b(t)\sin\theta \cos\theta + c(t)\sin^2\theta]. \tag{8.15}$$

When $\lambda > 0$ or $\lambda < 0$, we shall have $d\theta/dt \leqq 0$ or $d\theta/dt \geqq 0$, respectively. In other words, one has the following proposition.

2°. *As* t *increases, the radius-vector* $x = (x_1, x_2)$ *of any integral curve* $x_1 = x_1(t)$, $x_2 = x_2(t)$ *of system* (8.1) *rotates [more precisely, either does not change or rotates] clockwise (counterclockwise) for* $\lambda > 0$ *(respectively, for* $\lambda < 0$*).*

The difference

$$\Theta = \theta(0) - \theta(T)$$

will be called the *rotation angle* of the solution $x = (x_1, x_2)$ to system (8.1).

The angle coordinate $\theta(t)$ of this solution is determined, for any t, by its initial condition $\theta_0 = \theta(0)$, and so

$$\theta = \theta(t, \theta_0; \lambda) .$$

Consequently,

$$\Theta = \Theta(\theta_0; \lambda) .$$

It is easy to see that Θ is a continuous function of

θ_0 and λ, and satisfies

$$\Theta(\theta_0 + 2\pi;\lambda) = \Theta(\theta_0;\lambda) .$$

It follows that the functions of λ

$$\Theta_M(\lambda) = \max \Theta(\theta_0;\lambda) , \qquad \Theta_\mu(\lambda) = \min \Theta(\theta_0;\lambda) , \qquad (0 \leqq \theta_0 \leqq 2\pi)$$

exist and are continuous.

From now on we shall consider everything on the half line $\lambda > 0$, the case $\lambda < 0$ being similar.

According to a well-known, general lemma due to S.A. Chaplygin [1], one is able to assert that for $\lambda > 0$, the solution $\theta(t,\theta_0;\lambda)$ of equation (8.15) is a nondecreasing function not only of the argument t, but also of the argument λ. It follows that both the rotation angle $\Theta(\theta_0;\lambda)$ and the functions $\Theta_M(\lambda)$ and $\Theta_\mu(\lambda)$ are also nondecreasing and continuous.

We can now state the following important proposition due to V. A. Yakubovich [26,27].

THEOREM 8.3. *In order that* $\lambda > 0$ *belong to the* k-*th zone of stability of system* (8.1) (k = 1,2,...), *it is necessary and sufficient that*

$$(k - 1)\pi \leqq \Theta_\mu(\lambda) \leqq \Theta_M(\lambda) < k\pi \quad . \tag{8.16}$$

We should explain that in this formulation of the theorem, it is understood that the first number is reserved for the central zone (λ_{-1},λ_1), while the subsequent numbers refer to the zones that follow it on the right, so that under this arrangement the k-th zone of stability $(k \geqq 1)$ will be $(\lambda_{2k-2},\lambda_{2k-1})$.

PROOF. First of all, note that from the results of the preceding subsections of § 8 it follows almost at once that $\lambda > 0$ belongs to some zone of stability if and only if (8.16) holds for some integer k.

Indeed, suppose that for some integer $k \geqq 0$ one has

$$\Theta_\mu(\lambda) \leqq k\pi \leqq \Theta_M(\lambda) . \tag{8.17}$$

Then one can find θ_0 such that $\Theta(\theta_0;\lambda) = k\pi$. The solution $x = (x_1,x_2)$ with initial conditions $x_{10} = \cos\theta_0$, $x_{20} = \sin\theta_0$ will clearly have the property $x(T) = \rho x(0)$, i.e.,

$$x_1(T) = \rho x_1(0), \quad x_2(T) = \rho x_2(0), \tag{8.18}$$

where $\rho > 0$ or $\rho < 0$ depending on whether k is even or odd.

Therefore, when (8.17) holds, λ cannot belong to a zone of stability.

Conversely, if $\lambda > 0$ belongs to no zone of stability, then one can find a solution $x = (x_1,x_2)$ having property (8.18) for some real ρ. The rotation angle of this solution is an integer multiple of π, and so the relation (8.17) is valid for some integer $k > 0$.

We see that if $\lambda > 0$ belongs to some zone of stability, then (8.16) is fulfilled for some integer $k \geqq 0$. Letting λ vary continuously inside the zone, we become convinced that (8.16) holds with the same k for all λ in a fixed zone of stability.

If $\lambda > 0$ is sufficiently small, then $\Theta_M(\lambda)$ will be also. This shows that for values $\lambda > 0$ belonging to the first zone of stability, inequality (8.16) with $k = 1$ applies, i.e., we shall have

$$0 < \Theta_\mu(\lambda) < \Theta_M(\lambda) < \pi \quad \text{for} \quad 0 < \lambda < \lambda_1 . \tag{8.19}$$

Now take advantage of the fact that $\Theta_\mu(\lambda)$ and $\Theta_M(\lambda)$ are continuous, nondecreasing functions of λ, for $\lambda > 0$.

If $\lambda_1 = \lambda_2$, then, as we know, all solutions $x = x(t)$ to equation (8.1) for $\lambda = \lambda_1 = \lambda_2$ are antiperiodic: $x(t + T) = - x(t)$. Taking into account (8.19), we conclude that

$$\Theta_\mu(\lambda) = \Theta_M(\lambda) = \pi \quad \text{for} \quad \lambda = \lambda_1 = \lambda_2.$$

This further implies that

$$\pi < \Theta_\mu(\lambda) < \Theta_M(\lambda) < 2\pi \quad \text{for} \quad \lambda_2 < \lambda < \lambda_3.$$

If, however, $\lambda_1 < \lambda_2$, then for $\lambda = \lambda_1$ system (8.1) has a unique antiperiodic solution, up to a constant factor. Moreover, since $\rho_1(\lambda_1) = \rho_2(\lambda_1) = 1$, this solution is the unique solution to (8.1) having a rotation angle equal to π. Therefore, in this case (8.19) yields

$\Theta_\mu(\lambda_1) < \Theta_M(\lambda_1) = \pi$.

Thus, when λ is larger than but close to λ_1, we have

$\Theta_\mu(\lambda) < \pi < \Theta_M(\lambda) < 2\pi$.

Given any λ inside the interval (λ_1, λ_2), system (8.1) has two linearly independent solutions, $x^{(1)}(t;\lambda)$ and $x^{(2)}(t;\lambda)$ such that

$$x^{(j)}(T;\lambda) = \rho_j(\lambda) x^{(j)}(0;\lambda) \quad (j = 1,2;\ \rho_1(\lambda) = \rho_2^{-1}(\lambda) < 1).$$

With a suitable normalization, one can ensure that these solutions will depend continuously on λ.

Since for $\lambda > \lambda_1$ and λ close to λ_1, the rotation angle of these solutions is equal to π, then for all λ $(\lambda_1 < \lambda < \lambda_2)$ their rotation angle, being a multiple of π, will be precisely π.

On the other hand, every solution of (8.1) for $\lambda_1 < \lambda < \lambda_2$ having a rotation angle equal to a multiple of π, will differ from one of the solutions $x^{(j)}(t;\lambda)$ $(j = 1,2)$ only by a constant factor. Therefore,

$$\pi < \Theta_M(\lambda) < 2\pi \quad \text{for} \quad \lambda_1 < \lambda < \lambda_2,$$

whence

$$0 < \Theta_\mu(\lambda) \leqq \pi \quad \text{for} \quad \lambda_1 < \lambda < \lambda_2 .$$

Because the angle $\Theta_M(\lambda)$ does not decrease as λ increases, we see that for λ belonging to the second zone of stability inequality (8.16) can hold only for $k = 2$, i.e.,

$$\pi \leqq \Theta_\mu(\lambda) < \Theta_M(\lambda) < 2\pi \quad \text{for} \quad \lambda_2 < \lambda < \lambda_3 .$$

To analyze the behavior of the functions $\Theta_\mu(\lambda)$ and $\Theta_M(\lambda)$ on the intervals (λ_3, λ_4), (λ_4, λ_5), and so on, one proceeds similarly. The theorem is proved.

From Theorem 8.3 we immediately get the following test due to V. A. Yakubovich.

3°. *Let* $h_M(t)$ *and* $h_\mu(t)$ *be respectively the largest and the smallest eigenvalues of the matrix*

$$H(t) = \begin{pmatrix} a(t) & b(t) \\ b(t) & c(t) \end{pmatrix}$$

[*i.e.*, $h_M(t) = \frac{a+c}{2} + \{(\frac{a-c}{2})^2 + b^2\}^{1/2}$,

$h_\mu(t) = \frac{a+c}{2} + \{(\frac{a-c}{2})^2 - b^2\}^{1/2}$.]

If. given some integer $k \geqq 0$, *one can find values of* λ $(\lambda > 0)$ *satisfying*

$$(k - 1)\pi < \lambda\int_0^T h_\mu(t)dt, \quad \lambda\int_0^T h_M(t)dt < k\pi , \qquad (8.20)$$

then these values form a subinterval of the k*-th zone of stability of system* (8.1).

Indeed, according to (8.15) we have, for $\lambda > 0$,

$$\lambda h_\mu(t) \leqq \frac{d\theta}{dt} \leqq \lambda h_M(t),$$

and so the rotation angle $\Theta(\lambda)$ of any solution $x = x(t)$ of system (8.1) satisfies

$$\lambda\int_0^T h_\mu(t)dt \leqq \Theta(\lambda) \leqq \lambda\int_0^T h_M(t)dt .$$

It follows that (8.16) will hold true whenever (8.20) does.

We emphasize that the fundamental results in § 4 allowed M. G. Neigauz and V. B. Lidskii [16] to show that, in the general case $(n = 2m \geqq 2)$ of an equation (0.1) of positive type, the fact that inequalities (8.20) are fulfilled for some λ is also a sufficient condition for this value of λ to belong to one of the zones of stability of equation (0.1).

This conclusion can be quite simply obtained using Theorem 5.3.

5. Let $x^{(k)}(t)$ denote some nontrivial solution of system (8.1) for $\lambda = \lambda_k$ $(k = 1,2,\ldots)$, satisfying the condition

$$x^{(k)}(T) = \pm x^{(k)}(0) \quad (k = 1,2,\ldots) .$$

It is not hard to see, by applying Theorem 8.3, that the rotation angle Θ_k of this solution has the value

$$\Theta_k = [\frac{1 + k}{2}]\pi \quad (k = 1,2,\ldots) , \qquad (8.21)$$

where $[a]$ denotes the integral part of the number a.

This situation permits us to simplify the arguments

leading to the asymptotic formulas for λ_k as $k \longrightarrow \infty$. We borrowed the idea of using the values of the rotation angle for fundamental solutions of the boundary value problem in order to deduce asymptotic formulas for the characteristic numbers from the paper of P. D. Kalafati [5]. In his paper, this idea was applied to the Sturm-Liouville problem, whose differential equation was reduced to the form (8.1) with $b \equiv 0$.

THEOREM 8.4. *If the periodic functions* $a(t)$, $b(t)$, *and* $c(t)$ *are absolutely continuous and satisfy the conditions*

$$a(t) > 0, \quad \Delta(t) = a(t)c(t) - b^2(t) > 0 , \tag{8.22}$$

then for $k \longrightarrow \infty$

$$\lambda_k = [\tfrac{k+1}{2}]\pi\left(\int_0^T \sqrt{\Delta(t)}\,dt\right)^{-1} + \frac{1}{2}\int_0^T \frac{b(t)}{a(t)}\, d\,\frac{a(t)}{\sqrt{\Delta(t)}} + o(1). \tag{8.23}$$

If, in addition, the derivatives of the functions $a(t)$, $b(t)$, *and* $c(t)$ *are absolutely continuous, then one can replace* $o(1)$ *by* $O(1/k)$ *in this asymptotic formula.*

PROOF. In system (8.1) we pass to the new variable

$$\tau = \int_0^t \sqrt{\Delta(t)}\,dt .$$

Then this system takes on the form

$$\frac{dx_1}{d\tau} = \lambda(\beta x_1 + \gamma x_2), \quad \frac{dx_2}{d\tau} = -\lambda(\alpha x_1 + \beta x_2) , \tag{8.24}$$

where

$$\alpha(\tau) = \frac{a(t)}{\sqrt{\Delta(t)}}, \quad \beta(\tau) = \frac{b(t)}{\sqrt{\Delta(t)}}, \quad \gamma(\tau) = \frac{c(t)}{\sqrt{\Delta(t)}} \tag{8.25}$$

for $t = t(\tau)$.

Functions α, β, and γ have period

$$\Omega = \int_0^T \sqrt{\Delta(t)}dt . \tag{8.26}$$

Notice that

$$\alpha(\tau)\beta(\tau) - \gamma^2(\tau) \equiv 1 . \tag{8.27}$$

Furthermore, upon setting

$$u = \alpha x_1 + \beta x_2, \quad v = x_2 , \tag{8.28}$$

and using (8.25) and (8.27), we easily have

$$\frac{du}{d\tau} = \lambda v + \frac{d\alpha}{d\tau} x_1 + \frac{d\beta}{d\tau} x_2, \quad \frac{dv}{d\tau} = -\lambda u .$$

Eliminating x_1 and x_2 here by means of (8.28), we obtain

$$\frac{du}{d\tau} = \lambda v + \frac{1}{\alpha}\frac{d\alpha}{d\tau} u + \left(\frac{d\beta}{d\tau} - \frac{\beta}{\alpha}\frac{d\alpha}{d\tau}\right) v, \quad \frac{dv}{d\tau} = -\lambda u . \tag{8.29}$$

Now if one sets $u = r\cos\phi$, $v = r\sin\phi$, (8.29) easily yields

$$\frac{d\phi}{d\tau} = -\lambda - \frac{1}{\alpha}\frac{d\alpha}{d\tau}\sin\phi\cos\phi + \left(\frac{\beta}{\alpha}\frac{d\alpha}{d\tau} - \frac{d\beta}{d\tau}\right)\sin^2\phi =$$

$$= -\lambda - B(\tau) - A(\tau)\sin 2\phi + B(\tau)\cos 2\phi ,$$

where

$$A(\tau) = \frac{1}{2\alpha}\frac{d\alpha}{d\tau} , \quad B(\tau) = \frac{1}{2}\left(\frac{d\beta}{d\tau} - \frac{\beta}{\alpha}\frac{d\alpha}{d\tau}\right) . \tag{8.30}$$

Representing ϕ in the form

$$\phi = -\lambda\tau + \chi(\tau) , \tag{8.31}$$

we get

$$\frac{d\chi}{d\tau} = B(\tau) - A(\tau)\sin 2\phi + B(\tau)\cos 2\phi . \tag{8.32}$$

Therefore, one has a bound for $d\chi/d\tau$ which does not depend upon λ and the initial value $\phi_0 = \phi(0) = \chi(0)$; namely,

$$\left|\frac{d\chi}{d\tau}\right| \leqq \{A^2(\tau) + B^2(\tau)\}^{1/2} + |B(\tau)| ,$$

where the right-hand side is a function summable in the interval $(0,\Omega)$.

Notice that

$$\kappa = \int_0^\Omega B(\tau)d\tau = -\frac{1}{2}\int_0^\Omega \frac{\beta}{\alpha}\frac{d\alpha}{d\tau} d\tau = -\frac{1}{2}\int_0^\Omega \frac{b(t)}{a(t)}\, d\,\frac{a(t)}{\sqrt{\Delta(t)}} .$$

Using (8.31) and (8.32), we obtain

$$\chi(\Omega) - \chi(0) - \kappa = -\int_0^\Omega [A(\tau)\cos 2\chi(\tau) + B(\tau)\sin 2\chi(\tau)]\sin 2\lambda\tau d\tau +$$
$$+ \int_0^\Omega [-A(\tau)\sin 2\chi(\tau) + B(\tau)\cos 2\chi(\tau)]\cos 2\lambda\tau d\tau . \tag{8.33}$$

Assuming that $A(\tau)$ and $B(\tau)$ are absolutely continuous functions and integrating by parts in both the right-hand side integrals, we convince ourselves that this side is $O(1/\lambda)$.

If this condition on the functions $A(\tau)$ and $B(\tau)$ is not satisfied, then upon approximating them, in the metric of $L_1(0,T)$, by summable, absolutely continuous functions, we see that the first term in (8.32) tends to zero as $\lambda \longrightarrow \infty$ in any case.

Now take $\lambda = \lambda_k$, and let $x^{(k)} = (x_1, x_2)$ be, as above, a solution of system (8.1) satisfying

$$x^{(k)}(T) = \pm x^{(k)}(0) .$$

Then, according to what has been proved,

$$-\int_0^{\Omega} \frac{d}{d\tau} \arctan \frac{x_2}{x_1}\, d\tau = [\frac{k+1}{2}]\pi .$$

Since Ω is the period of the functions $\alpha(\tau)$ and $\beta(\tau)$, it is easy to see that the rotation angle of the vector $w = (u,v)$, as τ varies from 0 to Ω, is the same as that of the vector $x^{(k)}$, i.e.,

$$\phi(0) - \phi(\Omega) = \lambda_k \Omega + \chi(0) - \chi(\Omega) = [\frac{k+1}{2}]\pi . \tag{8.34}$$

According to what we showed earlier,

$$\chi(\Omega) - \chi(0) = \kappa + o(1) \quad \text{as} \quad \lambda \longrightarrow \infty ,$$

while if the functions a, b, and c have absolutely continuous derivatives, one has in addition

$$\chi(\Omega) - \chi(0) = \kappa + O(1/\lambda_k) \quad \text{as} \quad k \longrightarrow \infty .$$

Therefore, all the claims of the theorem are consequences of (8.34) and of the discussion above.

It follows from Theorem 8.4 that under the condition imposed above upon a, b, and c, the length of the zone of instability $[\lambda_{2k-1}, \lambda_{2k}]$ $(k = 1,2,\ldots)$ tends to zero as $k \longrightarrow \infty$, while the length of the zone of stability $(\lambda_{2k}, \lambda_{2k+1})$ $(k = 1,2,\ldots)$ tends to

$$\pi \left(\int_0^T \sqrt{\Delta(t)}\, dt \right)^{-1}$$

as $k \longrightarrow \infty$.

We should add that P. D. Kalafati [5] considered the case $b \equiv 0$ and other boundary conditions, and obtained a formula of the type (8.23) with $O(1)$ rather than $o(1)$.

In our case, we get such a formula without requiring that the coefficients a, b, and c be absolutely continuous. In fact, it follows from (8.30) and (8.32) that

$$|\chi(\Omega) - \chi(0) - \kappa| \leqq \frac{1}{2}\int_0^{\Omega} \frac{1}{\alpha}\left|\frac{d\alpha}{d\tau}\right| d\tau + \frac{1}{2}\int_0^{\Omega} \alpha\left|\frac{d}{d\tau}\,\frac{\beta}{\alpha}\right| d\tau .$$

Recalling (8.25) and (8.26), and using again (8.34), we obtain that for any $k = 1,2,\ldots$

$$|\lambda_k - \kappa - [\tfrac{k+1}{2}]\pi(\int_0^T \sqrt{\Delta(t)}dt)^{-1}| \leq \frac{1}{2}\int_0^T \left|d \ln \frac{a(t)}{\sqrt{\Delta(t)}}\right| + \frac{1}{2}\int_0^T \frac{a(t)}{\sqrt{\Delta(t)}}\left|d\,\frac{b(t)}{a(t)}\right| . \tag{8.35}$$

The first right-hand side integral in (8.35) makes sense whenever: 1) the function

$$\frac{a(t)}{\sqrt{\Delta(t)}} \tag{8.36}$$

has bounded variation in the interval $(0,T)$ and is bounded from below by a positive constant.

The second right-hand side integral in (8.35) and the quantity κ make sense whenever: 2) the function (8.36) is continuous, while the function $b(t)/a(t)$ has bounded variation in the interval $(0,T)$.

Suppose that the functions a, b, and c satisfy conditions 1) and 2). Then suitably approximating them with absolutely continuous functions $\tilde{a}$, $\tilde{b}$, and $\tilde{c}$, and using the fact that the bound (8.35) holds true for the system of type (8.1) having coefficients $\tilde{a}$, $\tilde{b}$ and $\tilde{c}$, we conclude that the same bound (8.35) is valid under our assumptions too.

Inequality (8.35) becomes particularly simple when applied to the equation

$$\frac{d^2y}{dt^2} + \lambda^2 p(t)y = 0 \qquad (p(t) = p(t+T) \geqq 0, \quad \int_0^T p(t)dt > 0).$$

Here

$$a(t) = p(t), \quad b(t) \equiv 0, \quad c(t) \equiv 1, \quad \kappa = 0,$$

and (8.35) gives

$$|\lambda_k - [\tfrac{k+1}{2}]\pi(\int_0^T \sqrt{p(t)}dt)^{-1}| < \frac{1}{2}(\int_0^T \sqrt{p(t)}dt)^{-1} \times \int_0^T |d \ln p(t)| \quad (k = 1,2,\ldots)$$

This inequality is valid whenever the function $p(t)$ is of bounded variation in $(0,T)$ and is bounded from below by a positive number.

To conclude this section, let us mention that in the most general case of system (8.1) with summable coefficients satisfying conditions (0.3) and (0.4), one can show that as soon as λ_k exists for all $k = 1,2,\ldots,$ one has

$$\lim_{k\to\infty} \frac{k}{\lambda_k} = \frac{2}{\pi}\int_0^T \sqrt{\Delta(t)}\,dt .$$

If the right-hand side integral is positive, then the λ_k-s do indeed exist for all $k = 1,2,\ldots$.

A necessary and sufficient condition for system (8.1) to have only a finite number of zones of stability, or for its zones of stability to extend to infinity on only one side, will be given elsewhere. In both these cases, $\Delta(t) = 0$ almost everywhere.

§ 9. THE ZONES OF STABILITY OF A SYSTEM OF SECOND-ORDER DIFFERENTIAL EQUATIONS

1. The results of the previous sections allow us to draw several conclusions concerning the first zone of stability of the differential equation (0.2)

$$\frac{d^2y}{dt^2} + \mu P(t)y = 0 ,$$

where $y = (y_1,\ldots,y_m)$ is an m-dimensional vector-function, and

$$P(t) = \|p_{jk}(t)\|_1^m$$

is a Hermitian, periodic, matrix-function $P(t + T) = P(t)$, summable over the interval $(0,T)$.

Setting $\lambda = \sqrt{\mu}$, $dy/dt = -\lambda z$, we reduce equation (0.2) to the system

$$\frac{dy}{dt} = \lambda z, \quad \frac{dz}{dt} = -\lambda P(t)y . \tag{9.1}$$

This system can be considered as a particular case of a system (0.1), where x is the direct sum of the vectors y and z:

$$x = y \dot{+} z, \quad H(t) = \begin{pmatrix} P(t) & 0 \\ 0 & I_m \end{pmatrix}. \tag{9.2}$$

Obviously, in order that the matrix $H(t)$ belong to the class $P_n(T)$ it is necessary and sufficient that the matrix $P(t)$ belong to the class $P_m(T)$.

Assuming first that the condition

$$P(t) \in P_m(T) \tag{9.3}$$

is satisfied, let us clarify what conclusions can be drawn for equation (0.2) on the basis of the results obtained for equation (0.1) in the previous sections.

We notice at the onset that for $\mu < 0$ the parameter in system (9.1) takes purely imaginary values, and so, according to Theorem 4.1, all the multipliers of this system have modulus different from 1.

Therefore, if $\mu < 0$, then each ($\not\equiv 0$) solution of equation (0.2) becomes unbounded as t tends to infinity in at least one direction.

This proposition should attributed actually to A. M. Lyapunov (see [18], Ch. III, § 52, and also [12]), who considered equation (0.2) under the assumption that the matrix-function $P(t)$ is real and continuous.

Let us remark that equation (0.2) can be reduced to system (9.1) as soon as $\mu \neq 0$. When $\mu = 0$, equation (0.2) has unbounded solutions which change linearly according to $y = \eta_0 + \eta_1 t$ $(\eta_0, \eta_1 \in E_m)$, while system (9.1) has only bounded solutions for $\lambda = 0$, and in fact these degenerate into constant vectors.

Therefore, when $P(t) \in P_m(T)$, all the solutions to equation (0.2) may be bounded only when $\mu > 0$.

A point $\mu > 0$ is called a *point of strong stability of equation* (0.2) if the corresponding point $\lambda = \sqrt{\mu} > 0$ is a point of strong stability of system (9.1).

The set of all points of strong stability of equation (0.2) decomposes into a union of open intervals situated on the half axis $\mu > 0$. The latter are called the *zones of stability of equation* (0.2). According to the order in which they appear on the half axis $\mu > 0$, these zones are called the *first*, the *second*, and so on. [Notice that in these definitions we do not assume that (9.3) is satisfied.]

The following theorem is a straightforward consequence of Theorem 6.1.

THEOREM 9.1. *Let* μ_1 *be the first characteristic number of the boundary value problem*

$$\frac{d^2y}{dt^2} + \mu P(t)y = 0, \quad y(0) = -y(T), \; y'(0) = -y'(T). \tag{9.4}$$

Then the open interval $(0,\mu_1)$ *belongs to the first zone of stability of* (0.2). *Moreover,* $(0,\mu_1)$ *coincides with the first zone of stability of equation* (0.2) *whenever the matrix-function* $P(t)$ *is real.*

Denote by $p_M(t)$ and $p_\mu(t)$ the largest and, respectively, the smallest eigenvalue of the matrix $P(t)$.*) Then

$$p_\mu(t)I_m \leq P(t) \leq p_M(t)I_m \; .$$

According to Theorem 3.3, if one substitutes $p_M(t)I_m$ or $p_\mu(t)I_m$ for the matrix $P(t)$ in (9.4), the μ_1 will decrease only or increase only, respectively, whence the next theorem.

THEOREM 9.2. *The first zone of stability of equation* (0.2) *is no smaller than the first zone of stability of the scalar equation*

$$\frac{d^2y}{dt^2} + \mu p_M(t)y = 0 \tag{9.5}$$

and, in the case of a real $P(t)$, *no larger than the first zone of stability of the scalar equation*

$$\frac{dy^2}{dt^2} + \mu p_\mu(t)y = 0 \; . \tag{9.6}$$

2. Combining Theorem 9.2 with the Lyapunov test for μ to belong to the first zone of stability, we obtain the <u>next generalization</u> of the latter.

*) Warning: do not confuse the two μ-s.

TEST A. *The number* μ *belongs to the first zone of stability of equation* (0.2) *whenever*

$$0 < \mu < \frac{4}{T} \left(\int_0^T p_M(t)\,dt\right)^{-1} .$$

We have reported this test previously in our note [9].

As an easy consequence of test I_n (§ 7), one obtains a test which is simpler in terms of computability and, in many cases, stronger.

TEST B. *The number* μ *belongs to the first zone of stability of equation* (0.2) *whenever*

$$0 < \mu < \frac{4}{TM(Q)} , \tag{9.7}$$

where $M(Q)$ *is the largest eigenvalue of the matrix*

$$Q = \left\| \int_0^T |p_{jk}(t)|\,dt \right\|_1^m .$$

Indeed, according to the notation of § 7, no. 2,

$$Q = \int_0^T P_a(t)\,dt .$$

Therefore, if $H(t)$ is defined by equality (9.2), then

$$C = J_a \int_0^T H_a(t)\,dt = \begin{pmatrix} 0 & TI_m \\ Q & 0 \end{pmatrix} ,$$

whence

$$\det(C - \lambda I_n) = \det(\lambda^2 I_m - TQ) .$$

It follows that the largest eigenvalue $M(C)$ of the matrix C satisfies

$$M^2(C) = TM(Q) . \tag{9.8}$$

The number $\mu > 0$ will belong to the first zone of stability of equation (0.2) if and only if $\lambda = \sqrt{\mu}$ belongs to the central zone of stability of system (9.1). According to test I_n (§ 7), the latter means that

$$(0 <)\ \lambda^2 < 4/M^2(C) .$$

As (9.8) shows, this inequality is equivalent for $\mu = \lambda^2$ to inequality (9.7).

3. Let us compare tests *A* and *B*. To this end, notice that for a Hermitian matrix A, the number M(A) is just the norm $|A|$ of A. And since $|A + B| \leq |A| + |B|$ for any two matrices A and B of the same order, one has

$$M(A + B) \leq M(A) + M(B) .$$

Therefore, if all elements of the matrix P(t) are nonnegative

$$p_{jk}(t) \geq 0 \qquad (0 \leq t \leq T \; ; \; j,k = 1,\ldots m) ,$$

then

$$M(Q) = M(\int_0^T P(t)dt) \leq \int_0^T M(P(t))dt = \int_0^T p_M(t)dt . \qquad (9.9)$$

In this way, we are led to one of the assertions made in the next proposition.

Test B is stronger than test A whenever one of the following conditions is fulfilled:

1) *All the elements of the matrix* P(t) *are nonnegative functions.*

2) *The matrix* P(t) *is Jacobi, i.e.,*

$$p_{jk}(t) \equiv 0 \quad \text{for} \quad |j - k| > 1 .$$

3) $m = 2$.

When condition 1) is fulfilled, the proposition follows from inequality (9.9).

Condition 3) can be considered as a particular case of condition 2).

We have therefore to consider only the case when P(t) is a Jacobi matrix.

We should remark that in the most general case one can always write, instead of (9.9), that

$$M(Q) = M(\int_0^T P_a(t)dt) \leq \int_0^T M(P_a(t))dt = \int_0^T p_a(t)dt ,$$

where $p_a(t)$ is the largest eigenvalue of the matrix $P_a(t)$.

If the matrix P(t) satisfies (9.3), as we assumed up

to now, then all its diagonal elements are nonnegative functions. Therefore, if $P(t)$ is also Jacobi, then the matrix $P_a(t)$ can differ from $P(t)$ only by the arguments (in the case of a real matrix-function - only by the signs) of the elements lying on the two diagonals adjacent to the principal one. But, as one can easily see, changing the arguments of these elements in a Jacobi matrix has no effect on its eigenvalues.

Thus, $p_a(t) \equiv p_M(t)$ and one has again

$$M(Q) \leq \int_0^T p_M(t)dt ,$$

which completes the proof of the proposition.

Let us mention that in studying the torsional oscillations of crank shafts one deals precisely with the case when condition 2) is fulfilled (see the paper of N. E. Kochin [7]).

4. Without giving complete proofs because of lack of space, we point out a number of possible generalizations and developments of the previous results.

First of all, let us remark that Theorem 5.3 allows us to establish a series of tests determining whether μ belongs to the zones of stability of equation (0.2) which succeed to the first one. Thus, for example, on the basis of this theorem (and of Theorem 4.2) one can assert that if any two zones of stability, having the same number $k > 1$, of the two equations (9.5) and (9.6), do intersect, then their intersection is included in one of the zones of stability of equation (0.2).

Using only this, one may obviously obtain various analytic tests for determining whether a given μ is a point of strong stability of equation (0.2). These tests are based on special tests [26,13] for μ to belong to a zone of stability having a specific number of the scalar equation

$$\frac{d^2y}{dt^2} + \mu p(t)y = 0 .$$

The tests appearing in the note by M. G. Neigauz and V. B. Lidskii [16] may be obtained in precisely this way. Those are tests for the point $\mu = 1$ to belong to some zone of stability of equation (0.2), and use no knowledge of the matrix-function

$P(t)$ except for the functions $p_\mu(t)$ and $p_M(t)$.

Incidentally, there are other results in [16] that may be deduced from Theorem 5.3 too.

5. Essential for applications is the fact that a number of fundamental propositions concerning the zones of stability of equation (0.2) retain their validity if one drops condition (9.3).

We do so, and in order to simplify the formulations of the results obtained here, we assume that there is no constant vector $\eta \neq 0$ $(\eta \in E_m)$ such that

$$P(t)\eta \equiv 0 \quad \text{(almost everywhere)} . \tag{9.10}$$

In making this assumption, we have not disregarded anything essential.

Indeed, if the linear set L of all vectors $\eta \in E_m$ satisfying condition (9.10) does not reduce to zero, then any vector $\eta \neq 0$ of this set will give a nonzero solution $y = \eta$ of equation (0.2).

Then any other solution $y = y(t)$ of (0.2) can be decomposed as $y = \eta + y_1$, where $\eta \in L$, and $y_1 = y_1(t)$ is a solution of (0.2) which is orthogonal to the vector η for all t, i.e., $(y_1,\eta) \equiv 0$. It is easy to see that the problem of finding the general solution y_1 to system (0.2) which is orthogonal to η reduces to solving a differential system of the same type, but made up of only $m - d$ equations, where $d = \dim L$.

Notice that the assumption

$$P(t)\eta \not\equiv 0 \quad \text{for} \quad \eta \neq 0 \tag{9.11}$$

does not at all exclude the possibility that the matrix $P(t)$ be degenerate for all t.

If (9.11) holds true, then it turns out that, given any real μ, precisely m multipliers of system (9.1) lie inside the unit circle, and precisely m - outside.

Considering the multipliers of system (0.2) as functions of the parameter μ from the upper half plane $\operatorname{Im} \mu > 0$, we can subdivide them into multipliers of the first and second kind according to whether the given multiplier lies inside or outside the unit circle. [Now, when we assume that $P(t) \notin P_m(T)$, this subdivision may be not possible if one proceeds from system (9.1) and considers the multipliers as functions of the parameter λ.] Subsequently, the theorem of § 4, which loses its meaning in our case and considers the behavior of the multipliers of one or the other kind as functions of λ $(\mathrm{Im}\,\lambda \geqq 0)$, can be reinstated by now considering the multipliers as functions of the parameter μ $(\mathrm{Im}\,\mu \geqq 0)$.

Then, by studying the zones of stability of equation (0.2) on the half axis $\mu > 0$, one can make the following statement with no difficulty.

Theorem 9.1 remains valid too when condition $P(t) \in P_m(T)$ *is not satisfied, but the condition of positivity in the mean is fulfilled:*

$$\int_0^T P(t)dt > 0 \ . \qquad (9.12)$$

One should add to this that given any summable, Hermitian, matrix-function $P(t)$ $(0 \leqq t \leqq T)$, the characteristic numbers of the boundary value problem (9.4) are all real, and if condition (9.12) is satisfied too, then among these numbers one finds an infinity of positive ones.

We remark that the existence of a zone of stability of equation (0.2) having $\mu = 0$ as the left endpoint is a straightforward consequence of Theorem 6.4, when condition (9.12) is fulfilled.

As soon as Theorem 9.1 is established under the general assumption (9.12), it is not hard to show, using the method of § 7, that test B is also valid under this assumption.

However, we should mention that the result contained in this test can be strengthened in many cases if one applies the test not directly to equation (0.2), but rather to the equation resulting from it in the way indicated below.

A Hermitian matrix-function $P(t)$ has an infinite number of representations of the form

$$P(t) = P^{+}(t) - P^{-}(t) \qquad (0 \leqq t < \infty) \ , \qquad (9.13)$$

where

$$P^{+}(t) \geqq 0, \quad P^{-}(t) \geqq 0 \quad (0 \leqq t < \infty) .$$

Moreover, if $P(t)$ is summable, then one can choose $P^{\pm}(t)$ to be summable too. For example, this always happens in the so-called *orthogonal* decomposition (9.13), characterized by the condition

$$P^{+}(t)P^{-}(t) \equiv P^{-}(t)P^{+}(t) \equiv 0 .$$

If one replaces the Hermitian matrix-function $P(t)$ by some other Hermitian matrix-function $P_1(t) \geqq P(t) \quad (0 \leqq t \leqq T)$, the positive characteristic numbers of the boundary value problem (9.4) can only decrease.

Therefore, if condition (9.12) is fulfilled, then the first zone of stability of equation (0.2) is not smaller than the first zone of stability of the equation

$$\frac{d^2y}{dt^2} + \mu P^{+}(t)y = 0 . \tag{9.14}$$

[Strictly speaking, this assertion follows from the above considerations only for a real, symmetric, matrix-function $P(t)$. To prove it in the general case of a Hermitian $P(t)$, one uses a theorem of the type 5.3.]

We indicated this result for the case of the scalar equation (0.2), i.e., when $m = 1$, in paper [13].

Combining the last assertion (where we choosed the orthogonal decomposition (9.13)) and Theorem 9.2, we can state that *the first zone of stability of equation* (0.2) *is not smaller than the first zone of stability of the scalar equation*

$$\frac{d^2y}{dt^2} + \mu p_M^{+}(t)y = 0 .$$

Here

$$p_M^{+}(t) = \frac{1}{2}(p_M(t) + |p_M(t)|) \quad (0 \leqq t < \infty) .$$

Returning to test $\mathcal{B}$, let us emphasize that in applying it, we have to keep in mind the following possibility: the interval of values of μ belonging to the first zone of stability of equation (9.14) produced by this test may be larger than the corresponding interval produced by this test for equation (0.2), althought the relationship between the first zones of stability of these equations always exhibits the opposite character.

For example, this always happens in the scalar case $m = 1$.

§ 10. THE CRITICAL FREQUENCIES OF AN ε-PARAMETRICALLY PERTURBED SYSTEM

1. Let us assume that the motion of some given, oscillating system S having m degrees of freedom, is described by the equation

$$\frac{d^2y}{dt^2} + P_0 y = 0 , \tag{10.1}$$

where $y = (y_1, \ldots, y_m)$ is the vector made up of the generalized coordinates y_j $(j = 1, \ldots m)$ of the system S, and P_0 is a constant, real, symmetric matrix of order m whose associated quadratic form is positive $(P_0 > 0)$.

Let $\eta^{(j)}$ $(j = 1, \ldots m)$ be a complete orthonormal system of real eigenvectors of the matrix P_0, i.e.,

$$P_0 \eta^{(j)} = \omega_j^2 \eta^{(j)} \quad (j = 1, \ldots m) ,$$

where

$$(\eta^{(j)}, \eta^{(k)}) = \delta_{jk} \quad (j,k = 1, \ldots, m) \tag{10.2}$$

and

$$0 < \omega_1 \leqq \omega_2 \leqq \ldots \leqq \omega_m .$$

The functions

$$y_{\pm j}(t) = e^{\pm i\omega_j t} \eta^{(j)} \quad (j = 1, \ldots m)$$

form a complete system of $2m$ linearly independent solutions of equation (10.1).

Consider the real, symmetric matrix $P_0^{1/2} > 0$ whose square equals P_0; $P_0^{1/2}$ is defined by the equalities

$$P_0^{1/2} \eta^{(j)} = \omega_j \eta^{(j)} \quad (j = 1, \ldots m) .$$

Then the general solution $y(t)$ of system (10.1) can be expressed as

$$y(t) = \cos(tP_0^{1/2})y(0) + P_0^{-1/2}\sin(tP_0^{1/2})y'(0) , \tag{10.3}$$

where

$$\cos(tP_0^{1/2}) = \sum_{n=0}^{\infty} \frac{(-1)^n t^{2n} P_0^n}{(2n)!} ,$$

$$P_0^{-1/2}\sin(tP_0^{1/2}) = \sum_{n=0}^{\infty} \frac{(-1)^n t^{2n} P_0^n}{(2n+1)!} .$$

Equation (10.1) is equivalent to the system

$$\frac{dy}{dt} = z, \quad \frac{dz}{dt} = - P_0 y . \tag{10.4}$$

From (10.3) it is not hard to conclude that the matrizant of system (10.4) is the J-orthogonal matrix-function

$$U_0(t) = \begin{pmatrix} \cos(tP_0^{1/2}) & P_0^{-1/2}\sin(tP_0^{1/2}) \\ P_0^{1/2}\sin(tP_0^{1/2}) & \cos(tP_0^{1/2}) \end{pmatrix} .$$

The eigenvalues of $U_0(t)$ are the numbers

$$\rho_j(t) = e^{i\omega_j t} \quad (j = 1,\dots m) \tag{10.5}$$

and

$$\rho_{-j}(t) = e^{-i\omega_j t} \quad (j = 1,\dots,m) . \tag{10.6}$$

Denoting by $\xi^{(j)}$ the direct sum of the vectors $\eta^{(j)}$ and $i\omega_j\eta^{(j)}$:

$$\xi^{(j)} = \eta^{(j)} \dot{+} i\omega_j\eta^{(j)} \quad (j = \pm 1,\dots,\pm m; \ \omega_{-j} = - \omega_j) ,$$

we can easily see that

$$U_0(t)\xi^{(j)} = \rho_j(t)\xi^{(j)} \quad (j = \pm 1,\dots,\pm m) .$$

Moreover, by (10.2) we get that

$$(J\xi^{(j)},\xi^{(k)}) = 0 \quad \text{for} \quad j \neq k \quad (j,k = \pm 1,\dots,\pm m; \ J = J_{2m}),$$

and

$$i(J\xi^{(j)},\xi^{(j)}) = 2\omega_j \quad (j = \pm 1,\dots,\pm m) .$$

It follows that an eigenvalue in the collection (10.6) (the collection (10.5)) is an eigenvalue of the first (second) kind if and only if it does not equal to any of the eigenvalues from the other collection.

Consequently, the J-orthogonal matrix $U_0(t)$ is of stable type for those, and only for those, values of t for which none of the numbers in (10.5) equals any number in (10.6), i.e.,

$$t(\omega_j + \omega_k) \neq 0 \pmod{2\pi} \quad (j,k = 1,\ldots,m) . \tag{10.7}$$

2. Now imagine that the system S is parametrically perturbed (excited) or, more precisely, that we allow some of its parameters (dimensions, masses, inertia momenta of rotating parts, capacitances of capacitors, self-inductances of circuits, and so on) to vary according to some arbitrary, periodic law with period $T = 2\pi/\omega$. Suppose, in addition, that the characteristics of the perturbation themselves depend continuously upon a small, scalar parameter $\varepsilon > 0$ in such a way that the amplitudes of the variation of the parameters of the system tend to zero as $\varepsilon \longrightarrow 0$ (but the period T does not depend upon ε).

In this case we shall say that system S is *ε-parametrically perturbed (excited)*.

The motion of the system S under an ε-parametric perturbation is usually described (in the linearized version) by the differential equation

$$\frac{d^2y}{dt^2} + P(\omega t;\varepsilon)y = 0 , \tag{10.8}$$

where $P(t;\varepsilon)$ is a continuous, symmetric, matrix-function of the arguments t and ε $(-\infty < t < \infty,\ 0 \leqq \varepsilon \leqq \varepsilon_0)$, satisfying

$$P(t + 2\pi;\varepsilon) = P(t;\varepsilon) \quad (-\infty < t < \infty, \quad 0 \leqq \varepsilon \leqq \varepsilon_0) \tag{10.9}$$

and

$$P(t;0) \equiv P_0 . \tag{10.10}$$

For the discussion below, it suffices to assume that the matrix-function $P(t;\varepsilon)$ is, for any fixed ε $(0 \leqq \varepsilon \leqq \varepsilon_0)$, summable with respect to t over the interval $(0,2\pi)$, and is continuous in the mean with respect to ε at $\varepsilon = 0$, i.e.,

$$\lim_{\varepsilon \to 0} \int_0^{2\pi} |P_0 - P(t;\varepsilon)|dt = 0 . \qquad (10.11)$$

Let us give the following definition.

A frequency ω is said to be *critical* for the given ε-parametric perturbation (i.e., for the given function $P(t;\varepsilon)$) of the system S if there is no $\varepsilon_\omega > 0$ such that the motion of S is stable for $0 < \varepsilon < \varepsilon_\omega$, i.e., such that all the solutions to equation (10.8) are bounded for $0 < \varepsilon < \varepsilon_\omega$.

THEOREM 10.1. *Independently of the ε-parametric perturbation of a system S with proper (natural) frequencies ω_j ($j = 1,\dots,m$) (i.e., independently of the choice of the matrix-function $P(t;\varepsilon)$ satisfying conditions* (10.9), (10.10), *and* (10.11)), *the critical frequencies of S must be among the numbers*

$$\omega_{j,k,N} = \frac{\omega_j + \omega_k}{N} \qquad (1 \leq j \leq k \leq m,\ N = 1,2,\dots) .$$

PROOF. Let $U_{\omega,\varepsilon}$ be the monodromy matrix of the system

$$\frac{dy}{dt} = z, \quad \frac{dz}{dt} = -P(\omega t;\varepsilon)y ,$$

equivalent to equation (10.8). In virtue of (10.11),

$$\lim_{\varepsilon \to 0} \int_0^{T} |P_0 - P(\omega t;\varepsilon)|dt = 0 \qquad (T = \frac{2\pi}{\omega}) ,$$

and hence, according to Lemma 5.1, $U_{\omega,\varepsilon}$ tends to the monodromy matrix of system (10.4) when $\varepsilon \longrightarrow 0$, if the latter is considered to be periodic with period T. Therefore,

$$\lim_{\varepsilon \to 0} U_{\omega,\varepsilon} = U_0(\frac{2\pi}{\omega}) .$$

If $\omega > 0$ is different from all the numbers $\omega_{j,k,N}$, then the value $t = T$ satisfies condition (10.7), i.e.,

$$\frac{2\pi}{\omega}(\omega_j + \omega_k) \neq 0 \pmod{2\pi} ,$$

and, as we proved earlier, the J-orthogonal matrix $U_0(\frac{2\pi}{\omega})$ will be of stable type. By Theorem 1.2, it then follows that one can find $\varepsilon_\omega > 0$ such that the matrix $U_{\omega,\varepsilon}$ will be also of stable type for all $0 < \varepsilon < \varepsilon_\omega$. Consequently, all the solutions to system (10.8) will be bounded for $0 < \varepsilon < \varepsilon_\omega$.

The theorem is proved.

We notice also that one can supplement Theorem 10.1 with the claim that *each* of the numbers $\omega_{j,k,N}$ does actually become a critical frequency ω of the system S for a suitable ε-parametric perturbation of this system, i.e., for a suitable choice of the matrix-function $P(t;\varepsilon)$.

The equations of motion of a mechanical system S having m degrees of freedom will have the form (10.1) if its kinetic and, respectively, potential energy, T and Π, have the following expressions in the generalized coordinates y_j $(j = 1,\ldots,m)$:

$$T = \frac{1}{2}\sum_{j=1}^{m}\left(\frac{dy_j}{dt}\right)^2, \quad \Pi = \frac{1}{2}\sum_{j,k=1}^{m} p_{jk}y_jy_k \quad (P_0 = \|p_{jk}\|_1^m).$$

It is always possible to reduce T and Π to this form by a suitable choice of the generalized coordinates y_j $(j = 1,\ldots,m)$ as linear combinations of the original coordinates q_j $(j = 1,\ldots,m)$, provided that T and Π are positive quadratic forms with respect the original coordinates:

$$T = \frac{1}{2}\sum_{j,k=1}^{m} a_{jk}\dot{q}_j\dot{q}_k, \quad \Pi = \frac{1}{2}\sum_{j,k=1}^{m} c_{jk}q_jq_k \quad \left(\dot{q}_j = \frac{dq_j}{dt}\right).$$

Therefore, Theorem 10.1 is applicable to any mechanical system having kinetic and potential energies of these forms, that is, practically speaking, to ε-parametric perturbations of any mechanical system having a finite number of degrees of freedom, and which performs small oscillations around a stable equilibrium position.

For systems S with one degree of freedom, and having, as such, a single natural frequency ω_1, this theorem was known, although it could be that it was never formulated in such a precise way. In this particular case, the possible values of the critical frequencies form a simple sequence

$$\omega_N = 2\omega_1/N \quad (N = 1,2,\ldots)$$

beginning with twice the frequency of the unperturbed system.

REFERENCES

1. Chaplygin, S.A.: *Foundations of a new method for approximative integration of differential equations*, Collected Works, Vol. 1, Gostekhizdat, 1948, p. 347. (Russian)

2. Chebotarev, N. G. and Meiman, I. N.: *The Routh-Hurwitz problem for polynomials and entire functions*, Trudy Mat. Inst. im. V. A. Steklova, Vol. 26, Izdat. Akad. Nauk SSSR (1949), 1-331. (Russian)

3. Gantmaher, F. R.: *The Theory of Matrices*, 2nd ed., "Nauka", Moscow, 1966; English. transl. of 1st ed., Chelsea, New York, 1959.

4. Gantmaher, F. R. and Krein, M. G.: *Oscillation Matrices and Kernels and Small Oscillations of Mechanical Systems*, 2nd ed. GITTL, Moscow, 1950; German transl., Akademie-Verlag, Berlin, 1960.

5. Kalafati, P. D.: *On a certain asymptotic formula*, Nauchn. Zap. Nikolaevsk. Gos. Pedag. Inst. im. V. G. Belinskogo, No. 3, Radyan. Shkola, Kiev (1951), 92-94. (Russian)

6. Karaseva, T. M.: *On the expansion of an arbitrary function into the eigenfunctions of a boundary value problem*, Zap. Khar'kovskogo Mat. Obshch., Vol. 21 (1949), 59-75. (Russian)

7. Kochin, N. E.: *On torsional oscillations of crankshafts*, Collected Works, Vol. II, Izdat. Akad. Nauk SSSR, 1949, pp. 507-535. (Russian)

8. Kovalenko, K. R. and Krein, M. G.: *On some investigations of A. M. Lyapunov on differential equations with periodic coefficients*, Dokl. Akad. Nauk SSSR 75, No. 4 (1950), 459-498. (Russian)

9. Krein, M. G.: *A generalization of some investigations of A. M. Lyapunov on linear differential equations with periodic coefficients*, Dokl. Akad. Nauk SSSR 73, No. 3 (1950), 445-448. (Russian)

10. Krein, M. G.: *On an application of the fixed-point principle in the theory of linear transformations of spaces with an indefinite metric*, Uspekhi Mat. Nauk 5, No. 2 (1950), 180-199; English transl., Amer. Math. Soc. Transl. (2) 1 (1955), 27-35.

11. Krein, M. G.: *A contribution to the theory of entire matrix-functions of exponential type*, Ukr. Mat. Zh., 3, No. 2 (1951), 163-173. (Russian)

12. Krein, M. G.: *On the application of an algebraic proposition in the theory of monodromy matrices*, Uspekhi Mat. Nauk 6, No. 1 (1951), 171-177. (Russian)

13. Krein, M. G.: *On certain problems on the maximum and minimum of characteristic values and on the Lyapunov zones of stability*, Priklad. Mat. i Mekh., 15, No. 3 (1951), 323-348;

English transl., Amer. Math. Soc. Transl. (2) 1 (1955), 163-187.

14. Krein, M. G.: *On weighted integral equations, the distribution functions of which are not monotonic*, Memorial Volume to D. A. Grave, GTTI, Moscow, 1940, pp. 88-103. (Russian)

15. Krein, M. G.: *On linear completely continuous operators in functional spaces with two norms*, Sb. Trudov Inst. Mat. Akad. Nauk Ukrain. SSR 9 (1947), 104-129. (Ukrainian)

16. Lidskii, V. B. and Neigauz, M. G.: *On the boundedness of the solutions to linear systems of differential equations with periodic coefficients*, Dokl. Akad. Nauk SSSR 77, No. 2 (1951), 189-192. (Russian)

17. Levin, B. Ya.: *Distributions of Zeros of Entire Functions*, GITTL, M scow, 1956; English transl., Transl. Math. Monographs, vol. 5, Amer. Math. Soc., Providence, R.I., 1964.

18. Lyapunov, A. M.: *The general problem of stability of motion*, 2nd ed., Gl. Red. Obshchetekh. Lit., Leningrad, Moscow, 1935; Liapunoff, A. M.: *Problème général de la stabilité du mouvement*, Ann. Fac. Sci. Toulouse 2 (1907), 203-474; reprinted Ann. Math. St dies, no. 17, Princeton Univ. Press, Princeton, N. J., 1947.

19. Lyapunov, A. M.: *On a problem concerning linear differential equations of second order with periodic coefficients*, Soobshch. Khar'kov. Mat. Obshch., 2nd series, 5, No. 3 (1896), 190-254. (Russian)

20. Lyapunov (Liapunoff), A. M.: *Sur une èquation diffèrentielle linèaire du second ordre*, C. R., t. 128, No. 15 (1899), 910-913.

21. Lyapunov (Liapunoff), A. N.: *Sur une équation transcendante et les équations différentielles linéaires du second ordre ā coefficients périodiques*, C. R., t. 128, No. 18 (1899), 1085-1088.

22. Lyapunov, A. N.: *Sur une série dans la théorie des equations différentielles linéaires du second ordre ā coefficients périodiques*, Zap. Akad. Nauk Fiz. Mat. Otdel., Ser. 8, vol. 13, No. 2 (1902), 1-70.

23. Makarov, S. M.: *Investigation of the characteristic equation of a linear system of two equations of first order with periodic coefficients*, Priklad. Mat. i Mekh., 15, No. 3 (1951), 373-378. (Russian)

24. Mal'tsev (Mal'cev), A. I.: *Foundations of Linear Algebra*, OGIZ, Moscow, 1948; English transl., Freeman, San Francisco, CA., 1963.

25. Nemytskii, V. V. and Stepanov, V. V.: *Qualitative Theory of Differential Equations*, 2nd ed., Gostekhizdat, 1949; English transl., Princeton Univ. Press, Princeton, N. J., 1960.

26. Yakubovich, V. A.: *On the boundedness of the solutions of*

the equation $y'' + p(t)y = 0$, $p(t + \omega) = p(t)$, Dokl. Akad. Nauk SSSR 74, No. 5 (1950), 901-903. (Russian)

27. Yakubovich, V. A.: *Tests of stability for systems of two equations of canonical form with periodic coefficients*, Dokl. Akad. Nauk SSSR 78, No. 2 (1951), 221-224. (Russian)

28. Yakubovich, V. A.: *Stability problems for a system of two linear differential equations with periodic coefficients*, Auto-synopsis of Candidate Dissertation, Leningrad State Univ., 1953. (Russian)

29. Zhukovskii, N. E.: *A condition for the boundedness of the integrals of the equation* $d^2y/dx^2 + py = 0$, Collected Works, vol. 1, 1948, pp. 246-253. (Russian)

ON CERTAIN NEW STUDIES IN THE PERTURBATION THEORY FOR SELFADJOINT OPERATORS*

M. G. Krein

FIRST LECTURE

In the last decade, a number of significant results have been obtained in the perturbation theory for selfadjoint and nonselfadjoint operators. In particular, several important results have been gained in the theory of nonselfadjoint perturbations of selfadjoint operators. We should mention, for example, the various theorems concerning the completeness of root vectors which have been proven by Soviet mathematicians, starting with the well-known work of M. V. Keldysh.

In the present series of lectures, we shall discuss some investigations into the theory of selfadjoint perturbations of selfadjoint operators with arbitrary spectra. The motivating force behind these investigations comes entirely from physicists, who advanced many new, and sometimes most paradoxical ideas. At this time, mathematicians have joined in the effort of elaborating these ideas. As a result, a rather "formidable" theory has been constructed, which cannot be covered in a few lectures. This is why the aim of the present series of lectures is to explain (not always with complete proofs) the fundamental propositions of this theory, and to formulate a number of open problems.

In all probability, these lecture notes would not have

---

*Translation of First Summer Math. School (Kanev, 1963), part I, "Naukova Dumka," Kiev (1964), 103-187.

come to light, had they been not written down and elaborated upon by a group of attentive listeners, composed of I. S. Iokhvidov (the group's leader), M. L. Gorbachuk, V. I. Gorbachuk, V. I. Kolomytsev, and L. P. Nizhnik. After their efforts, the task of bringing these lectures into the present form yielded no particular difficulty. The author expresses his heartfelt and profound gratitute to all these people, and especially to I. S. Iokhvidov and to the Gorbachuks husband and wife.

INTRODUCTION
(a review of results)

1. The elementary part of the theory has analogues in linear algebra, though these disappear as we proceed further into the depth of the theory.

Let G_m be the usual m-dimensional space, and let H be a Hermitian operator defined in it:

$$H = \sum_{j=1}^{m} \lambda_j(\cdot,\phi_j)\phi_j \quad ,$$

where ϕ_j $(j = 1,\ldots,m)$ is an orthonormal system of eigenvectors of H. In other words,

$$H = \sum_{j=1}^{m} \lambda_j E_j \quad ,$$

where E_j $(j = 1,\ldots,m)$ are the rank-one, mutually orthogonal projections onto the eigendirections of H. Let $\Phi(\lambda)$ $(-\infty < \lambda < \infty)$ be some complex-valued function. Then

$$\Phi(H) = \sum_{j=1}^{m} \Phi(\lambda_j)E_j \quad .$$

The trace of the operator $\Phi(H)$ is given by

$$\mathrm{Sp}\,\Phi(H) = \sum_{j=1}^{m} \Phi(\lambda_j) \quad . \tag{1}$$

For the operator H we introduce the distribution

function of its eigenvalues, written $n(\lambda)$. Namely, $n(\lambda)$ is the number of eigenvalues of H which are no larger than λ:

$$n(\lambda) = \int_{-\infty}^{\infty} \sum_{j=1}^{m} \delta(\mu - \lambda_j)d\mu \quad .$$

Assuming that the function Φ is continuous, one can rewrite (1) in the form

$$\mathrm{Sp}\,\Phi(H) = \int_{-\infty}^{\infty} \Phi(\lambda)dn(\lambda) \quad .$$

Now suppose that we are given two Hermitian operators in G_m: the "perturbed" one, $\tilde{H}$, and the "initial" one, H. Then:

$$\mathrm{Sp}\{\Phi(\tilde{H}) - \Phi(H)\} = \int_{-\infty}^{\infty} \Phi(\lambda)d\{\tilde{n}(\lambda)-n(\lambda)\} \quad .$$

Since $\tilde{n}(\lambda)-n(\lambda)$ vanishes for large enough values of $|\lambda|$, one can integrate by parts, under the assumption that function $\Phi(\lambda)$ is absolutely continuous, to obtain

$$\mathrm{Sp}\{\Phi(\tilde{H}) - \Phi(H)\} = \int_{-\infty}^{\infty} [n(\lambda)-\tilde{n}(\lambda)]\Phi'(\lambda)d\lambda \quad .$$

Denoting $n(\lambda)-\tilde{n}(\lambda)$ by $\xi(\lambda)$, we arrive at the "trace formula"

$$\mathrm{Sp}\{\Phi(\tilde{H}) - \Phi(H)\} = \int_{-\infty}^{\infty} \xi(\lambda)\Phi'(\lambda)d\lambda \quad . \tag{2}$$

We call $\xi(\lambda) = \xi(\lambda;\tilde{H},H)$ the *spectral shift function*. In the finite dimensional case we have considered, $\xi(\lambda)$ is an integer-valued function. It is not hard to see that the function $\xi(\lambda)$ is nonnegative whenever $\tilde{H} \geq H$, where inequality is understood as inequality of the corresponding quadratic forms.

2. The notion of trace has a precise meaning also for some classes of operators in a Hilbert space G.

Let A be a compact selfadjoint operator in G, and let $\lambda_j(A)$ be its eigenvalues. The operator A is said to have an absolutely convergent trace if the series $\sum\lambda_j(A)$ is absolutely convergent. When this is so, the trace of A is defined to be the sum of this series:

$$\operatorname{Sp} A = \sum \lambda_j(A) \quad .$$

A compact nonselfadjoint operator A is said to have a trace if each of its Hermitian components

$$A_R = (A + A^*)/2 \quad \text{and} \quad A_I = (A - A^*)/2i$$

has a trace. Then the trace $\operatorname{Sp} A$ of the operator A is defined as the sum $\operatorname{Sp} A_R + \operatorname{Sp} A_I$.

The operators having a trace are called *trace class* (or *nuclear*). The set of trace class operators is denoted by I_1. An equivalent definition of the class I_1 will be given in the second lecture.

Let there be given two selfadjoint, not necessarily bounded operators H and $\tilde{H}$ in the Hilbert space G. If the difference $\Phi(\tilde{H}) - \Phi(H) \in I_1$, then $\operatorname{Sp}\{\Phi(\tilde{H}) - \Phi(H)\}$ makes sense. Now one may ask if it would be possible to associate some function $\xi(\lambda)$ to the ordered pair $H, \tilde{H}$ such that the "trace formula" (2) would hold for a sufficiently large class of absolutely continuous functions $\Phi(\lambda)$.

This question was first posed by the physicist I. M. Lifshits [47,48], in connection with certain problems arising in the theory of crystals. He showed that if the "perturbation" is of finite rank, i.e.,

$$\tilde{H} = H + \sum_{j=1}^{\nu} \varepsilon_j(\cdot, \chi_j)\chi_j \quad , \quad \varepsilon_j = \pm 1 \; ,$$

then such a function $\xi(\lambda)$ exists. The requirement that the spectral function of the operator H be smooth in a certain sense played an essential role in the reasoning of I. M. Lifshits. Moreover, it was assumed that the operator H has no discrete spectrum. When $\tilde{H}$ has such a spectrum, formula (2) has a different form.

Subsequently, it was shown in M. G. Krein's work [35,36] that the trace formula (2) is valid under very general assumptions on the "closeness" between the operators H and $\tilde{H}$, and with no restrictions on the smoothness of the spectral function of H. The fulfillment of these general "closeness" conditions does not even exclude the possibility that H and $\tilde{H}$ be so "far apart" that their domains intersect only at zero.

The function $\xi(\lambda)$ is not always uniquely defined by the trace formula (2). This situation may occur when the collection of functions $\Phi(\lambda)$ for which $\Phi(\tilde{H}) - \Phi(H) \in I_1$ is not a rich enough collection. However, if this collection does contain the function $\Phi(\lambda) = \lambda$, i.e., $\tilde{H} = H + V$ with $V \in I_1$, then $\xi(\lambda)$ is uniquely determined. For operators H and $\tilde{H}$ semibounded below, the uniqueness of $\xi(\lambda)$ is a consequence of the natural requirement that $\xi(\lambda) \longrightarrow 0$ as $\lambda \longrightarrow -\infty$. When the latter is fulfilled, $\xi(\lambda) \equiv 0$ at the left of the spectra of H and $\tilde{H}$. Moreover, if that part of the spectrum of H lying to the left of some point a $(\leqq \infty)$ is discrete, and supposing that $\xi(\lambda)$ exists, then the spectrum of $\tilde{H}$ has the same property. At the same time, $\xi(\lambda) = n(\lambda)-\tilde{n}(\lambda)$ for $\lambda < a$, where $n(\lambda)$ $(\tilde{n}(\lambda))$ denotes the number of eigenvalues of the operator H $(\tilde{H})$ (counted with their multiplicities) which are no larger than λ, as we recall from the finite dimensional case. A reasonable normalization of the function $\xi(\lambda)$ which ensures its uniqueness up to an integer term is available in other cases too.

3. The spectral shift function appears in a new light when one considers the radial Schrödinger equation.

Let $G = L_2(0,\infty)$, and let

$$H = - \frac{d^2}{dr^2}\bigg|_h , \quad \tilde{H} = \left(- \frac{d^2}{dr^2} + V\right)\bigg|_h$$

where V is the operator representing multiplication by a function $v(r)$ $(0 \leqq r < \infty)$ satisfying

$$\int_0^\infty r|v(r)|dr < \infty , \tag{3'}$$

$$\int_1^\infty |v(r)|\, dr < \infty . \tag{3''}$$

Here the symbol h is appended to the differential expression in order to emphasize that we consider the operator to be defined by this differential expresion together with the boundary condition $\psi(0) = 0$ (this notation is connected with the notion of "hard" extension of symmetric operators).

The system

$$-\frac{d^2\psi}{dr^2} + v\psi - \lambda\psi = 0, \qquad \psi(0) = 0, \ \psi'(0) = 1 , \tag{4}$$

defines the generalized eigenfunctions of the operator $\tilde{H}$ (the condition $\psi'(0) = 1$ is added to normalize these eigenfunctions).

Condition (3') guarantees that the solution $\psi(r;\lambda)$ of system (4) exists for all λ, while (3") implies that $\psi(r;\lambda)$ has the following asymptotics for $\lambda > 0$:

$$\psi(r;\lambda) \sim A(k)\sin(kr + \eta(k)) \qquad (k = \sqrt{\lambda} > 0).$$

The functions $A(k)$ and $\eta(k)$ are called the *asymptotic amplitude* and *asymptotic phase* respectively.

It is well known that the spectrum of $\tilde{H}$ consists of the semiaxis $\lambda \geqq 0$, and, possibly, a sequence (finite or infinite) of negative eigenvalues $\lambda_j = -\kappa_j^2$, which may have zero as a unique limit point.

For the pair of operators H and $\tilde{H}$ introduced above, one can exhibit a sufficiently rich class of functions Φ such that the "trace formula" (2) holds.

Moreover, it turns out that

$$\begin{aligned} \xi(\lambda) &= -\eta(\sqrt{\lambda})/\pi \quad \text{for} \quad \lambda > 0, \\ \xi(\lambda) &= -\int_{-\infty}^{\lambda} \sum_j \delta(\lambda-\lambda_j)\,d\lambda \quad \text{for} \quad \lambda < 0. \end{aligned} \tag{5}$$

Here we compute the value of $\xi(\lambda)$ for $\lambda < 0$ according to the general rule indicated above for semibounded operators: $\xi(\lambda) = n(\lambda)-\tilde{n}(\lambda)$.

The connection between the spectral shift function $\xi(\lambda)$ and the asymptotic phase $\eta(k)$ was already clear to I. M.

Lifshits [48]. However, the first rigorous proof of formula (5), under the restriction $rv(r) \in L_1(0,\infty)$, is due to V. S. Buslaev and L. D. Faddeev [12]. In the form given above, which is definitive in a certain sense, this result has been obtained by V. A. Yavryan [57].

In the light of formula (5), Levinson's relation [1] acquires an interesting connotation, namely

$$\frac{1}{\pi}\eta(0+) = \int_{-\infty}^{0} \sum_j \delta(\lambda-\lambda_j)d\lambda = m ,$$

where m is the number of negative eigenvalues of the operator $\widetilde{H}$. This formula expresses the continuity of the function $\xi(\lambda)$ at the point $\lambda = 0$:

$$\xi(0+) = \xi(0-) ,$$

and is valid under the hypotheses that $rv(r) \in L_1(0,\infty)$ and that there is no virtual level at the point $\lambda = 0$ [i.e., the function $\psi(r;0)$ is not bounded. If $\psi(r;0)$ is bounded on the entire half line, then Levinson's formula should be modified: $\eta(0+) = \eta(m+\frac{1}{2})\pi$. In this case, the function $\xi(\lambda)$ has at $\lambda = 0$ a first-order discontinuity: $\xi(0+) = \xi(0-) + \frac{1}{2} = m + \frac{1}{2}$].

Moreover, all these results may be extended to the operators defined by the radial Schrödinger equations having centrifugal terms

$$H = [-\frac{d^2}{dr^2} + \frac{\ell(\ell-1)}{r^2}]\Big|_h \quad \text{and} \quad \widetilde{H} = [-\frac{d^2}{dr^2} + \frac{\ell(\ell-1)}{r^2} + V]\Big|_h$$

4. In a finite dimensional space G_m, two Hermitian operators H and $\widetilde{H}$ are unitarily equivalent if and only if $\xi(\lambda) \equiv 0$. In an infinite dimensional Hilbert space it could happen that the spectral shift function $\xi(\lambda)$ is not identically zero and even has a constant sign, but the operators H and $\widetilde{H}$ are unitarily equivalent, i.e., $W^{-1}\widetilde{H}W = H$, where W is a unitary operator. The paradoxical situation that two operators H and $\widetilde{H}$ may be unitarily equivalent even in the case of a positive perturbation ($\widetilde{H} = H + V$, $V > 0$) was observed for the first time by physicists. It it easy to see that when there

exists at least one unitary operator W_0 transforming $\tilde{H}$ into H, then there are infinitely many such operators, given by $W = W_0C$, where C is an arbitrary unitary operator commuting with H.

It turns out that, among all these operators W, one can single out (under the assumption that H and $\tilde{H}$ are somehow "close") two special unitary operators $W_\pm = W_\pm(\tilde{H},H)$, called *wave operators*, which themselves define the so-called S-*operator* (or *scattering operator*)

$$S = W_+^{-1}W_- \quad . \tag{6}$$

The wave operators $W_\pm$ were introduced for the first time by Møller [5] in 1945, who defined them by the formula

$$W_\pm = \lim_{t\to\pm\infty} e^{it\tilde{H}}e^{-itH} \quad . \tag{7}$$

The precise meaning of formula (7) was elucidated only in subsequent papers by several authors [20,14,30,25,10,28].

In the framework of abstract operator theory, the rigorous treatment of wave operators begins with the work of M. Rosenblum [55] and T. Kato [32,33]. We shall discuss the appropriate results in the fifth lecture, but already at this point we give the following definition.

Let

$$H = \int_{-\infty}^{\infty} \lambda dE(\lambda)$$

be the spectral resolution of the operator H. An element $f \in G$ is called *absolutely continuous* relative to the operator H if the function $(E(\lambda)f,f)$ is absolutely continuous, i.e., it has a derivative almost everywhere and

$$(E(\lambda)f,f) = \int_{-\infty}^{\lambda} \frac{d}{d\mu}(E(\mu)f,f)d\mu \quad , \quad (-\infty < \lambda < \infty).$$

Denote by $A = A_H$ the set of all elements of G which are absolutely continuous relative to H. Then one can show that the subspace A_H reduces the operator H. Consequently, the subspace $A_H^\perp = G \ominus A_H$ (the orthogonal complement of A_H), also reduces H.

The subspace $A_H^{\perp}$ consists of all elements of $f \in G$ such that the function $(E(\lambda)f,f)$ is singular (i.e., it has a derivative equal to zero almost everywhere).

In particular, $A_H^{\perp}$ will contain all the eigenspaces of H, corresponding to its various eigenvalues.

The restrictions H_a and H_s of the operator H to A_H and $A_H^{\perp}$, respectively, are called the absolutely continuous and the singular parts of H.

Concerning problems arising in physics, the selfadjoint operator H (the unperturbed Hamiltonian) usually has only an absolutely continuous spectrum $(H = H_a)$, while the selfadjoint operator $\tilde{H}$ (the perturbed Hamiltonian) has, in addition to its absolutely continuous spectrum in A_H, a spectrum consisting of isolated eigenvalues.

For the sake of simplicity, we assume here that $H = H_a$. Then the wave operators $W_{\pm}(\tilde{H},H)$ are defined by formula (7) as strong limits. Therefore, the fact that they do exist means that the limits

$$\lim_{t\to\pm\infty} (e^{it\tilde{H}}e^{-itH}f) \quad (= W_{\pm}f)$$

exist, in the sense of norm convergence, for any $f \in G$.

It turns out that when the wave operators $W_{\pm}$ exist, they map G isometrically into $A_{\tilde{H}}$, and

$$\tilde{H}_a W_{\pm} = W_{\pm}H \quad .$$

When the wave operators $W_{\pm}$ map the entire space G isometrically (unitarily) onto the entire space $A_{\tilde{H}}$, they are said to be *complete*. In this case, the $W_{\pm}$ realise a unitary equivalence between the Hamiltonian H and the absolutely continuous part $\tilde{H}_a$ of the Hamiltonian $\tilde{H}$: $H = W_{\pm}^{-1}\tilde{H}_a W_{\pm}$.

It turns out that by using the language of abstract operator theory, one can produce several criteria for the existence of complete wave operators, and these criteria have simple formulations. For example, the Kato-Rosenblum theorem [32,33,55]. guarantees the existence of complete wave operators $W_{\pm}$ when $\tilde{H} = H + V$ and V is a trace class operator.

This last criterion is, however, unnecessarily restrictive and unapplicable in real physical situations.

A broad range of applications fall under the criterion (see Lecture 6) guaranteing existence and completeness of the wave operators W under the assumption that the condition

$$R_z^p(\tilde{H}) - R_z^p(H) \in \mathcal{I}_1 \quad (z \in \rho(H) \cap \rho(\tilde{H})) \tag{8}$$

holds for some natural number p. [We denote by $\rho(H)$ the set of all regular points of the operator H, i.e., the set of all complex points z such that the resolvent of H, $R_z(H) = (H-zI)^{-1}$ exists. It can be shown easily that if condition (8) is satisfied for at least one point z_0 which is regular for both the operators H and $\tilde{H}$ (for example, for some nonreal point z_0), then it is satisfied for all such points.]

When complete wave operators exist, the scattering operator S defined by formula (6) becomes particularly "valuable." In this case the operator S is unitary and commutes with H, i.e., $S\mathcal{D}(H) = \mathcal{D}(H)$ and $SHf = HSf$ $(f \in \mathcal{D}(H))$. [$\mathcal{D}(H)$ denotes the domain of the operator H.]

In order to investigate more properties of S, first consider the case when the operator H has a *simple spectrum*, i.e., there is a *generating vector* u for H. Using the generating vector u, each element f can be represented in the form

$$f = \int_{-\infty}^{\infty} f_u(\lambda)\, dE(\lambda)u \quad .$$

Moreover, the formula $f \longmapsto f_u(\lambda)$ defines a unitary map from the entire space G onto the entire space $L_2(\sigma)$, where $\sigma(\lambda) = (E(\lambda)u,u)$ $(-\infty < \lambda < \infty)$, and under this map, the element Hf is transformed into the function $\lambda f_u(\lambda)$ $(f \in \mathcal{D}(H))$, i.e.,

$$Hf = \int_{-\infty}^{\infty} \lambda f_u(\lambda)\, dE(\lambda)u \quad .$$

Since the operator S commutes with H, it also *diagonalizes* in the representation $f \longmapsto f_u(\lambda)$. In other words, one can associate to S a function $s(\lambda)$ such that

$$Sf = \int_{-\infty}^{\infty} s(\lambda) f_u(\lambda) dE(\lambda) u \quad . \tag{9}$$

This equality defines the function $s(\lambda)$ only on the spectrum $\sigma(H)$ of H, and the unitarity of S implies the unitarity of $s(\lambda)$, i.e., $|s(\lambda)| = 1$ $(\lambda \in \sigma(H))$. It turns out that, under sufficiently general assumptions,

$$s(\lambda) = e^{-2\pi i \xi(\lambda)} \qquad (\lambda \in \sigma(H)). \tag{10}$$

This equality emphasizes the importance of the possibility to define the function $\xi(\lambda)$ independently, up to an integer term, by means of the trace formula.

When the operator H has a multiple spectrum, and especially when it has a spectrum of infinite multiplicity, the relation between the function $\xi(\lambda)$ and the operator S becomes complicated.

Without making things precise (see Lecture 6), let us point out merely that in this case the fact that S and H commute implies, instead of (9), that

$$S = \int_{\sigma} \mathcal{S}(\lambda) dE(\lambda) \quad , \tag{11}$$

where $\mathcal{S}(\lambda)$ $(\lambda \in \sigma(H))$ is now a certain operator (called the *scattering suboperator*) acting in the elementary space G_λ of generalized eigenvectors.

M. Sh. Birman and the author [8,9] showed that under very general conditions that ensure the existence of a trace formula (2), complete wave operators, and hence the scattering operator S, exist, and

$$\mathcal{S}(\lambda) = I + T_\lambda \quad ,$$

where T_λ is a trace class operator in G_λ, and

$$\det \mathcal{S}(\lambda) = e^{-2\pi i \xi(\lambda)} \quad . \tag{12}$$

In the next lecture, the notion of a determinant $\det(I + A)$ will be made precise for any trace class operator A.

One of the facts which is implicit in the above discussion is that the operator $S(\lambda)$ can be expressed as

$$S(\lambda) = e^{-2\pi i K(\lambda)} ,$$

where $K(\lambda)$ is a trace class operator in G_λ satisfying

$$\mathrm{Sp}\, K(\lambda) = \xi(\lambda).$$

The rigorous proof of relation (12) in the particular case of the Schrödinger operator in the space $L_2(\mathbb{R}^3)$ can be found in the paper of Buslaev [11] (with a reference to L. D. Faddeev's work).

Further on (in Lecture 6), we shall give (without proof) a formula relating the operator T_λ with the so-called asymptotic (scattering) amplitude in the case of the Schrödinger operator in $L_2(\mathbb{R}^3)$. This formula, remarkable in its simplicity, was discovered by physicists and then rigorously developed in papers of several mathematicians (A. Ya. Povzner [53, 54], T. Ikebe [27], and others).

Besides the nonstationary (time-dependent) approach to wave operators that we have discussed above, there is also a stationary (time-independent) approach. Although work has been already done on the problem of establishing a rigorous base in the frame of abstract operator theory for this important approach (S. T. Kuroda [42], M. Sh. Birman and S. B. Entina [7]), we shall not be concerned with it, due to a lack of time.

We remark only that it is precisely the stationary approach which, in the case of the Schrödinger operator in $L_2(\mathbb{R}^3)$, allows us to interpret the scattering operator S as the operator which transforms "incoming waves" into "outgoing" ones.

SECOND LECTURE

It is known that trace class (nuclear) operators play an important role in the theory of topological vector spaces. It turns out that they play an absolutely exceptional part in perturbation theory for selfadjoint operators in Hilbert space. Trace class operators form an ideal in the ring of all bounded operators. In this ring there exist other ideals consisting of compact operators, some of which will be considered later, as a comparison to the ideal of trace class operators. At the same time we shall discuss relatively new facets of the theory of compact operators, some of which are new even for the algebraic case. We remark from the beginning that not all the propositions presented below will be used in the next lectures, and that many of them have independent interest.

Compact operators

1. Let G be a separable Hilbert space, and denote by R the normed linear ring (Banach algebra) of all bounded operators acting in G. It is known that the norm in this ring is defined by the equality

$$||A|| = \sup_{x \in G} \frac{|Ax|}{|x|}$$

and is called the uniform norm. We denote by I_∞ the linear collection of all compact operators in R. If $A \in I_\infty$ and $X \in R$, then $AX \in I_\infty$ and $XA \in I_\infty$, and so I_∞ is a two-sided ideal in the ring R. J. Calkin has shown that I_∞ is the unique, non-trivial, closed (in the uniform norm) ideal of the ring R (see [45]). In what follows we shall become acquainted with other ideals of R which, although not closed in the uniform norm, are indeed closed in other norms.

Let λ be an eigenvalue of some operator $A \in R$, and consider $S_\lambda = \{\phi | A\phi = \lambda\phi\}$ - the eigenspace of A corresponding to eigenvalue λ. The dimension α of the subspace S_λ is called the *proper (geometric) multiplicity* of the eigenvalue λ: $\dim S_\lambda = \alpha$. Let L_λ be the set of all root vectors corresponding to the eigenvalue λ: $L_\lambda = \{\phi \mid (A-\lambda I)^p\phi = 0$ for some natural $p = p(\phi)\}$. The dimension ν of the linear manifold L_λ is called the *algebraic multiplicity* of λ: $\dim L_\lambda = \nu$. If A is compact, then, when $\lambda \neq 0$, the numbers α and ν are finite, and, obviously, $\alpha \leqq \nu$. From now on, the eigenvalues of a compact operator A will be indexed in the order of decreasing absolute values

$$|\lambda_1(A)| \geqq |\lambda_2(A)| \geqq \ldots ,$$

and each eigenvalue $\lambda_j(A)$ is repeated according to its algebraic multiplicity.

Now let us define the so-called s-*numbers* (*singular numbers*), introduced for the first time by E. Schmidt for integral operators. Let $A \in I_\infty$. Then $A^* \in I_\infty$ (I_∞ is a selfadjoint ideal), and A^*A is a positive selfadjoint operator. Therefore, its eigenvalues $\lambda_j(A^*A) = s_j^2(A)$ $(j = 1,2,\ldots)$ are nonnegative. The arithmetic square root $s_j(A)$ of the number $\lambda_j(A^*A)$ is called the j-th s-number of the operator A.

If $A \neq 0$, then at least one of the numbers $s_j(A)$ is different from zero. If the operator A^*A has only a finite number m of nonzero eigenvalues, then $s_j(A) = 0$ for $j > m$. If A^*A has an infinite set of eigenvalues, then the family $\{s_j^2(A)\}$ consist only of those eigenvalues and does not "capture" the zero eigenvalues, if any, of the operator A^*A.

For example, we compute $s_1(A)$:

$$s_1^2(A) = \lambda_1(A^*A) = \max \frac{(A^*Ax,x)}{(x,x)} = \max \frac{(Ax,Ax)}{(x,x)} = \|A\|^2 ,$$

i.e., $s_1(A) = \|A\|$.

The remaining s-numbers can be calculated as the arithmetic square roots of the sequence of minimax values of the quadratic form (Ax,Ax) on the sphere $(x,x) = 1$.

One may easily verify the following properties of the

s-numbers. For any $A \in I_\infty$:

1) $s_j(A) = s_j(A^*)$;

2) $s_j(cA) = |c| s_j(A)$ $(c = \text{const})$;

3) $s_j(XA) \leqq s_j(A) \|X\|$ $(X \in R)$;

Properties 1) and 3) imply

3') $s_j(AX) \leqq s_j(A) \|X\|$.

If $X = U$ is a unitary operator, then together 3) and 3') yield

$$s_j(UA) = s_j(AU) = s_j(A) .$$

The first singular number is $s_1(A) = \|A\|$, and so it satisfies the triangle inequality

$$s_1(A + B) \leqq s_1(A) + s_1(B) .$$

This statement is not true for the other singular numbers. However, for each n, they satisfy the so-called Ky Fan inequality

$$\sum_{j=1}^{n} s_j(A + B) \leqq \sum_{j=1}^{n} s_j(A) + \sum_{j=1}^{n} s_j(B) .$$

Until recently, this result has been little known even for the algebraic case. It is not trivial even when A and B are self-adjoint.

The Ky Fan inequality shows that, for each fixed n, the sum $\sum_{j=1}^{n} s_j(A)$ is a norm on I_∞. This norm is topologically equivalent to the uniform one for any n.

2. Let us formulate a theorem relating the singular numbers and the eigenvalues of an operator $A \in I_\infty$. This result, in the algebraic case, is due to H. Weyl, but it remains valid for compact operators too [45].

H. WEYL'S THEOREM. *Let* $f(r)$ $(0 \leqq r < \infty, f(0) = 0)$ *be a continuous function such that the function* $\Phi(t) = f(e^t)$ $(-\infty < < t < \infty)$ *is convex. Then for any* $A \in I_\infty$

$$\sum_{j=1}^{n} f(|\lambda_j(A)|) \leq \sum_{j=1}^{n} f(s_j(A)) \quad (n = 1,2,\ldots) .$$

When the function $\Phi(t)$ *is strictly convex, then, under the assumption that that the right-hand side is finite, the equality*

$$\sum_{j=1}^{\infty} f(|\lambda_j(A)|) = \sum_{j=1}^{\infty} f(s_j(A))$$

holds if and only if the operator A *is normal (i.e.,* $A^*A = AA^*$*).*

An example of function satisfying the conditions of the theorem is given by $f(r) = r^p$ $(0 \leq r < \infty)$, for any $p > 0$. Therefore, Weyl's theorem implies, in particular, that

$$\sum_{j=1}^{n} |\lambda_j(A)|^p \leq \sum_{j=1}^{n} s_j^p(A) \quad (n = 1,2,\ldots) . \tag{1}$$

3. Let us introduce, for $p > 0$, the classes of operators

$$I_p = \{A \mid A \in I_\infty, \ \sum_{j=1}^{\infty} s_j^p(A) < \infty\} .$$

A consequence of property 1) for the s-numbers is that $A^* \in I_p$ and $cA \in I_p$ whenever $A \in I_p$. Moreover, using properties 3) and 3'), we conclude that $A \in I_p$ implies $AX, XA \in I_p$. Furthermore, one can show that $A + B \in I_p$ for all $A,B \in I_p$. In other words, I_p is a two-sided selfadjoint ideal in the ring R for all $p > 0$.

When $p \geq 1$, one can prove (von Neumann-Schatten) that I_p is a Banach space with the norm

$$||A||_p = \{\sum_{j=1}^{\infty} s_j^p(A)\}^{1/p} \quad (=\{Sp[(A^*A)^{p/2}]\}^{1/p}) .$$

This norm enjoys the following properties:

a) it is invariant under the adjoint involution, i.e., $||A||_p = ||A^*||_p$;

b) $||AX||_p \leq ||A||_p \, ||X||$, $X \in R$.

Using a), we also get that

$$\|XA\|_p \leqq \|A\|_p \; \|X\| \; .$$

If U is a unitary operator, then

$$\|AU\|_p = \|UA\|_p = \|A\|_p \; .$$

Properties a) and b) say that, for $p \geqq 1$, I_p is a symmetrically normed ideal of the ring R. We notice that I_1 is the smallest symmetrically normed ideal of R, i.e., I_1 is contained in any two-sided, selfadjoint ideal I of the ring which is a Banach space relative to a norm $\|\cdot\|_I$ satisfying conditions a) and b) (the von Neumann-Schatten Theorem, see [22]).

The classes I_1 and I_2 are called the class of *trace class* operators and the class of *Hilbert-Schmidt* operators, respectively. Accordingly, the norms $\|A\|_1 = \sum_{j=1}^{\infty} s_j(A)$ and $\|A\|_2 = \{\sum_{j=1}^{\infty} s_j^2(A)\}^{1/2}$ are called the *trace* and the *Hilbert-Schmidt* norms, respectively.

4. Now we study the class I_1 of trace class operators in more detail. This class can be characterized as follows.

THEOREM. *A necessary and sufficient condition for an operator* A $(\in R)$ *to be trace class is the existence of an absolutely convergent matrix trace* $\sum_{j=1}^{\infty}(A\phi_j, \phi_j)$ *for any orthonormal basis* $\{\phi_j\}$. *If this condition is fulfilled, then the sum of the series does not depend upon the choice of basis.*

The matrix trace $\sum_{j=1}^{\infty}(A\phi_j, \phi_j)$ is denoted by Sp:

$$\mathrm{Sp}\, A = \sum_{j=1}^{\infty}(A\phi_j, \phi_j).$$

V. B. LIDSKII'S THEOREM [45]. *If* $A \in I_1$, *then* $\sum_j |\lambda_j(A)| < \infty$ *and* $\mathrm{Sp}\, A = \sum_{j=1}^{\infty}\lambda_j(A)$.

Therefore, if an operator has a matrix trace in every matrix representation, then it has an absolutely convergent spectral trace, and these two traces coincide. The converse of the last statement is true only for compact selfadjoint operators.

The fact that a nonselfadjoint operator $A \in I_\infty$ has an absolutely convergent spectral trace $(\sum_j |\lambda_j(A)| < \infty)$ does not allow to conclude that A belongs to even one of the classes I_p.

5. Relation (1) with $p = 1$ shows that $\sum_j |\lambda_j(A)| \leq \sum_j s_j(A)$ for all operators $A \in I_\infty$. In particular, if $A \in I_1$, then we recover a bound that we have demostrated before:

$$|\mathrm{Sp}\, A| \leq \|A\|_1 .$$

This bound shows that $\mathrm{Sp}\, A$ is a linear continuous functional on I_1. This functional has the following property: if $A \in I_\infty$, $B \in R$, and $AB, BA \in I_1$, then $\mathrm{Sp}(AB) = \mathrm{Sp}(BA)$.

Now we give the general form of the linear continuous functionals on each of the spaces I_p $(1 \leq p \leq \infty)$.

1°. The general form of a linear continuous functional $F(X)$ on the space I_∞ is

$$F(X) = \mathrm{Sp}(AX) ,$$

where A is an arbitrary operator belonging to I_1. Also,

$$\|F\| = \sup_{X \in I_\infty} (|F(X)| / \|X\|_\infty) = \|A\|_1 .$$

Therefore, the space I_1 is the dual of the space I_∞: $I_1 = I_\infty^*$.

2°. The general form of a linear continuous functional $F(X)$ on the space I_1 is

$$F(X) = \mathrm{Sp}(AX) ,$$

where A is any operator from R. Also,

$$\|F\| = \sup_{X \in I_1} (|F(X)| / \|X\|_1) = \|A\| .$$

Thus, R is the dual of the space I_1: $R = I_1^*$, and hence R is the bidual of the space I_∞.

3°. The general form of a linear continuous functional $F(X)$ on the space I_p $(p > 1)$ is

$$F(X) = \mathrm{Sp}(AX) ,$$

where A is an arbitrary operator from I_q $(1/p + 1/q = 1)$. Also

$$\|F\| = \sup_{X \in I_p} (|F(X)| / \|X\|_p) = \|A\|_q .$$

In other words, the spaces I_p $(p \geq 2)$ are reflexive, meaning that $I_p^* = I_q$ and $I_q^* = I_p$.

In particular, $I_2^* = I_2$, i.e., the Hilbert-Schmidt operators form a self-dual space. In fact, I_2 becomes a Hilbert space when one defines the scalar product

$$(A,B) = \mathrm{Sp}(AB^*) \qquad (A,B \in I_2).$$

Notice that we have already used the fact that $X \in I_p$ and $Y \in I_q$ imply $XY \in I_1$ in the formulation of proposition 3°.

6. Let us state two propositions concerning functions of the operators $A \in I_1$.

I. *Let* $A \in I_1$ *and* $f(\lambda)$ *be a function holomorphic in a domain* G *which contains the spectrum of* A, *and satisfying* $f(0) = 0$. *Then* $f(A) \in I_1$.

PROOF. In G choose a contour Γ encircling the spectrum of the operator A. Recalling the definition of a function of an operator, we have

$$f(A) = \frac{1}{2\pi i} \oint_\Gamma (\lambda I - A)^{-1} f(\lambda)\, d\lambda = \frac{1}{2\pi i} \oint_\Gamma A(\lambda I - A)^{-1} \frac{f(\lambda)}{\lambda} d\lambda ,$$

the last equality being an easily verifiable consequence of the condition $f(0) = 0$. Since the resolvent

$$R_\lambda(A) = (A - \lambda I)^{-1}$$

of A is an R-valued, continuous, operator-function of $\lambda \in \Gamma$, we see that $AR_\lambda(A)$ $(\lambda \in \Gamma)$ is a I_1-valued, continuous, operator-function. Consequently,

$$f(A) = \frac{1}{2\pi i} \oint_\Gamma X_\lambda d\lambda ,$$

where X_λ is a I_1-valued, continuous, operator-function, which implies that $f(A) \in I_1$.

It is not hard to see that Proposition I remains valid

when one replaces I_1 by I_p $(1 \leq p \leq \infty)$. Using the same method, one proves the following somewhat more difficult proposition.

II. *Let* A_t $(a \leq t \leq b)$ *be a continuously differentiable,* I_1*-valued, operator-function, and let* $f(\lambda)$ *be a function holomorphic in a domain* G *which contains the spectra of all operators* A_t*, and satisfying* $f(0) = 0$. *Then*

$$\frac{d}{dt}\mathrm{Sp}\, f(A_t) = \mathrm{Sp}[f'(A_t)\frac{dA_t}{dt}] \; .$$

7. On the class I_1 one can define an important nonlinear functional, denoted by $\det(I + A)$. Namely, the following result holds true.

THEOREM. *If* $A \in I_1$, *then for any orthonormal basis* ϕ_j *the limit*

$$\lim_{n\to\infty} \det \; \|\delta_{jk} + (A\phi_j, \phi_k)\|_1^n = \det(I + A) \; .$$

exists and does not depend upon the choice of this basis.

As in the algebraic case, it turns out that

$$\det(I + A) = \prod_{j=1}^{\infty} (1 + \lambda_j(A)) \; . \tag{2}$$

One has the following bound for $\det(I + A)$:

$$\ln|\det(I + A)| \leq \sum_j \ln(1 + s_j(A)) \leq \|A\|_1 \; . \tag{3}$$

Indeed, (2) implies that $\det(I + A) \leq \prod_j (1 + |\lambda_j(A)|)$, whence

$$\ln|\det(I + A)| \leq \sum_j \ln(1 + |\lambda_j(A)|).$$

Now applying H. Weyl's theorem with the function $f(r) = \ln(1 + r)$, we get

$$\sum_j \ln(1 + |\lambda_j(A)|) \leq \sum_j \ln(1 + s_j(A)),$$

which establishes the first inequality in (3). Taking into account that $\ln(1 + s_j(A)) \leq s_j(A)$ $(j = 1,2,\ldots)$, we obtain the second inequality too. The bound (3) extends at once to the case of Fredholm determinants

$$D_A(\mu) = \det(I - \mu A) = \prod_j \left(1 - \frac{\mu}{\mu_j(A)}\right) ,$$

where $\mu_j(A) = 1/\lambda_j(A)$ are the characteristic numbers of the operator A. Namely,

$$|D_A(\mu)| \leq \prod_j (1 + |\mu| s_j(A)).$$

The following estimate, which is not trivial even in the algebraic case, is also valid (see [46] and [22]):

$$\|(I - \mu A)^{-1}\| \leq \frac{1}{|D_A(\mu)|} \prod_{j=1}^{\infty} (1 + |\mu| s_j(A)) ,$$

[In [46], this estimate was obtained with an additional factor of 2 in the right-hand side. The above form is derived in [22].]

THIRD LECTURE

1. Let us state two more propositions concerning the determinant $\det(I + A)$.

I. *If* $A \in I_1$, *then the operator* $(I + A)^{-1}$ *exists in* R *if and only if* $\det(I + A) \neq 0$.

In fact, since $\det(I + A) = \prod_j (1 + \lambda_j(A))$, we see that $\det(I + A) \neq 0$ if and only if -1 is not in the spectrum of A. According to Hilbert's Theorem, the last condition on A $(\in I_\infty)$ means that the operator $I + A$ has a bounded inverse.

II. *The determinant* $\det(I + A)$ *is a continuous functional on the space* I_1.

The proof of this fact is not completely trivial, but we omit it.

Now we formulate several simple propositions concerning operations with determinants.

1°. $\det(I + A^*) = \overline{\det(I + A)}$.

2°. $\det[(I + A)(I + B)] = \det(I + A)\det(I + B)$.

Here the left-hand side makes sense because $A + B + AB \in \mathcal{I}_1$ whenever $A,B \in \mathcal{I}_1$.

3°. $\det(I + AB) = \det(I + BA)$ whenever $A \in \mathcal{I}_\infty$, $B \in \mathcal{R}$, and $AB,BA \in \mathcal{I}_1$. In particular, this is true when at least one of the operators A,B is trace class and the other is in $\mathcal{R}$, or when $A \in \mathcal{I}_p$ and $B \in \mathcal{I}_q$ $(1/p + 1/q = 1)$. Property 3° implies

4°. If $S,S^{-1} \in \mathcal{R}$ and $A \in \mathcal{I}_1$, then

$\det[S^{-1}(I + A)S] = \det(I + A)$.

Indeed, $\det[S^{-1}(I + A)S] = \det(I + S^{-1}AS) = \det(I + ASS^{-1}) = \det(I + A)$.

Let us compute $\det(I + A)$ when A is a finite rank operator. In this case

$$Af = \sum_{j=1}^{n} c_j(f)\chi_j$$

for all $f \in \mathcal{G}$, where $\{\chi_j\}_1^n$ is any basis in the range of A. Choose a system $\{\phi_k\}_1^n$ biorthogonal to $\{\chi_j\}_1^n$: $(\phi_k,\chi_j) = \delta_{jk}$. Then $c_j(f) = (Af,\phi_j) = (f,A^*\phi_j) = (f,\psi_j)$, where we denote $\psi_j = A^*\phi_j$. Therefore,

$$A = \sum_{j=1}^{n} (\cdot,\psi_j)\chi_j$$

and so in this case

$$\det(I + A) = \det(I + \sum_{j=1}^{n} (\cdot,\psi_j)\chi_j) = \det \|\delta_{jk} + (\chi_k,\psi_j)\|_1^n . \quad (1)$$

To verify (1), one has to choose the $\{\chi_j\}_1^n$ to be orthonormal, and then extend them to an orthonormal basis for the whole space $\mathcal{G}$. Computing $\det(I + A)$ in this basis, we get (1). When one passes to another such basis, the matrix $\|\delta_{jk}+(\chi_k,\psi_j)\|_1^n$ undergoes a similarity transformation and the determinant preserves its value.

LEMMA. *Let* A_z *be an* I_1*-valued operator-function holomorphic in a domain* G *of the complex* z*-plane. Then* $\det(I + A_z)$ *is a holomorphic function in the same domain, and*

$$\frac{d}{dz}\ln\det(I + A_z) = \mathrm{Sp}[(I + A_z)^{-1}\frac{dA_z}{dz}]$$

at all points z *where* $\det(I + A_z) \neq 0$.

We omit the proof that $\det(I + A_z)$ is holomorphic. Now let z_0 be a point where $\det(I + A_{z_0}) \neq 0$. Then the same holds true in a neighborhood of z_0. The spectrum of the operator $B_{z_0} = I + A_{z_0}$ can have an accumulation point only at $\lambda = 1$. Encircle this spectrum by a contour Γ such that zero does not lie inside. Now the spectra of the operators $B_z = I + A_z$ are contained inside Γ for all z belonging to a small enough neighborhood of z_0. Choose a single-valued branch of the function $f(\lambda) = \ln(1 + \lambda)$ in the domain bounded by the contour Γ. Then (see Lemma 1, 2-nd Lecture) the operator function $C_z = f(A_z) = \ln(I + A_z)$ takes values in I_1. Since

$$\lambda_j(C_z) = \ln(1 + \lambda_j(A_z)),$$

we obtain

$$\ln\det(I + A_z) = \ln\prod_j(1 + \lambda_j(A_z)) = \sum_j \ln(1 + \lambda_j(A_z)) = $$
$$= \mathrm{Sp}\, C_z .$$

Therefore

$$\ln\det(I + A_z) = \mathrm{Sp}\ln(I + A_z) .$$

(Notice that the last equality may be taken as the definition of $\det(I + A_z)$.) It remains to apply Lemma 2 of the 2-nd Lecture in order to get

$$\frac{d}{dz}\ln\det(I + A_z) = \mathrm{Sp}[(I + A_z)^{-1}\frac{dA_z}{dz}] .$$

2. Perturbation determinants. Let A be a closed linear operator in G acting in some domain $\mathcal{D}(A)$: $A\mathcal{D}(A) \subset G$.

Recall that a point z is called *regular* for A if the operator $A - zI$ is one-to-one from $\mathcal{D}(A)$ onto the entire space G and its inverse $R_z(A) = (A - zI)^{-1}$ is a bounded operator. We shall denote by $\rho(A)$ the set of regular points for the operator A. This is always an open set, and its complement $\sigma(A)$ is called the *spectrum* of A.

A point z_0 is said to be *normal* for the operator A if it is either regular or an eigenvalue of finite algebraic multiplicity and having a normally splitting root linear manifold L_{z_0}. The latter means that G splits into a direct sum $G = L_{z_0} \dot{+} N_{z_0}$, where N_{z_0} is an invariant subspace for A, and z_0 is a regular point for the operator A_0 induced by A on N_{z_0}. The set $\tilde{\rho}(A)$ of all normal points of the operator A is open and the normal eigenvalues form an isolated subset in it.

We say that two linear operators A and B are *close* if the following conditions are satisfied:

1. $\mathcal{D}(A) = \mathcal{D}(B)$;

2. the intersection $\rho(A) \cap \rho(B)$ is not empty;

3. there exists at least one point $z \in \rho(A)$ such that $(B - A)R_z(A) \in I_1$.

Let us show that if condition 3 is satisfied at one point $\zeta \in \rho(A)$, then it is satisfied at all points $z \in \rho(A)$. Indeed, multiplying both sides of the Hilbert identity below by $B - A$ from the left

$$R_z(A) = R_\zeta(A) + (z - \zeta)R_z(A)R_\zeta(A) ,$$

we see that the two terms in the right-hand side are in I_1, and so the left-hand side is also.

Notice that one can interchange the roles of the operators A and B in condition 3. In fact, let $z \in \rho(A)\cap\rho(B)$. The operator $T = I + (B - A)R_z(A) = (B - zI)(A - zI)^{-1}$ is invertible, and $(I + T)^{-1} = I + (A - B)R_z(B) = I + T_1$, where $T_1 = (A - B)R_z(B)$. The relation $I = (I + T)(I + T_1)$ makes it clear that $T_1 \in I_1$ whenever $T \in I_1$.

One can show that if two close operators A and B

have a common normal point, then the connected components of the sets $\tilde{\rho}(A)$ and $\tilde{\rho}(B)$ which contain this point coincide.

Given two close operators A and B, we introduce the perturbation determinant

$$\Delta_{B/A}(z) = \det[(B - zI)(A - zI)^{-1}] = \det[I + (B - A)R_z(A)].$$

Function $\Delta_{B/A}(z)$ is holomorphic in $\rho(A) \cap \rho(B)$. It is also holomorphic in a neighborhood of any normal point z_0 that the operators A and B have in common, except, possibly, at z_0. The point z_0 itself is a zero or a pole of the determinant depending on whether the number $k = \nu(z_0,B) - \nu(z_0,A)$ is positive or negative. [Here $\nu(z_0,A)$ $(\nu(z_0,B))$ stands for the algebraic multiplicity of the point z_0 when z_0 is an eigenvalue of operator A (B), and equals zero when z_0 is a regular point of the same operator.] The order of the zero (pole) is k.

Below we list the formal rules governing the operations with perturbation determinants.

If the operators A, B, and C are mutually close and $\rho(A) \cap \rho(B) \cap \rho(C)$ is not empty, then

$$\Delta_{C/B}(z) \cdot \Delta_{B/A}(z) = \Delta_{C/A}(z) .$$

This results from rule 2° for the multiplication of determinants. In particular, upon setting $A = C$ we find that

$$\Delta_{A/B}(z) \cdot \Delta_{B/A}(z) = 1 .$$

Furthermore, let us compute the derivative of the logarithm of the perturbation determinant. Using the previous Lemma, we have

$$\frac{d}{dz} \ln \det(I + (B - A)R_z(A)) =$$

$$= \mathrm{Sp}[(I + (B - A)R_z(A))^{-1}(B - A)R_z^2(A)] =$$

$$= \mathrm{Sp}[(A - zI)(B - zI)^{-1}(B - A)R_z^2(A)] =$$

$$= \mathrm{Sp}[(B - zI)^{-1}(B - A)(A - zI)^{-1}] =$$

$$= \mathrm{Sp}[(A - zI)^{-1} - \mathrm{Sp}(B - zI)^{-1}] .$$

Thus,

$$\frac{d}{dz}\ln \Delta_{B/A}(z) = \mathrm{Sp}\,[\, R_z(A) - R_z(B)\,] . \tag{1}$$

The notion of perturbation determinant can be generalized to the case when the operators A and B are not close in the sense introduced above, namely, when one of the conditions 1 and 3, or both of them, is not satisfied, but instead A and B are resolvent-comparable. In other words, we assume that there is at least one point $z \in \rho(A) \cap \rho(B)$ such that $R_z(A)-R_z(B) \in I_1$ (and this is then true for all points of $\rho(A) \cap \rho(B)$; see the 4-th Lecture).

Under these circumstances then, by choosing an arbitrary point $z_0 \in \rho(A) \cap \rho(B)$, one sets

$$\widetilde{\Delta}_{B/A}(z) = \Delta_{B_0/A_0}(\zeta) ,$$

where $A_0 = R_{z_0}(A)$, $B_0 = R_{z_0}(B)$, and $\zeta = (z - z_0)^{-1}$.

One can show that, up to a constant factor, this generalized perturbation determinant $\widetilde{\Delta}_{B/A}(z)$ does not depend upon the choice of the point $z_0 \in \rho(A) \cap \rho(B)$.

Relation (1) remains valid for $z \in \rho(A) \cap \rho(B)$ if one replaces Δ by $\widetilde{\Delta}$.

3. Perturbation determinants for selfadjoint operators. Let H and $\widetilde{H}$ be selfadjoint operators such that $\widetilde{H} = H + V$, with V a trace class operator. Then

$$\Delta_{\widetilde{H}/H}(z) = \det[I + VR_z(H)] .$$

We remark that

$$\Delta_{\widetilde{H}/H}(\bar{z}) = \overline{\Delta_{\widetilde{H}/H}(z)} .$$

Indeed,

$$\Delta_{\widetilde{H}/H}(z) = \det(I + VR_{\bar{z}}(H)) = \det(I + R_{\bar{z}}(H)V) =$$

$$= \det(I + (VR_z(H))^*) = \overline{\det(I + VR_z(H))} = \overline{\Delta_{\widetilde{H}/H}(z)} \ .$$

In particular, the determinant $\Delta_{\widetilde{H}/H}(z)$ takes real values for real $z \in \rho(H) \cap \rho(\widetilde{H})$. Since $\|R_z(H)\| \leq |\mathrm{Im}\, z|^{-1}$, we have

$$|\Delta_{\widetilde{H}/H}(z)| \leq \exp(\|VR_z(H)\|_1) \leq \exp(\frac{\|V\|_1}{|\mathrm{Im}\, z|}) \ .$$

Exchanging the places of H and $\widetilde{H}$, we obtain the corresponding estimate for the inverse of the quantity $|\Delta_{\widetilde{H}/H}(z)|$. Finally,

$$\exp(-\frac{\|V\|_1}{|\mathrm{Im}\, z|}) \leq |\Delta_{\widetilde{H}/H}(z)| \leq \exp(\frac{\|V\|_1}{|\mathrm{Im}\, z|}) \ .$$

This estimate shows that $\Delta_{\widetilde{H}/H}(z)$ does not vanish for $\mathrm{Im}\, z \neq 0$ and converges uniformly to the value one as $|\mathrm{Im}\, z| \longrightarrow \infty$. Therefore, it makes sense to define the function $\ln \Delta_{\widetilde{H}/H}(z)$ and choose a single-valued branch for it in the upper half plane $(\mathrm{Im}\, z > 0)$, which tends to zero as $|\mathrm{Im}\, z| \longrightarrow \infty$. We use this in the theorem below.

One more preliminary remark. The resolvents $R_z(H)$ and $R_z(\widetilde{H})$ are bounded by

$$\|R_z(H)\| \leq \frac{1}{d(z;H)} \ , \qquad \|R_z(\widetilde{H})\| \leq \frac{1}{d(z;\widetilde{H})} \ ,$$

where $d(z;H)$ and $d(z;\widetilde{H})$ denote the distances from the point z to the spectra $\sigma(H)$ and $\sigma(\widetilde{H})$, respectively. If the spectrum $\sigma(H)$ (and thus the spectrum $\sigma(\widetilde{H})$) does not fill up the entire real axis, then for certain z, each of these distances is larger than $|\mathrm{Im}\, z|$. In this case, the estimate given above for $\Delta_{\widetilde{H}/H}(z)$ can be made more precise:

$$\exp(-\frac{\|V\|_1}{d(z;\widetilde{H})}) \leq |\Delta_{\widetilde{H}/H}(z)| \leq \exp(\frac{\|V\|_1}{d(z;H)}) \ .$$

This estimate applies for $z \in \rho(H) \cap \rho(\widetilde{H})$ too.

THEOREM. *Let* H *be a selfadjoint operator, let* $V = V^* \in I_1$, *and set* $\widetilde{H} = H + V$. *Then*

$$\ln \Delta_{\widetilde{H}/H}(z) = \int_{-\infty}^{\infty} \frac{\xi(\lambda)}{\lambda - z} dz \qquad (\operatorname{Im} z > 0) \tag{2}$$

where

$$\int_{-\infty}^{\infty} |\xi(\lambda)| d\lambda \leqq \|V\|_1 \quad \text{and} \quad \int_{-\infty}^{\infty} \xi(\lambda) d\lambda = \operatorname{Sp} V \ .$$

The function $\xi(\lambda)$ *is given for almost all* λ *by the formula*

$$\xi(\lambda) = \frac{1}{\pi} \lim_{\mu \downarrow 0} \arg \Delta_{\widetilde{H}/H}(\lambda + i\mu) \ ,$$

and has the following property: if the operator V *has precisely* p *positive* (q *negative) eigenvalues, then* $\xi(\lambda) \leqq p$ *(respectively,* $\xi(\lambda) \geqq -q$).

PROOF. Consider first the case of a rank-one perturbation

$$V = v(\cdot, \phi)\phi \qquad (|\phi| = 1) \ .$$

With no loss of generality, one can assume that $v > 0$ (if not, one can switch the roles of H and $\widetilde{H}$). Using formula (1), we find

$$\Delta(z) = \det(I + VR_z) = \det(I + R_z V) =$$

$$= \det(I + v(\cdot, \phi) R_z \phi) = 1 + v(R_z \phi, \phi) \ ,$$

where, for simplicity, we have used the notations

$$R_z = R_z(H), \quad \Delta(z) = \Delta_{\widetilde{H}/H}(z).$$

If

$$H = \int_{-\infty}^{\infty} \lambda dE_\lambda \ ,$$

then

$$\Delta(z) = i + v\int_{-\infty}^{\infty} \frac{d(E_\lambda \phi, \phi)}{\lambda - z} = 1 + v\int_{-\infty}^{\infty} \frac{d\sigma(\lambda)}{\lambda - z} \ ,$$

where

$$\int_{-\infty}^{\infty} d\sigma(\lambda) = \int_{-\infty}^{\infty} d(E_\lambda \phi, \phi) = (\phi, \phi) = 1 .$$

It follows that

$$\Delta(iy) = 1 - \frac{v + o(1)}{iy} \qquad (y \uparrow \infty).$$

Moreover,

$$\operatorname{Im} \Delta(z) = \int_{-\infty}^{\infty} \frac{\operatorname{Im} z \, d\sigma(\lambda)}{(\lambda - z)^2} , \quad \text{and so} \quad \frac{\operatorname{Im} \Delta(z)}{\operatorname{Im} z} > 0 .$$

Choose that branch of the function $\ln \Delta(z) = \ln|\Delta(z)| + i \arg \Delta(z)$ satisfying $0 < \arg \Delta(z) < \pi$. Then

$$\ln \Delta(iy) = - \frac{v + o(1)}{iy} \tag{3}$$

because $\ln(1 + \varepsilon) = \varepsilon + o(\varepsilon)$. Isolating the imaginary part, we get that

$$\arg \Delta(iy) = \frac{v + o(1)}{y} .$$

Since the analytic function $F(z) = \ln \Delta(z)$ has a bounded, non-negative imaginary part, and since it behaves asymptotically on the imaginary axis as in (3), it can be expressed in terms of the boundary values of its imaginary part by the Poisson integral:

$$F(z) = \frac{1}{\pi} \int_{-\infty}^{\infty} \frac{\operatorname{Im} F(\lambda + i0)}{\lambda - z} \, d\lambda .$$

Therefore, we have established the representation (2) with $\xi(\lambda) = \frac{1}{\pi} \arg \Delta(\lambda + i0)$. Since $0 < \arg \Delta(z) < \pi$, we have $0 \leqq \xi(\lambda) \leqq 1$ (here $p = 1$ and $q = 0$).

Comparing relations (2) and (3) with the formula

$$\int_{-\infty}^{\infty} \frac{\xi(\lambda)}{\lambda - iy} \, d\lambda = \frac{i}{y}\left(\int_{-\infty}^{\infty} \xi(\lambda) d\lambda + o(1)\right) \qquad \text{as } y \uparrow \infty ,$$

we see that

$$\int_{-\infty}^{\infty} \xi(\lambda)d\lambda = \int_{-\infty}^{\infty} |\xi(\lambda)|d\lambda = v = \mathrm{Sp}\,V = \|V\|_1 .$$

Obviously, when $v < 0$ we have $-1 \leqq \xi(\lambda) \leqq 0$ and

$$\int_{-\infty}^{\infty} \xi(\lambda)d\lambda = - \int_{-\infty}^{\infty} |\xi(\lambda)|d\lambda = v = \mathrm{Sp}\,V = - \|V\|_1 .$$

This completes the proof of the theorem for rank-one perturbations.

FOURTH LECTURE

Now let us prove the theorem in the general case, when V is an arbitrary selfadjoint operator in I_1:

$$V = \sum_j v_j(\cdot,\phi_j)\phi_j, \quad (\phi_j,\phi_k) = \delta_{jk}, \quad \sum_j |v_j| = \|V\|_1 \leqq \infty.$$

Introduce the notations

$$H_0 = H, \quad H_n = H_0 + \sum_{j=1}^{n} v_j(\cdot,\phi_j)\phi_j \qquad (n = 1,2,\ldots) .$$

Then $H_n - H_{n-1} = v_n(\cdot,\phi_n)\phi_n$ is a rank-one operator, and referring to the first part of the proof,

$$\ln \Delta_{H_n/H_{n-1}}(z) = \int_{-\infty}^{\infty} \frac{\xi_n(\lambda)}{\lambda - z} d\lambda .$$

In virtue of the theorem concerning multiplication of perturbation determinants,

$$\ln \Delta_{H_n/H_0}(z) = \sum_k \ln \Delta_{H_k/H_{k-1}}(z) ,$$

whence

$$\ln \Delta_{H_n/H_0}(z) = \int_{-\infty}^{\infty} \frac{\sum_{k=1}^{n} \xi_k(\lambda)}{\lambda - z} d\lambda .$$

At the same time, each function $\xi_k(\lambda)$ has the following properties:

$$0 \leqq \pm\xi_k(\lambda) \leqq 1, \quad \int_{-\infty}^{\infty} |\xi_k(\lambda)|\, d\lambda = |v_k| \quad (= \pm\,\mathrm{Sp}(H_k - H_{k-1})) . \quad (1)$$

Consequently, the series $\xi_1(\lambda) + \xi_2(\lambda) + \ldots$ converges absolutely in the metric of $L_1(-\infty,\infty)$ to some function $\xi(\lambda)$, and

$$\int_{-\infty}^{\infty} \frac{\sum_{k=1}^{n} \xi_k(\lambda)}{\lambda - z} d\lambda \longrightarrow \int_{-\infty}^{\infty} \frac{\xi(\lambda)}{\lambda - z} d\lambda \qquad (n \longrightarrow \infty).$$

Let V_n denote the operator

$$V_n = \sum_{j=1}^{n} v_j(\cdot,\phi_j)\phi_j \ .$$

Then

$$\Delta_{H_n/H_0}(z) = \det[I + V_n R_z(H_0)] \ .$$

Since

$$\|V - V_n\|_1 = \|\sum_{n+1}^{\infty} v_k(\cdot,\phi_k)\phi_k\|_1 = \sum_{n+1}^{\infty} |v_k| \longrightarrow 0$$
$$(n \longrightarrow \infty),$$

we have

$$\Delta_{H_n/H_0}(z) \longrightarrow \det[I + VR_z(H_0)] = \Delta_{H/H_0}(z) = \Delta(z) \ .$$

Therefore,

$$\ln \Delta(z) = \int_{-\infty}^{\infty} \frac{\xi(\lambda)}{\lambda - z} d\lambda \ , \qquad (2)$$

where $\xi(\lambda) = \sum_{k=1}^{\infty} \xi_k(\lambda)$ in the metric of $L_1(-\infty,\infty)$. Moreover, the L_1-norm of the function $\xi(\lambda)$ is equal to

$$\int_{-\infty}^{\infty} |\xi(\lambda)| \, d\lambda = \lim_{n\to\infty} \int_{-\infty}^{\infty} \Big|\sum_{k=1}^{n} \xi_k(\lambda)\Big| \, d\lambda \leqq \lim_{n\to\infty} \int_{-\infty}^{\infty} \sum_{k=1}^{n} |\xi_k(\lambda)| \, d\lambda =$$
$$= \lim_{n\to\infty} \sum_{k=1}^{n} |v_k| = \|V\|_1 \ .$$

while

$$\int_{-\infty}^{\infty} \xi(\lambda) d\lambda = \sum_{k=1}^{\infty} \int_{-\infty}^{\infty} \xi_k(\lambda) d\lambda = \sum_{k=1}^{\infty} v_k = \mathrm{Sp}\, V \ .$$

The representation (2) determines the function $\xi(\lambda)$ uniquely via the inversion formula

$$\xi(\lambda) = \frac{1}{\pi} \lim_{\mu \downarrow 0} \Delta(\lambda + i\mu) \ .$$

Finally, the last claim of the theorem and concerning the numbers p and q is an immediate consequence of the bounds (1). The theorem's proof is now complete.

As a corollary, we obtain the following statement.

Again, let $\tilde{H} = H + V$, $V \in I_1$. *Then to the pair* $H, \tilde{H}$ *there corresponds a unique function* $\xi(\lambda;\tilde{H},H) \in L_1(-\infty,\infty)$ *such that the trace formula*

$$\mathrm{Sp}[\Phi(\tilde{H}) - \Phi(H)] = \int_{-\infty}^{\infty} \Phi'(\lambda)\xi(\lambda)d\lambda \tag{3}$$

holds true for any rational function $\Phi(\lambda)$ *with poles in* $\rho(H) \cap \rho(\tilde{H})$ *and having a pole of order no larger than one at* $\lambda = \infty$.

Indeed, any such $\Phi(\lambda)$ is a linear combination of functions

$$1, \quad , \frac{1}{\lambda - z}, \quad \frac{1}{(\lambda - z)^2}, \ldots \quad (z \in \rho(H) \cap \rho(\tilde{H})) .$$

Formula (3) for $\Phi(\lambda) \equiv 1$ is trivial. For $\Phi(\lambda) = \lambda$, (3) was established in the previous theorem. Now let $\Phi(\lambda) = \frac{1}{\lambda - z}$. We have to check that

$$\mathrm{Sp}[(\tilde{H} - zI)^{-1} - (H - zI)^{-1}] = - \int_{-\infty}^{\infty} \frac{\xi(\lambda)}{(\lambda - z)^2} d\lambda . \tag{4}$$

According to the last theorem

$$\ln \det[(\tilde{H} - zI)(H - zI)^{-1}] = \int_{-\infty}^{\infty} \frac{\xi(\lambda)}{\lambda - z} d\lambda .$$

Differentiating both sides with respect to z, and using formula (1) from the 3-rd Lecture, we obtain the desired result (4).

Since

$$\mathrm{Sp}[R_z(\tilde{H}) - R_z(H)] = \mathrm{Sp}[- R_z(\tilde{H})VR_z(H)] ,$$

and the I_1-valued function $R_z(\tilde{H})VR_z(H)$ can be differentiated any number of times in I_1, formula (4) can also be differentiated with respect to z any number of times. This proves the validity of formula (3) for any function

$$\Phi(\lambda) = \frac{1}{(\lambda - z)^k} \qquad (k = 1,2,\dots) .$$

The uniqueness of $\xi(\lambda)$ is made plain by integrating both sides of (4) with respect to z, which gives

$$\ln \Delta_{\widetilde{H}/H}(z) = \int_{-\infty}^{\infty} \frac{\xi(\lambda)}{\lambda - z}\, d\lambda + C .$$

The constant C equals 0 because $\Delta(z) \longrightarrow 1$ as $z \longrightarrow \infty$ along the imaginary axis.

We comment also that the inverse problem is always solvable. That is to say, given any function $\xi(\lambda) \in L_1(-\infty,\infty)$, one can find, and even in an infinity of ways, pairs of operators $H,\widetilde{H}$ for which $\xi(\lambda)$ is the spectral shift function.

As we have remarked in the introduction (1-st Lecture), the fact that $\widetilde{H} > H$ (i.e., $V > 0$) implies $\xi(\lambda) \geqq 0$ in the algebraic case. It is not difficult to verify that this remains true as one passes to the general formula (3) established above, i.e., $V > 0$ implies $\xi(\lambda) \geqq 0$.

Moreover, the analogy with the algebraic case is preserved when the operators H and $\widetilde{H}$ are semibounded, i.e., $H,\widetilde{H} > aI$. Now the resolvent $R_\lambda(H)$ exists for all real $\lambda < a$ and so

$$\xi(\lambda) = \frac{1}{\pi} \arg \det(I + VR_\lambda(H)) .$$

Since the function

$$\det(I + VR_\lambda(H)) = \det(I + R_\lambda(H)V) = \det(I + R_\lambda^*(H)V^*) = \overline{\det(I + VR_\lambda(H))}$$

is real and holomorphic for $\lambda < a$, and tends to 1 as $\lambda \longrightarrow \infty$, one has that

$$\xi(\lambda) = \frac{1}{\pi} \arg \det(I + VR_\lambda(H)) = 0$$

for $\lambda < a$.

Let the operator H have a "spectral hatch", i.e., there is an interval of the real line $(a,b) \subset \rho(H)$. Then according to the well-known theorem of H. Weyl concerning compact perturbations of selfadjoint operators, the perturbed operator

$\widetilde{H} = H + V$ $(V \in I_1)$ may have only a discrete spectrum $\mu_k \in \sigma(\widetilde{H})$ in the interval (a,b):

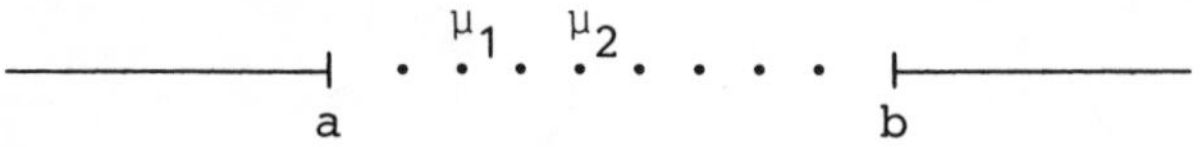

For the trace class perturbations that we consider, this result is also a direct consequence of the fact that, according to the Lemma from the 3-rd Lecture, $\det(I + VR_z(H))$ is a holomorphic function in $\rho(H)$ and, in particular, is holomorphic in the interval (a,b). Consequently, its zeros (the points of the spectrum of the operator H lying in the interval (a,b)) can have accumulation points only at the endpoints a and b of this interval.

Let μ_j be one of the eigenvalues of the perturbed operator in (a,b). Then

$$\frac{1}{\pi}\,[\arg \Delta(\mu_j + 0) - \arg \Delta(\mu_j - 0)] = -k_j \tag{5}$$

where k_j is the multiplicity of the eigenvalue μ_j (as a zero of the function $\Delta(z)$). Since the determinant $\Delta(z)$ does not vanish between two adjacent points $\mu_j < \mu_{j+1}$ and is real, $\arg \Delta(\mu_j + 0) = \arg \Delta(\mu_{j+1} - 0)$, whence

$$\frac{1}{\pi}\,[\arg \Delta(\mu_{j+1} - 0) - \arg \Delta(\mu_j - 0)] = -k_j\,.$$

Summing over j from 1 to some number n, we find

$$\frac{1}{\pi}\,[\arg \Delta(\mu_n + 0) - \arg \Delta(\mu_1 - 0)] = -\sum_{j=1}^{n} k_j\,,$$

i.e.,

$$\xi(\mu_n + 0) - \xi(\mu_1 - 0) = -\sum_{j=1}^{n} k_j\,. \tag{6}$$

If the perturbation V is of finite rank, then the estimate $-q \leqq \xi(\lambda) \leqq p$ together with formula (6) imply

$$\sum_j k_j \leqq p + q.$$

We do not mention numerous other consequences of formula (6).

2. The following problems do arise in connection with the trace formula (3) for trace class perturbations.

1) How can one characterize the class F of all functions $\Phi(\lambda)$ having the property that $\Phi(\tilde{H}) - \Phi(H) \in I_1$ for all pairs of selfadjoint operators H and $\tilde{H}$ which differ by a trace class operator: $\tilde{H} - H \in I_1$?

2) Will the trace formula (3) hold true for all functions belonging to F?

As formula (3) itself makes clear, the class F_1 of all functions for which this formula holds cannot be larger than the set of all functions $\Phi(\lambda)$ having a bounded derivative on the whole axis. It would be remarkable if one could succeed in showing that the class F_1 coincides with this set.

The following theorem shows that the class F_1 is, in any case, sufficiently rich.

THEOREM. *Let* $\tilde{H} = H + V$ $(V \in I_1)$. *Then the trace formula* (3) *holds true for all functions of the form*

$$\Phi(\lambda) = \text{const} + \int_{-\infty}^{\infty} \frac{e^{it} - 1}{it} \, d\omega(t) , \tag{7}$$

where $\omega(t)$ *is a complex-valued function of bounded variation on the whole axis:*

$$\int_{-\infty}^{\infty} |d\omega(t)| < \infty .$$

We remark that the validity of the representation (7) is equivalent to the function $\Phi(\lambda)$ having a continuous derivative which is a Fourier-Stieltjes transform:

$$\Phi'(\lambda) = \int_{-\infty}^{\infty} e^{i\lambda t} d\omega(t) ,$$

where $\omega(t)$ is an arbitrary function of bounded variation. In particular, when $d\omega(t) = \delta(t)dt$ (where $\delta(t)$ is Dirac's function), we obtain $\Phi(\lambda) = \lambda$. If $d\omega(t) = \frac{1}{2} e^{-|t|}dt$, then $\Phi(\lambda) = \text{arctg}\,\lambda$.

Notice that the superposition of two real function from the class (7) considered above does not belong, in general, to the same class. It would be interesting to derive, starting with the class (7), a linear class of functions closed under the operation of superposition of any two of its real functions.

Now we shall explain, without giving full proof of the theorem, why $\Phi(\tilde{H}) - \Phi(H) \in I_1$ when $\Phi(\lambda)$ has the form (7).

First, let us prove that

$$e^{it\tilde{H}} - e^{itH} = (e^{it\tilde{H}}e^{-itH} - I) \in I_1 . \tag{8}$$

Since the operator $e^{it\tilde{H}}e^{-itH}$ is unitary, the expression $(e^{it\tilde{H}}e^{-itH} - I)f$ makes sense for all $f \in G$. For $f \in \mathcal{D}(H) = \mathcal{D}(\tilde{H})$, we have

$$\frac{d}{dt}(e^{it\tilde{H}}e^{-itH} - I)f = ie^{it\tilde{H}}Ve^{-itH}f .$$

The operator $e^{it\tilde{H}}Ve^{-itH}$ is in I_1. Moreover, one can show that this operator is a continuous operator-function of t relative to the trace norm. Therefore,

$$(e^{it\tilde{H}}e^{-itH} - I)f = i\int_0^t e^{is\tilde{H}}Ve^{-isH}f\,ds .$$

Since f runs over a dense set in G,

$$e^{it\tilde{H}}e^{-itH} - I = i\int_0^t e^{is\tilde{H}}Ve^{-isH}\,ds \in I_1 . \tag{9}$$

This implies that the operator (8) is trace class. Furthermore,

$$\|e^{it\tilde{H}}e^{-itH} - I\|_1 \leq |t|\,\|V\|_1 ,$$

whence

$$\left\|\frac{e^{it\tilde{H}} - e^{itH}}{it}\right\|_1 \leq \|V\|_1 .$$

Then

$$\int_{-\infty}^{\infty} \frac{e^{it\tilde{H}} - e^{itH}}{it}\,d\omega(t) \in I_1$$

because

$$\left\| \int_{-\infty}^{\infty} \frac{e^{it\widetilde{H}} - e^{itH}}{it} \, d\omega(t) \right\|_1 = \|V\|_1 \int_{-\infty}^{\infty} |d\omega(t)| \ .$$

In other words,

$$\Phi(\widetilde{H}) - \Phi(H) \in I_1 \ .$$

3. In all our previous considerations, we assumed that the difference $V = \widetilde{H} - H$ was trace class. It turns out however that the trace formula holds true under more general assumptions.

<u>Definition.</u> Two operators H and $\widetilde{H}$ are called *resolvent-comparable* if they have a common regular point z where

$$R_z(\widetilde{H}) - R_z(H) \in I_1 \ . \tag{10}$$

It is not difficult to check that if condition (10) is satisfied at some point $z \in \rho(H) \cap \rho(\widetilde{H})$, then it is satisfied at all points of $\rho(H) \cap \rho(\widetilde{H})$.

It is worthwhile mentioning that this condition may be fulfilled even in the case when

$$\mathcal{D}(H) \cap \mathcal{D}(H) = \{0\} \ .$$

Since the Cayley transform $U(H)$ of the operator H satisfies

$$U(H) = (H - iI)(H + iI)^{-1} = I + 2i(H + iI)^{-1} \ ,$$

condition (10) is equivalent to the fact that the Cayley transforms of the operators H and $\widetilde{H}$ differ by a trace class operator: $U(\widetilde{H}) - U(H) \in I_1$. This observation allows us to prove the following theorem.

THEOREM [36]. *Let the selfadjoint operators* H *and* $\widetilde{H}$ *be resolvent comparable. Then there exists a function* $\xi(\lambda)$ *such that*

$$(1 + \lambda^2)^{-1} \xi(\lambda) \in L_1(-\infty, \infty)$$

and

$$\operatorname{Sp}\{\Phi(\tilde{H}) - \Phi(H)\} = \int_{-\infty}^{\infty} \Phi'(\lambda)\xi(\lambda)d\lambda$$

for all rational functions $\Phi(\lambda)$ *which have poles in* $\rho(H) \cap \rho(\tilde{H})$ *and are regular at infinity. The function* $\xi(\lambda)$ *is uniquely defined, up to an additive constant, and can be obtained from the formula*

$$\Delta(\lambda) = \frac{1}{\pi} \lim_{\mu \downarrow 0} \arg \Delta(\lambda + i\mu) + \text{const} \quad (\textit{almost everywhere}),$$

where $\Delta(z) = \tilde{\Delta}_{\tilde{H}/H}(z)$ *is the generalized perturbation determinant, and one chooses any single-valued harmonic branch for* $\arg \Delta(z)$ $(\operatorname{Im} z > 0)$.

If the operator H is semibounded (and then so is the operator $\tilde{H}$), then, conforming to the formula $\xi(\lambda) = n(\lambda;H) - n(\lambda;\tilde{H})$ for the algebraic case, it is natural to define the function $\xi(\lambda)$ uniquely, requiring $\xi(\lambda)$ to vanish identically at the left of the spectra of the operators H and $\tilde{H}$. In the general case of resolvent-comparable operators H and $\tilde{H}$, one can acquire uniqueness of the definition for $\xi(\lambda)$, up to an integer term, by imposing the additional constraint

$$\operatorname{Sp}'\{\operatorname{arctg} \tilde{H} - \operatorname{arctg} H\} = \int_{-\infty}^{\infty} \frac{\xi(\lambda)}{1 + \lambda^2} d\lambda . \tag{11}$$

Here the notation Sp' indicates that the trace should be understood in the following sense.

The point is that

$$\operatorname{arctg} H = \frac{1}{2i} \ln\{(H - iI)(H + iI)^{-1}\} = \frac{1}{2i} \ln U(H)$$

and so, formally,

$$\operatorname{arctg} \tilde{H} - \operatorname{arctg} H = \frac{1}{2i} \ln U(\tilde{H})U^{-1}(H).$$

By assumption, $T = U(\tilde{H}) - U(H) \in I_1$. Consequently, $U(\tilde{H})U^{-1}(H) = I + T_1$, where $T_1 = TU^{-1}(H) \in I_1$, and so, according to Proposition I of the 2-nd Lecture, one can define the operator $\ln U(\tilde{H})U^{-1}(H)$ is such a way as to ensure that it is trace class.

Now we take in (11)

$$\mathrm{Sp}'\{\mathrm{arctg}\, \tilde{H} - \mathrm{arctg}\, H\} = \frac{1}{2i}\, \mathrm{Sp}\, \ln U(\tilde{H})U^{-1}(H) \quad .$$

FIFTH LECTURE

In 1945, while attempting to lay the foundation, in the framework of Hamiltonian formalism, for the notion of the scattering operator introduced by Heisenberg [26], Møller [5] introduced the so-called wave operators:

$$W_{\pm} = \lim_{t \to \pm\infty} e^{it\tilde{H}} e^{-itH} \quad .$$

In the algebraic case, the expression appearing on the right has no meaning, because in this case $\exp(it\tilde{H})\exp(-itH)$ is an almost periodic polynomial with operator (matrix) coefficients. In the infinite dimensional case, the notion of wave operator was rigorously derived for the first time by K. Friedrichs, in 1948 [20]. In his work, the operators H and $\tilde{H}$ were assumed to admit a certain special representation, and the perturbation contained a small parameter $(\tilde{H} = H + \varepsilon V)$. Friedrichs's research was continued in the work of O. A. Ladyzhenskaya and L. D. Faddeev [17], who aimed at eliminating the small parameter ε. Wave operators for the Schrödinger operator in $L_2(\mathbb{R}^3)$ were studied, under various assumptions concerning the potential $v(x)$, by a number of authors.

A new stage in the theory of wave operators began with the work appearing in 1957. (M. Rosenblum [55] and T. Kato [32,33]). These authors succeeded in establishing a criterion for the existence of wave operators and some of their important properties, these results being formulated in the language of abstract operator theory in Hilbert space. At present time, the "abstract" approach has been further developed. In particular [8 ,9], it was seen to be possible and expedient to study the scattering suboperator $S(\lambda)$ on the energy shell by abstract methods. Some of the results obtained in this way were new even for the three-dimensional quantum scattering problem, with which mathematicians concerned themselves for a long time. We remark also that, by use of the abstract theory of wave operators, M. Sh. Birman [2 ,5] obtained a series of general results in the spectral theory of elliptic boundary value problems.

1. Fundamental notions and notations. Let G be a separable Hilbert space. If M is a subspace of G, we shall denote by $M^{\perp}$ its orthogonal complement:

$$G = M \oplus M^{\perp} .$$

Every vector $f \in G$ can be represented in the form

$$f = g + h, \quad f \in M, \quad g \in M^{\perp} .$$

Therefore, $g = P_M f$, where $P_M = P_M^* = P_M^2$ is the orthogonal projection onto the subspace M.

An operator W is said to be *partially isometric* if there exists a projection $P = P_M$ such that

$$|Wf| = |Pf|$$

for all $f \in G$. This means that for $f \in M$, $|Wf| = |f|$, while for $f \in M^{\perp}$, $Wf = 0$. In other words, W is partially isometric if G splits into a direct orthogonal sum of subspaces M and $M^{\perp}$ in such a way that W is isometric on M and equals to zero on $M^{\perp}$. We call the subspaces M and $N = WM$ the *initial* and the *final domains* of the operator W, respectively. Since

$$(Wf,Wf) = (Pf,Pf)$$

for all $f \in G$, $W^*W = P$ $(= P_M)$. In fact, the last relation may be taken as the definition of partially isometric operators.

It is not hard to show that the operator W^* is partially isometric whenever W is. Moreover, $W^*N = M$ and $WW^* = P_N$. If one restricts the operators W and W^* to their initial domains M and N, respectively, then they turn out to be inverses of each other.

2. We recall also several definitions given in the 1-st Lecture.

Let H be a selfadjoint operator:

$$H = \int_{-\infty}^{\infty} \lambda dE(\lambda) .$$

An element $f \in G$ is called *absolutely continuous* relative to H if $(E(\lambda)f,f)$ is absolutely continuous as a function of λ, i.e.,

$$(E(\lambda)f,f) = \int_{-\infty}^{\infty} \frac{d(E(\mu)f,f)}{d\mu} d\mu .$$

An element $f \in G$ is called *singular* relative to H if

$$\frac{d(E(\lambda)f,f)}{d\lambda} = 0$$

almost everywhere. In particular, any eigenvector of the operator H is a singular vector.

It is not hard to show that the set $A(H) = A$ of absolutely continuous elements relative to the operator H is a subspace and that

$$G = A \oplus A^{\perp},$$

where $A^{\perp}$ is precisely the set of all elements which are singular relative to H. These two subspaces reduce the operator H, i.e., $H = H_a \oplus H_s$, where H_a and H_s are selfadjoint operators acting in A and $A^{\perp}$, respectively. The operators H_a and H_s are called the *absolutely continuous* and *singular parts* of the operator H, respectively.

If $\{A_n\}$ is a sequence of operators from R having a strong limit A, we shall write

$$A = \text{s-lim } A_n .$$

Recall that $A = \text{s-lim } A_n$ and $B = \text{s-lim } B_n$ imply

$$AB = \text{s-lim } A_n B_n .$$

3. Let H_0 and H_1 be initial and perturbed self-adjoint operators in G, respectively:

$$H_k = \int_{-\infty}^{\infty} \lambda dE_k(\lambda) \qquad (k = 0,1) .$$

Introduce the simplified notation

$$A_k = A(H_k), \quad P_k = P(H_k) = P_{A_k} \qquad (k = 0,1) .$$

<u>Definition (T. Kato).</u> If the strong limit

$$\operatorname*{s-lim}_{t\to\infty} (e^{itH_1} e^{-itH_0} P_0) = W_+$$

exists, then $W_+ = W_+(H_1,H_0)$ is called a *wave operator.*

The wave operator W_- is defined similarly:

$$W_- = W_-(H_1,H_0) = \operatorname*{s-lim}_{t\to-\infty} (e^{itH_1} e^{-itH_0} P_0) .$$

Let us establish a number of properties of the wave operator W_+ (the operator W_- has analogous properties).

1°. For any $t \in (-\infty,\infty)$

$$e^{itH_1} W_+ = W_+ e^{itH_0} . \tag{1}$$

Indeed, let τ be an arbitrary real number. Then

$$e^{i\tau H_1} W_+ = \operatorname*{s-lim}_{t\to\infty} (e^{i(t+\tau)H_1} e^{-i(t+\tau)H_0} e^{i\tau H_0} P_0) =$$
$$= \operatorname*{s-lim}_{t\to\infty} (e^{i(t+\tau)H_1} e^{-i(t+\tau)H_0} P_0 e^{i\tau H_0}) = W_+ e^{i\tau H_0} ,$$

as claimed.

Applying both sides of equality (1) to an element $f \in G$ and taking the scalar products of the results with $g \in G$, we get

$$\int_{-\infty}^{\infty} e^{it\lambda} d(E_1(\lambda)W_+f,g) = \int_{-\infty}^{\infty} e^{it\lambda} d(W_+E_0(\lambda)f,g) .$$

Using the uniqueness theorem for the inversion of the Fourier-Stieltjes integral, and by taking into account that the spectral functions are normalized in a unique way, we conclude from the last inequality that

2°. $E_1(\lambda)W_+ = W_+E_0(\lambda)$.

THEOREM 1. *The operator* W_+ *is partially isometric with initial domain* A_0 *and final domain contained in* A_1: $W_+A_0 \subset A_1$.

PROOF. We have, for any vector $f \in G$,

$$|W_+f| = \lim_{t\to\infty} |e^{itH_1}e^{-itH_0}P_0f| = |P_0f|$$

which proves the partial isometry of W_+.

The initial domain of the operator W_+ is clearly $P_0G = A_0$. Let us investigate the final domain. For $f \in G$ arbitrary,

$$(E_1(\lambda)W_+f,W_+f) = |E_1(\lambda)W_+f|^2 = |W_+E_0(\lambda)f|^2 =$$

$$= |P_0E_0(\lambda)f|^2 = |E_0(\lambda)P_0f|^2 = (E_0(\lambda)g,g),$$

where $g = P_0f$. Therefore, W_+f is an absolutely continuous element relative to the operator H_1. The theorem is proved.

The wave operator W_+ is said to be *complete* if $W_+A_0 = A_1$. In this case W_+ transforms the absolutely continuous parts of the operators H_1 and H_0 into each other unitarily. Namely, we have

$$e^{itH_1}W_+f = W_+e^{itH_0}f \ . \tag{2}$$

In particular, when $f \in \mathcal{D}(H_{0a}) = A_0$, one can differentiate equality (2) with respect to t. Then, taking $t = 0$, we get

$$H_1W_+f = W_+H_0f \ ,$$

or

$$W_+^*H_1W_+f = H_0f.$$

Because $f \in \mathcal{D}(H_{0a}) = A_0$ here, the last equality says that

$$W_+^*H_{1a}W_+f = H_{0a}f \ ,$$

as stated.

THEOREM 2 (on the multiplication of wave operators). *Let* H_0, H_1 *and* H_2 *be selfadjoint operators and assume that the wave operators* $W_+(H_1,H_0)$ *and* $W_+(H_2,H_1)$ *exist. Then the wave operator* $W_+(H_2,H_0)$ *exists also, and*

$$W_+(H_2,H_0) = W_+(H_2,H_1)W_+(H_1,H_0) .$$

PROOF. Consider

$$W_+(H_2,H_1)W_+(H_1,H_0) = \underset{t\to\infty}{\text{s-lim}} \{e^{itH_2}e^{-itH_1}P_1e^{itH_1}e^{-itH_0}P_0\} . \quad (3)$$

One has, for $t \longrightarrow \infty$,

$$(I - P_1)e^{itH_1}e^{-itH_0}P_0 \xrightarrow{s} (I - P_1)W_+(H_1,H_0) = 0 .$$

Multiplying this relation on the left by the unitary operator $e^{itH_2}e^{-itH_1}$, we obtain

$$e^{itH_2}e^{-itH_1}(I - P_1)W_+(H_1,H_0) \xrightarrow{s} 0 \quad (t \longrightarrow \infty).$$

Adding this last relation to (3), we complete the proof of the theorem.

Notice, by the way, that $W_\pm(H,H) = P(H)$ $(= P_{A_H})$.

THEOREM 3. *In order that the operator* $W_+(H_1,H_0)$ *be complete, it is necessary and sufficient that the operator* $W_+(H_0,H_1)$ *exists.*

PROOF OF SUFFICIENCY. Let both the operators $W_+(H_1,H_0)$ and $W_+(H_0,H_1)$ exist. Then, by Theorem 2,

$$A_1 = W_+(H_1,H_1)G = W_+(H_1,H_0)W_+(H_0,H_1)G \subseteq$$

$$\subseteq W_+(H_1,H_0)A_0 \subseteq A_1 .$$

Therefore, one can replace all the inclusions by equalities, and we see that the operator $W_+(H_1,H_0)$, and even more so, the operator $W_+(H_0,H_1)$ are complete.

PROOF OF NECESSITY. Let the operator $W_+(H_1,H_0)$ be complete. We have to show that the operator $W_+(H_0,H_1)$ exists, i.e., that the expression $\exp(itH_0)\exp(-itH_1)f$ has a limit (as $t \longrightarrow \infty$) when $f \in A_1$. But $W_+(H_1,H_0)A_0 = A_1$ by assumption, whence $f = W_+(H_1,H_0)g$ for some $g \in A_0$, or

$$f = \lim_{t\to\infty} f_t , \quad f_t = e^{itH_1}e^{-itH_0}g .$$

Consequently,

$$e^{itH_0}e^{-itH_1}f = e^{itH_0}e^{-itH_1}(f - f_t + f_t) =$$

$$= e^{itH_0}e^{-itH_1}(f - f_t) + g \longrightarrow g \quad (t \longrightarrow \infty).$$

We see that $W_+(H_0,H_1)$ exists and $f = W_+(H_1,H_0)g$, $g = W_+(H_0,H_1)f$. The last two equalities show that the operators $W_+(H_0,H_1)$ and $W_+(H_1,H_0)$ are inverses of each other on the corresponding domains, i.e.,

$$W_+(H_1,H_0) = W_+^*(H_0,H_1).$$

THEOREM 4 (on the S-operator). *Suppose that there exist complete wave operators* $W_\pm(H_1,H_0)$. *Then*

$$S = S(H_1,H_0) = W_+^*(H_1,H_0)W_-(H_1,H_0)$$

is a unitary opeartor on A_0, *which commutes with* H_0.

PROOF. The first claim is a consequence of the fact that $W_-(H_1,H_0)$ takes A_0 isometrically onto A_1, while $W_+^*(H_1,H_0)$ does the same for A_1 and A_0. It remains to check that S and H_0 commute. We know already that

$$H_1W_+ = W_+H_0 , \quad H_1W_- = W_-H_0 .$$

The first relation gives $H_1 = W_+H_0W_+^*$, which substituted into the second relation shows that

$$W_+H_0W_+^*W_- = W_-H_0 .$$

Multiplying from the left by W_+^*, we finally obtain

$$H_0 W_+^* W_- = W_+^* W_- H_0 ,$$

i.e.,

$$H_0 S = S H_0 .$$

The theorem is proved.

From the 1-st Lecture we already know that the operator $S(H_1, H_0)$ is called the *scattering operator* for the ordered pair of selfadjoint operators H_0, H_1. Let us agree on the convention that the expression "the complete scattering operator $S(H_1, H_0)$ exists" means that there exist complete wave operators $W_\pm(H_1, H_0)$ and $S(H_1, H_0) = W_+^*(H_1, H_0) W_-(H_1, H_0)$.

The following result is a direct consequence of theorems 2 and 3.

THEOREM 5 (on the multiplication of scattering operators). *Let* H_0, H_1 *and* H_2 *be selfadjoint operators, and assume that there exist complete scattering operators* $S_{10} = S(H_1, H_0)$ *and* $S_{21} = S(H_2, H_1)$. *Then the complete scattering operator* $S_{20} = S(H_2, H_0)$ *exists also, and*

$$S_{20} = S_{21}^{+} \cdot S_{10} ,$$

where

$$S_{21}^{+} = W_+(H_0, H_1) S_{21} W_+(H_1, H_0) .$$

PROOF. By hypothesis, there exist complete wave operators $W_\pm(H_1, H_0)$ and $W_\pm(H_2, H_1)$. Then, according to Theorem 2, the wave operators

$$W_\pm(H_2, H_0) = W_\pm(H_2, H_1) W_\pm(H_1, H_0)$$

exist and are complete. Consequently, the complete scattering operator S_{20} exists too, and we have

$$S_{20} = W_+^*(H_2, H_0) W_-(H_2, H_0) =$$

$$= W_+^*(H_1, H_0) W_+^*(H_2, H_1) W_-(H_2, H_1) W_-(H_1, H_0) =$$

$$= W_+^*(H_1,H_0)W_+^*(H_2,H_1)W_-(H_2,H_1)W_+(H_1,H_0)W_+^*(H_1,H_0)W_-(H_1,H_0) =$$

$$= W_+(H_0,H_1)S_{21}W_+(H_1,H_0)S_{10}.$$

4. The framework of ideas and theorems 1-4 discussed above are taken from papers by T. Kato [32,33] (see also the review paper by T. Kuroda [38]).

We should like to add the following "physical" interpretation of the scattering operator to this discussion.

Suppose that $H_k = H_{ka}$ $(k = 0,1)$. Then

$$W_+(H_0,H_1) = \underset{t\to\infty}{\text{s-lim}}\,(e^{itH_0}e^{-itH_1}) .$$

Applying both members of this equality to some vector ψ_0, and then multiplying the result from the left by the unitary operator e^{-itH_0}, we get

$$e^{-itH_0}W_+(H_0,H_1)\psi_0 - e^{-itH_1}\psi_0 \longrightarrow 0 \quad (t \longrightarrow \infty) . \tag{4}$$

A solution ψ of the Schrödinger equation

$$i\frac{d}{dt}\psi = H\psi ,$$

may be written in the form

$$\psi = e^{-itH}\psi_0 ,$$

where ψ_0 is the initial state. The vector $\phi = e^{-itH_1}\psi_0$ describes the perturbed state of the system at the moment t, while $e^{-itH_0}\psi_0$ describes the free state at the same moment. Relation (4) signifies that, as $t \longrightarrow \infty$, the perturbed state behaves as a free state, the latter corresponding, however, to a different initial state, namely

$$\psi_+ = W_+(H_0,H_1)\psi_0 .$$

Similarly, when $t \longrightarrow -\infty$, the perturbed state behaves like the free state with initial state

$$\psi_- = W_-(H_0,H_1)\psi_0 .$$

The scattering operator S is defined by the equality

$$\psi_+ = S\psi_- .$$

Indeed, this equality is equivalent to the following one:

$$W_+(H_0,H_1) = SW_-(H_0,H_1) .$$

Multiplying this by $W_-(H_1,H_0) = W_-^*(H_0,H_1)$, we obtain

$$S = W_+(H_0,H_1)W_-(H_1,H_0) = W_+^*(H_1,H_0)W_-(H_1,H_0),$$

i.e., we recover the original definition for the operator S.

SIXTH LECTURE

In this final lecture we will have to restrict ourselves to stating a series of propositions and problems and to giving short explanations of them.

1. Criteria for the existence of complete wave operators and the invariance principle. We mentioned in the introductory lecture that in 1957, M. Rosenblum [55] and T. Kato [33] showed that a sufficient condition for the existence of complete wave operators $W_\pm(H_1,H_0)$ is that the closure $\overline{H_1 - H_0}$ of the difference $H_1 - H_0$ be a trace class operator (M. Rosenblum proved this result for operators having an absolutely continuous spectrum only; the general case is due to T. Kato). Their result has been generalized, in two directions, to the case of unbounded perturbations, by S. T. Kuroda [39,40].

Subsequently, M. Sh. Birman and M. G. Krein [8] showed that a sufficient condition for the existence of complete wave operators $W_\pm$ for a given pair of operators H_0 and H_1 is the resolvent-comparability of the latter, i.e., the condition

$$R_z(H_1) - R_z(H_0) \in I_1 \quad (z \in \rho(H_0) \cap \rho(H_1)) . \tag{1}$$

This criterion is stronger than both criteria introduced by Kuroda.

[We must mention here at once (in the favour of Kuroda's analysis) that I. V. Stankevich [56] has developed Kuroda's criteria and has obtained a criterion for the existence of complete wave operators which is not covered by those we shall discuss below.]

Even before the formulation of the Birman-Krein criterion, M. Sh. Birman showed that, for semibounded (for example, from below) H_0 and H_1, complete wave operators exists as soon as the condition

$$R_z^p(H_1) - R_z^p(H_0) \in I_1 \quad (z \in \rho(H_0) \cap \rho(H_1)) \tag{2}$$

is satisfied for some natural number p [3]. [In paper [3], without loss of generality, positive definite H_0 and H_1 were considered, which allows one to take $z = 0$ in condition (2). We recall that if condition (2) is fulfilled at some point z that is regular for both H_0 and H_1, then it is fulfilled at all such points z.] At the same time, he proved that given $G_k = (H_k - aI)^{-1}$ $(k = 0,1)$, where a is a point lying on the left of the spectra of H_0 and H_1, complete wave operators exist for G_0, G_1, and one has the relation $W_\pm(G_1,G_0) = W_\pm(H_1,H_0)$.

Let us mention that in the process of establishing criterion (1), it was shown simultaneously that complete wave operators exist for $G_k = \operatorname{arctg} H_k$ $(k = 0,1)$, and that $W_\pm(G_1,G_0) = W_\pm(H_1,H_0)$.

Apparently, it was in these papers by M. Sh. Birman and the author that, for the first time, certain functions of the operators H_0 and H_1, rather than H_0 and H_1 themselves, appeared in criteria for the existence of complete wave operators and, at the same time, particular cases of the invariance principle for wave operators were dicovered. In subsequent papers, M. Sh. Birman [4,6] laid the foundation for the invariance principle under sufficiently general assumptions, concerning both the operators H_0 and H_1, and the function $\Phi(\lambda)$ figuring in the basic equality

$$W_{\pm}(\Phi(H_1),\Phi(H_0)) = W_{\pm}(H_1,H_0) .$$

In the spring of 1963, the author gratefully received a preprint of T. Kato's paper [34], where the results indicated above were strengthened more. With the intent of formulating T. Kato's result, let us follow him and give the next definition.

Definition. A function $\Phi(\lambda)$ $(-\infty < \lambda < \infty)$ is called *admissible* if one can find a finite system of points $-\infty = a_0 < a_1 < \dots < a_{n-1} < a_n = \infty$ such that in each open interval $J_k = (a_{k-1},a_k)$ $(k = 1,2,\dots,n)$ $\Phi(\lambda)$ is continuously differentiable, strictly monotonic, and its derivative $\Phi'(\lambda)$ has locally a bounded variation (i.e., it has bounded variation in the vicinity of any point of J_k).

We say that the intervals $J_k = (a_{k-1},a_k)$ are associated to the function $\Phi(\lambda)$.

KATO'S THEOREM (a criterion for the existence of complete wave operators). *Complete wave operators* $W_{\pm}(H_1,H_0)$ *exist for the selfadjoint operators* H_0 *and* H_1 *whenever given* N (> 0), *one can construct an admissible function* $\Phi_N(\lambda)$ $(-\infty < < \lambda < \infty)$, *monotonic in the interval* $(-N,N)$, *and such that*

$$\Phi_N(H_1) - \Phi_N(H_0) \in I_1 .$$

Let us show that this criterion contains as a particular case the following one.

Complete wave operators $W_{\pm}(H_1,H_0)$ *exist for the selfadjoint operators* H_0 *and* H_1 *provided condition* (2) *is satisfied for some natural number* p.

Indeed, consider the function

$$\Phi_h(\lambda) = i[(h - i\lambda)^{-p} - (h + i\lambda)^{-p}] .$$

Then for any selfadjoint operator H:

$$\Phi_h(H) = (-1)^p i^{-p+1}[R^p_{ih}(H) - R^p_{-ih}(H)] .$$

Putting here $H = H_0$ and $H = H_1$, and taking into account (2), we have easily that

$$\Phi_h(H_1) - \Phi_h(H_0) \in I_1 .$$

On the other hand,

$$\Phi_h'(\lambda) = -p[(h - i\lambda)^{-p-1} + (h + i\lambda)^{-p-1}] =$$

$$= -2p \operatorname{Re}(h - i\lambda)^{-p-1} = \frac{2p}{h^{p+1}} \operatorname{Re}[1 - \frac{i\lambda}{h}]^{-p-1} .$$

The last expression has a constant sign on each finite interval $(-N,N)$ as soon as h is large enough, i.e., the function $\Phi_h(\lambda)$ is monotonic. Moreover, $\Phi_h(\lambda)$ is rational and real on the real line, and so it changes sign only a finite number of times. Also, since $\Phi_h(\pm\infty) = 0$, the function $\Phi_h(\lambda)$ has a bounded variation on the entire real line. Therefore, all the conditions of Kato's theorem are satisfied.

Now we state the invariance principle for wave operators due to M. Sh. Birman, in the stronger version that was given by T. Kato [34].

THE INVARIANCE PRINCIPLE. *Let the operators* H_0 *and* H_1 *satisfy the conditions of the previous theorem. Then for any admissible function* $\Phi(\lambda)$ *for which the operators* $\Phi(H_0)$ *and* $\Phi(H_1)$ *exist, there exist complete wave operators* $W_\pm(\Phi(H_1),\Phi(H_0))$ *too. Moreover, let* Δ *be any interval where* $\Phi(\lambda)$ *is monotonic. Then one has*

$$W_\pm(\Phi(H_1),\Phi(H_0))E_0(\Delta) = W_\pm(H_1,H_0)E_0(\Delta) , \tag{3}$$

if $\Phi(\lambda)$ *is increasing on* Δ, *and*

$$W_\pm(\Phi(H_1),\Phi(H_0))E_0(\Delta) = W_\mp(H_1,H_0)E_0(\Delta) , \tag{4}$$

if $\Phi(\lambda)$ *is decreasing on* Δ.

Here the interval Δ may be finite, semiinfinite, or the entire real axis. In the last case one has, for $\Phi(\lambda)$ monotonically increasing,

$$W_{\pm}(\Phi(H_1),\Phi(H_0)) = W_{\pm}(H_1,H_0) \ .$$

In general, relation (3) shows that the last equality will be valid each time one can find, for some admissible function $\Phi(\lambda)$, a system $\{\Delta_k\}$ of open intervals covering the entire absolutely continuous spectrum of H_0 and such that $\Phi(\lambda)$ is increasing on each interval Δ_k. To make the last statement, we took advantage of the fact that the absolutely continuous subspaces of the operators H_0 and $H_0' = \Phi(H_0)$, A_0 and A_0', coincide when $\Phi(\lambda)$ is admissible.

In order to understand better the relations (3) and (4), we mention the following property of wave operators, which is easy to infer, but is still important.

THE LOCAL DEPENDENCE PROPERTY OF WAVE OPERATORS. *Let* H_0, H_1 *be selfadjoint operators (with spectral functions* $E_0(\lambda)$ *and* $E_1(\lambda)$*) such that the wave operators* $W_{\pm}(H_1,H_0)$ *exist. Then for any interval* Δ, *one has*

$$W_{\pm}(H_1,H_0)E_0(\Delta) = W_{\pm}(H_1(\Delta),H_0(\Delta)) \ , \tag{5}$$

where $H_0(\Delta) = H_0E_0(\Delta)$ *and* $H_1(\Delta) = H_1E_1(\Delta)$.

Indeed, as we remember from the 5-th Lecture,

$$W_{\pm}(H_1,H_0)E_0(\Delta) = E_1(\Delta)W_{\pm}(H_1,H_0),$$

and so

$$W_{\pm}(H_1,H_0)E_0(\Delta) = E_1(\Delta)W_{\pm}(H_1,H_0)E_0(\Delta) =$$

$$= \operatorname*{s-lim}_{t\to\pm\infty} (E_1(\Delta)e^{itH_1}e^{-itH_0}E_0(\Delta)) =$$

$$= \operatorname*{s-lim}_{t\to\pm\infty} (e^{itH_1(\Delta)}e^{-itH_0(\Delta)}) = W_{\pm}(H_1(\Delta),H_0(\Delta)) \ .$$

The corresponding property of the operator S is a direct consequence of (5):

$$S(H_1,H_0)E_0(\Delta) = S(H_1(\Delta),H_0(\Delta)) \ . \tag{6}$$

2. The scattering suboperator. We now give the rigorous definition of the notion of a scattering suboperator and describe some of its properties. To this end we need the concept of an absolutely continuous sum (direct integral) of Hilbert spaces.

Let $G^{(1)}, G^{(2)}, \ldots, G^{(\infty)}$ be a complete collection of Hilbert spaces of Hilbert spaces of dimensions $1,2,\ldots,$, respectively ($G^{(\infty)}$ is a separable Hilbert space). Further, let σ be some measurable set on the real line such that to each point $\lambda \in \sigma$ there is associated a space G_λ from the above collection, in such a way that the function $n(\lambda) = \dim G_\lambda$ is measurable on σ.

Now let $f(\lambda)$ $(\lambda \in \sigma)$ be a vector-function satisfying $f(\lambda) \in G_\lambda$ for all λ. We call such functions *nomadic*. We say that a nomadic function $f(\lambda)$ is *measurable* if $(f(\lambda), g)$ is a measurable function for all $g \in G^{(m)}$, where $m = \dim G_\lambda$.

Consider nomadic measurable functions $f(\lambda)$ satisfying the condition

$$\int_\sigma \|f(\lambda)\|^2 d\lambda < \infty .$$

Then the set $G^{\oplus}$ of all such vector-functions is a Hilbert space with the scalar product

$$(f,g) = \int_\sigma (f(\lambda), g(\lambda)) d\lambda .$$

The space $G^{\oplus}$ is called the *absolutely continuous orthogonal sum* (direct integral) of the given Hilbert spaces and one writes

$$G^{\oplus} = \int_\sigma \oplus\, G_\lambda d\lambda .$$

In particular, if all spaces are one-dimensional ($G_\lambda \equiv G^{(1)}$), then $G^{\oplus}$ is just the usual space $L_2(\sigma)$ of square-summable, scalar functions on the set σ. When $G_\lambda \equiv G^{(n)}$ for all $\lambda \in \sigma$ (n fixed), we get the space $L_2^{(n)}(\sigma)$ of n-dimensional, square-summable, vector-functions. Finally, when

$G \equiv G^{(\infty)}$ for all $\lambda \in \sigma$, we obtain the space of infinite dimensional, square-summable, vector-functions on σ.

In the space $G^{\oplus}$ we define the operator Λ to be multiplication by λ. If the element $f \in G^{\oplus}$ is a nomadic function $f = \{f(\lambda)\}$ satisfying

$$\int_{\sigma} \lambda^2 \|f(\lambda)\|^2 d\lambda < \infty ,$$

then we set $\Lambda f = \{\lambda f(\lambda)\}$.

It is easy to see that Λ is a symmetric operator. Moreover, Λ is selfadjoint because

$$(\Lambda - zI)^{-1} f = \{ \frac{f(\lambda)}{\lambda - z} \}$$

for all z $(\operatorname{Im} z \neq 0)$.

Let us explain the form of all bounded operators K in $G^{\oplus}$ which commute with Λ. It turns out [18] that $K\Lambda = \Lambda K$ if and only if one can associate to K a family of bounded operators $\{K(\lambda)\}$ $(K(\lambda)G_{\lambda} \subset G_{\lambda})$, such that

$$Kf = \{K(\lambda)f(\lambda)\}$$

and

$$\operatorname{ess\,sup}_{\lambda\in\sigma} \|K(\lambda)\| \quad (= \|K\|) < \infty.$$

In the theory of spectral types [52], one proves the following statement.

Let $H = H_a$ *be a selfadjoint operator acting in a Hilbert space* G. *Then there exists a unitary transformation of* G *onto some space*

$$G^{\oplus} = \int_{\sigma} \oplus G_{\lambda} d\lambda ,$$

where σ *is the spectrum of the operator* H, *and such that* H *is taken into the operator* Λ *in* $G^{\oplus}$.

Now we shall apply all these facts in the theory of wave operators $W_{\pm}$.

Suppose we are given two selfadjoint operators H_0 and H_1 such that complete wave operators $W_\pm(H_1,H_0)$ exist, and hence the scattering operator S exists too. Then one can identify the absolutely continuous subspace A_0 of H_0 with a space

$$G^{\oplus} = \int_\sigma^{\oplus} G_\lambda d\lambda$$

in such a way that the operator H_{0a} becomes the operator Λ. Recall that the operators S and H_{0a} commute. Therefore, $S = \{S(\lambda)\}$, where, as one can easily see, the $S(\lambda)$ are unitary operators acting in the corresponding spaces G_λ. The operators $S(\lambda)$ are called, after Pauli, the *scattering suboperators*.

In the 1-st Lecture, starting from the spectral resolution

$$H_0 = \int_{-\infty}^{\infty} \lambda dE_0(\lambda)$$

and the fact that S commutes with $H_0 = H_{0a}$, we wrote

$$S = \int_\sigma S(\lambda) dE_0(\lambda) \ , \qquad \sigma = \sigma(H_0) \ .$$

At that point, we used this symbolic expression to denote precisely the fact that $S = \{S(\lambda)\}$, where $S(\lambda)$ $(\lambda \in \sigma)$ are the scattering suboperators introduced above.

As a direct consequence of relation (6), we get the property of local dependence of the suboperator $S(\lambda)$ upon H_0 and H_1: the values of $S(\lambda)$ on some interval Δ of variation of λ are completely determined by the values taken by the spectral functions $E_k(\lambda)$ $(k = 0,1)$ on Δ.

M. Sh. Birman and the author [8] established the following result.

THEOREM (on the scattering suboperator). *Let* H_0 *and* H_1 *be resolvent-comparable selfadjoint operators. Then for almost all* $\lambda \in \sigma$ $(= \sigma(H_{0a}))$, *the scattering suboperator* $S(\lambda) = S(\lambda; H_1, H_0)$ *has the form*

$$S(\lambda) = I_\lambda + T_\lambda \ , \tag{7}$$

where I_λ *and* T_λ *are the identity operator and a trace class operator in* G_λ, *respectively. Moreover,*

$$S(\lambda) = e^{-2\pi i\xi(\lambda)} \tag{8}$$

for almost all $\lambda \in \sigma$, *where* $\xi(\lambda)$ *is the spectral shift function for the pair* H_0, H_1 : $\xi(\lambda) = \xi(\lambda; H_1, H_0)$.

In particular, one can apply the theorem to the case where $H_1 = H_0 + V$, V being a trace class operator.

Now let H_0 and H_1 be semibounded from below operators satisfying for some natural number p

$$R_z^p(H_1) - R_z^p(H_0) \in I_1.$$

Pick an arbitrary number a to the left of the spectra of H_0 and H_1, and write $H_k' = (H_k - aI)^{-p}$ $(k = 0,1)$. We have $H_1' - H_0' \in I_1$, and so one can apply the theorem to the operators H_0' and H_1'. But then, as one can easily prove, the conclusion of the theorem holds true for the initial operators H_0 and H_1 too. [At the same time, here the function $\xi(\lambda) = \xi(\lambda; H_1, H_0)$ is uniquely determined by the following requirement: it should vanish to the left of the spectra of H_0 and H_1, and the trace formula (see (3), the 4-th Lecture) should hold, for example, for all rational functions $\Phi(\lambda)$ having nonreal zeros and a pole of order $\geq p$ at infinity.]

This remark may be generalized immediately. Namely, let $\Phi(\lambda)$ be some admissible function that is univalent on the spectra of the operators H_0 and H_1, and let the operators $\Phi(H_0)$ and $\Phi(H_1)$ be resolvent-comparable. Then the first statement of the theorem holds true for the operators H_0 and H_1, and so does the second, provided a suitable definition of the function $\xi(\lambda)$ is adopted.

More precisely, our theorem is applicable to the operators $H_k' = \Phi(H_k)$ $(k = 0,1)$, because they are resolvent-comparable. The requirement that the function $\Phi(\lambda)$ be univalent on the spectra of H_0 and H_1 guarantees that $H_k = \Psi(H_k')$ $(k = 0, 1)$, where $\Psi(\lambda)$ is an admissible function. Therefore, if one assumes, with no loss of generality, that $\Phi(\lambda)$ is an increasing function on the spectra of H_0 and H_1, then one can deduce from the previous theorems that there exist complete wave operators $W_\pm(H_1,H_0)$ and that $S(H_1,H_0) = S(H_1',H_0')$. Consequently, $S(\lambda;H_1,H_0) = S(\Psi(\lambda);H_1,H_0)$, which gives at once the representation (7).

It would be interesting to generalize the theorem on scattering suboperators to include operators H_0 and H_1 satisfying the conditions of Kato's first theorem. In this case, in order that the second statement of the theorem make sense, it would be necessary to generalize the trace formula beforehand.

As we noticed in the introductory lecture, representation (7) and formula (8) imply that the following representation is possible for almost all $\lambda \in \sigma$:

$$S(\lambda) = e^{-2\pi i K(\lambda)} ,$$

where $K(\lambda)$ is a trace class operator in G_λ and

$$\mathrm{Sp}\, K(\lambda) = \xi(\lambda) .$$

Moreover, if $H_1 = H_0 + V$ with $V \in I_1$, then the operator-function $K(\lambda)$ can be chosen to satisfy

$$\int_\sigma \|K(\lambda)\| d\lambda \leq \|V\|_1 ,$$

while if H_1 and H_0 are resolvent-comparable, then $K(\lambda)$ can be chosen to satisfy

$$\int_\sigma \frac{\|K(\lambda)\|}{1 + \lambda^2} d\lambda \leq \|U_1 - U_0\| ,$$

where $U_k = (H_k - iI)(H_k + iI)^{-1}$ $(k = 0,1)$.

In [8,9] it was shown also under which conditions one can assert that the operator phase $K(\lambda)$ $(\lambda \in \sigma)$ is a nonnegative trace class operator. In general, these conditions include, among others, the requirement that $\mathcal{D}(H_0) = \mathcal{D}(H_1)$ and that the perturbation $V = H_1 - H_0$ be nonnegative. If one additionally sets $H_1(\varepsilon) = H_0 + \varepsilon V$ $(0 \leq \varepsilon \leq 1)$, then the operator phase

$K_\varepsilon(\lambda)$ corresponding to the pair $H_0, H_1(\varepsilon)$ will be a nondecreasing function of ε.

[In the case of operators H_0 and H_1 semibounded from below which satisfy (2), a sufficient condition for the existence of a nonnegative trace class phase $K(\lambda)$ is that the difference $R_a(H_0) - R_a(H_1)$ be a nonnegative operator for some a lying to the left of the spectra of H_0 and H_1 (and then for all such a).]

3. It is not difficult to obtain conditions expressing the resolvent-comparability of Schrödinger operators.

Let $G = L_2(\mathbb{R}^m)$, $H_0 = -\Delta = -\sum_{i=1}^m \partial^2/\partial x_i$, $H_1 = H_0 + V$, where V is the operator of multiplication by a real "potential" $v(x)$. It turns out that the condition

$$v(x) \in L_1(\mathbb{R}^m) \cap L_2(\mathbb{R}^m) \tag{9}$$

is sufficient to guarantee that H_1, together with H_0, is selfadjoint, and that these operators are resolvent comparable.

M. Sh. Birman has shown that for $m = 3$ condition (9) can be weakened and replaced by

$$v(x) \in L_1(\mathbb{R}^3) \quad \text{and} \quad \sup_{x \in \mathbb{R}^3} \int_{|x-y|=1} v^2(y)\,dy < \infty . \tag{10}$$

When $m = 1$ it suffices to assume that

$$v(x) \in L_1(\mathbb{R}^1) = L_1(-\infty,\infty).$$

In addition, let us consider the situation

$$G = L_2(0,\infty), \quad H_0 = -\frac{d^2}{dr^2}\bigg|_h , \quad H_1 = \left(-\frac{d^2}{dr^2} + v(r)\right)\bigg|_h .$$

Here the operators H_0 and H_1 will be resolvent-comparable as soon as

$$\int_0^1 r|v(r)|\,dr + \int_1^\infty |v(r)|\,dr < \infty.$$

We have already encountered this condition in the 1-st Lecture. If one translates it into the language of operator theory, then it expresses (as does, indeed, the condition $v(x) \in L_1(-\infty,\infty)$ in the previous case) precisely the fact that

$$(H_0 + I)^{-1/2}|v|(H_0 + I)^{-1/2} \in I_1 .$$

In general, this condition is more rigid than the resolvent-comparability of H_0 and H_1.

It seems that, at present, it is not clear which analytic conditions, imposed upon the potential $v(x)$, adequately express the resolvent-comparability of the pairs H_0 and H_1 considered above.

4. Let us pause and discuss in more detail the important case

$$G = L_2(\mathbb{R}^3), \quad H_0 = -\Delta, \quad H_1 = -\Delta + v(x) .$$

If, for example, conditions (10) are fulfilled, then H_0 and H_1 are resolvent-comparable. Therefore, the corresponding wave operators $W_\pm(H_1,H_0)$ and scattering operator S exist, and so the suboperators $S(\lambda)$ do too. We try to calculate $S(\lambda)$.

To this end, we pass to the so-called p-representation, i.e., we associate to each function $f(x) \in L_2(\mathbb{R}^3)$ its Fourier transform

$$F(p) = \frac{1}{(2\pi)^{3/2}} \int_{\mathbb{R}^3} e^{ipx} f(x)\,dx$$

The map $f(x) \longmapsto F(p)$ is unitary according to Plancherel's theorem. Moreover, it takes

$$-\Delta f(x) \longmapsto p^2 F(p) .$$

Comparing to the map $H_0 \longmapsto \Lambda$ that we are looking for, we see that one must take

$$p^2 = p_1^2 + p_2^2 + p_3^2 = |p|^2 = \lambda .$$

Now in the p-space introduce spherical coordinates $\omega, |p|$, where ω is the unit vector defining the point of intersection between the radius-vector of the point p and the unit sphere Ω, and $|p| = \sqrt{\lambda}$ is the length of the radius-vector p. From now on we can write

$$F(p) = F(\omega, \sqrt{\lambda}) \ .$$

The map $f \longmapsto F$ being unitary, we have

$$\|f\|^2 = \|F\|^2 = \int_0^\infty \int_\Omega |F(\omega,\sqrt{\lambda})|^2 \lambda d\omega d(\sqrt{\lambda}) =$$

$$= \int_0^\infty [\int_\Omega |\frac{1}{\sqrt{2}} F(\omega,\sqrt[4]{\lambda})\sqrt{\lambda}|^2 d\omega] d\lambda \ .$$

Therefore, the operator $H_0 = -\Delta$ has only an absolutely continuous spectrum, and this fills up the semiaxis $(0,\infty)$. Setting

$$G^{\oplus} = \int_0^\infty \oplus \, G_\lambda d\lambda \ , \quad G_\lambda \cong L_2(\Omega) \ ,$$

the map

$$f \longmapsto f(\lambda) = \frac{\sqrt[4]{\lambda}}{\sqrt{2}} F(\omega,\sqrt{\lambda})$$

becomes a unitary transformation from $L_2(\mathbb{R}^3)$ onto $G^{\oplus}$ and takes the operator H_0 into the operator Λ.

Now let us find the operator T_λ which acts in $L_2(\Omega)$ and gives the scattering suboperator $S(\lambda) = I_\lambda + T_\lambda$. Since the operator T_λ is trace class and, even more, a Hilbert-Schmidt operator, it will be realized by a kernel $T_\lambda(\omega,\omega')$, i.e.,

$$T_\lambda F(\omega) = \int_\Omega T_\lambda(\omega,\omega') F(\omega') d\omega' \quad (F \in L_2(\Omega)) \ .$$

Under broad enough assumptions concerning the potential $v(x)$, one succeeds in showing that

$$T_\lambda(\omega,\omega') = \frac{i\sqrt{\lambda}}{2\pi} A_\lambda(\omega,\omega') \ , \tag{11}$$

where $A_\lambda(\omega,\omega')$ is the so-called *asymptotic (scattering) amplitude* for the perturbed equation

$$(-\Delta + v)\psi - \lambda\psi = 0 \tag{12}$$

Let us explain that the free (unperturbed) equation

$$-\Delta\psi - \lambda\psi = 0$$

admits, as the simplest generalized solution, the function

$$\psi_\omega^{(0)}(x) = e^{i\sqrt{\lambda}\omega x} .$$

As for the eigenfunctions of the perturbed equation (12), they have the form

$$\psi_\omega(x) = e^{i\sqrt{\lambda}\omega x} + \frac{e^{i\sqrt{\lambda}|x|}}{|x|} A_\lambda(\omega,\frac{x}{|x|}) + o(\frac{1}{|x|}) . \tag{13}$$

Thus, it turns out that the scattering suboperators are defined by the asymptotics (13) of the eigenfunctions of the perturbed equation.

The formal procedure leading to formula (11) was proposed by physicists [21,50]. It became completelly well-founded, under certain assumptions on the potential (which ensure, in particular, that the asymptotic formula (13) holds true), following work by A. Ya. Povzner [53,54] and T. Ikebe [27], who proved the completeness of the system of generalized eigenfunctions (13) supplemented by the usual (i.e., belonging to $L_2(\mathbb{R}^3)$) eigenfunctions of equation (12).

At the same time, let us point out the the trace class property of the operator T_λ defined by the scattering amplitude via formula (11), as well as other properties which are consequences of the general facts discussed at no.3, were apparently not known before the works [8,9] appeared.

5. To conclude, let us formulate a number of problems.

I. Consider a differential operator with constant,

scalar or matrix coefficients acting on scalar or vector-valued functions, respectively. What conditions should such an operator satisfy in order to ensure that, after perturbing it by a local operator, the operator T_λ (or $S(\lambda)$) can be calculated from the corresponding asymptotics of the solution to the perturbed equation ?

II. We have seen that one of the invariants of the scattering operator (or, equivalently, of the operator phase $K(\lambda)$) can be calculated directly by the formula

$$\mathrm{Sp}\, K(\lambda) = \xi(\lambda) ,$$

where the definition of the function $\xi(\lambda)$ is completely independent of $K(\lambda)$.

Can one obtain something similar for other invariants of the operator-function $K(\lambda)$?

III. The previous problem can be generalized as follows. We know that the complete wave operators $W_\pm(H_1,H_0)$ provide a unitary equivalence between the operators H_{1a} and H_{0a} ($H_{1a}W_\pm = W_\pm H_{0a}$). Is it no possible, by some intrinsic properties, to distinguish the operators $W_\pm$ amond all those operators W which both map A_0 unitarily onto A_1 and transform H_{0a} unitarily into H_{1a} ?

Thus we would obtain also some intrinsic characteristic of the operator S.

IV. It would be nice to clarify whether or not the invariance principle holds in the stronger from, when no restrictions other than the existence of complete wave operators $W_\pm(H_1,H_0)$ is imposed upon the given operators H_0 and H_1.

Finally, let us remark that we have not touched upon at all the problem of the analyticity of the scattering suboperator $S(\lambda)$, a subject that is related in quantum mechanics to causality problems [31,19]. The lattest works by Lax and Phillips [43,44] give us hope that this profound question will also be made tractable in the framework of abstract operator theory, and that in this way its connection with the theory of stationary stochastic processes will come to light.

Let us emphasize also that the abstract schemes discussed here do not encompass the scattering problem for many particles (the multi-channel scattering problem). Nevertheless, there has been serious progress recently in the theory concerned with this more difficult problem. Here one should mention, first of all, the fundamental paper of L. D. Faddeev [16] on the scattering theory for three-particle systems and then the interesting paper of A. L. Chistyakov [13] on the multi-channel scattering problem, as well as the preceding papers [29] and [24].

REFERENCES

1. Agranovich, Z. S. and Marchenko, V. A.: *The Inverse Problem of Scattering Theory*, Khar'kov Gosud. Univ., 1960; English transl., Gordon and Breach, New York, 1963.

2. Birman, M. Sh.: *Perturbation of spectrum of a singular elliptic operator under variation of boundary and boundary conditions*, Dokl. Akad. Nauk SSSR 137, No. 4 (1961), 761-763; English transl., Soviet Math. Dokl. 2 (1961), 326-328.

3. Birman, M. Sh.: *Conditions for the existence of wave operators*, Dokl. Akad. Nauk SSSR 143, No. 3 (1962), 506-509; English transl., Soviet Math. Dokl. 3 (1962), 408-411.

4. Birman, M. Sh.: *A test for the existence of wave operators*, Dokl. Akad. Nauk SSSR 3 (1962), 1008-1009; English transl., Soviet Math. Dokl. 3 (1962), 1747-1748.

5. Birman, M, Sh.: *Perturbations of the continuous spectrum of a singular elliptic operator under changes of boundary and boundary conditions*, Vestnik Leningrad. Univ., No. 1 (1962), 22-55. (Russian)

6. Birman, M. Sh.: *Existence conditions for wave operators*, Izv. Akad. Nauk SSSR Ser. Mat., 27 (1963), 883-906; English transl., Amer. Math. Soc. Transl. (2) 54 (1966), 91-117.

7. Birman, M. Sh. and Entina, S. B.: *A stationary approach in the abstract theory of scattering*, Dokl. Akad. Nauk SSSR 155, No. 3 (1964), 506-508; English transl., Soviet Math. Dokl. 5 (1964), 432-435.

8. Birman, M. Sh. and Krein, M. G.: *On the theory of wave operators and scattering operators*, Dokl. Akad. Nauk SSSR 144, No. 3 (1962), 475-478; English transl., Soviet Math. Dokl. 3 (1962), 740-744.

9. Birman, M. Sh. and Krein, M. G.: *Some problems in the theory of wave operators and scattering operators*, Proceedings of the Joint Soviet- American Symposium on Partial Differential

Equations, Novosibirsk, 1963.

10. Brownell, F. H.: *A note on Cook's wave-matrix theorem*, Pacific J. Math. 12 (1962), 47-52.

11. Buslaev, V. S.: *Trace formula for Schrödinger's operator in three-space*, Dokl. Akad. Nauk SSSR 143, No. 5 (1962), 1067-1070; English transl., Soviet Physics Dokl. 7, No. 4 (1962), 293-297.

12 Buslaev, V. S. and Faddeev, L. D.: *Formulas for traces for a singular Sturm-Liouville differential operator*, Dokl. Akad. Nauk SSSR 132, No. 1 (1960), 13-16; English transl., Soviet Math. Dokl. 1 (1960), 451-454.

13. Chistyakov, L. A.: *On multi-channel scattering*, Uspekhi Mat. Nauk 18, No. 5 (1963), 201-208. (Russian)

14. Cook, J. M.: *Convergence to the Møller wave matrix*, J. Math. Phys. 36 (1957), 82-87.

15. Faddeev, L. D.: *The inverse problem in the quantum theory of scattering*, Uspekhi Mat. Nauk 14, No. 4 (1959), 57-119; English transl., J. Math. Phys. 4, No. 1 (1963), 72-104.

16. Faddeev, L. D.: *Mathematical Aspects of the Three-Body Problem in Quantum Scattering Theory*, Trudy Mat. Inst. im Steklova 69 (1963), 1-222; English transl., Israel Program for Scientific Translations, Jerusalem, 1965.

17. Faddeev, L. D. and Ladyzhenskaya, O. A.: *On continuous spectrum perturbation theory*, Dokl. Akad. Nauk SSSR 120, No. 6 (1958), 1187-1190. (Russian)

18. Fomin, S. V. and Naimark, M. A.: *Continuous direct sums of Hilbert spaces*, Uspekhi Mat. Nauk 10, No. 2 (1955), 111-142. (Russian)

19. Fourès, Y. and Segal, I. E.: *Causality and analyticity*, Trans. Amer. Math. Soc. 78 (1955), 385-405.

20. Friedrichs, K. O.: *On the perturbation of continuous spectra*, Comm. Pure Appl. Math. 1 (1948), 361-406.

21. Gell-Mann, M. and Goldberger, M. L.: *The formal theory of scattering*, Phys. Rev. 91, No. 2 (1953), 398-408.

22. Gohberg, I. C. and Krein, M. G.: *Introduction to the Theory of Linear Nonselfadjoint Operators*, "Nauka", Moscow, 1965; English transl., Transl. Math. Monographs, vol. 18, Amer. Math. Soc., Providence, R. I. 1970.

23. Gohberg, I. C. and Krein, M. G.: *The basic propositions on defect numbers, root numbers and indices of linear operators*, Uspekhi Mat. Nauk 12, No. 3 (1957), 43-118; English transl., Amer. Math. Soc. Transl. (2) 13 (1960), 185-264.

24. Grawert, G. and Petzold, I.: *Formale Mehrkanal-Streutheorie*, Zeitshrift für Naturforsch. 15 a, No. 4 (1960), 311-319.

25. Hack, M. N.: *On the convergence of Møller operators*, Nuovo Cimento 9 (1958), 731-733.

26. Heisenberg, W.: *Die "beobachtbaren Grössen" in der Theorie der Elementarteilchen, I* Zeitschrift für Phys. 120 (1943), 513; *II*, *ibid.* 120 (1943), 673.

27. Ikebe, T.: *Eigenfunctions expansions associated with the Schrödinger operator and their applications to scattering theory*, Arch. Rational Mech. Anal. 5 (1960), 1-34.

28. Jauch, J. M.: *Theory of the scattering operator*, Helv. Phys. Acta 31 (1958), 127-158.

29. Jauch, J. M.: *Theory of the scattering operator. II. Multichannel scattering*, Helv. Phys. Acta 31 (1958), 661-684.

30. Jauch, J. M. and Zinnes, I. I.: *The asymptotic condition for simple scattering systems*, Nuovo Cimento 11 (1959), 553-567.

31. van Kampen, N. G.: *Note on the analytic continuation of the S-matrix*, Phil. Mag. 42 (1951), 851-855.

32. Kato, T.: *On finite-dimensional perturbation of selfadjoint operators*, J. Math. Soc. Japan 9 (1957), 239-249.

33. Kato, T.: *Perturbation of continuous spectra by trace class operators*, Proc. Japan Acad. 33 (1957), 260-264.

34. Kato, T.: *Wave operators and unitary equivalence*, Pacific J. Math. 15 (1965), 171-180.

35. Krein, M. G.: *On the trace formula in perturbation theory*, Mat. Sb. 33 (1953), 597-626. (Russian)

36. Krein, M. G.: *On perturbation determinants and a trace formula for unitary and selfadjoint operators*, Dokl. Akad. Nauk SSSR 144, No. 2 (1962), 268-271; English transl., Soviet Math. Dokl. 3 (1962), 707-710.

37. Krein, M. G.: *Criteria for completeness of the system of root vectors of a dissipative operator*, Uspekhi Mat. Nauk 14, No. 3 (1959), 145-152; English transl., Amer. Math. Soc. Transl. (2) 26 (1963), 221-229.

38. Kuroda, S. T.: *On the existence and the unitary property of the scattering operator*, Nuovo Cimento 12 (1959), 431-454.

39. Kuroda, S. T.: *Perturbation of continuous spectra by unbounded operators. I.* J. Math. Soc. Japan 11 (1959), 247-262.

40. Kuroda, S. T.: *Perturbation of continuous spectra by unbounded operators. II.* J. Math. Soc. Japan 12 (1960), 243-257.

41. Kuroda, S. T.: *On a paper of Green and Lanford*, J. Math. Phys. 3 (1962), 933-935.

42. Kuroda, S. T.: *Finite-dimensional perturbation and a representation of scattering operator*, Pacific J. Math. 13 (1963), 1305-1318.

43. Lax, P. D. and Phillips, R. S.: *The wave equation in exterior domains*, Bull. Amer. Math. Soc. 68 (1962), 47-49.

44. Lax, P. D. and Phillips, R. S.: *Scattering Theory*, Academic Press, New York, 1967.

45. Lidskii, V. B.: *Nonselfadjoint operators with a trace*, Dokl. Akad. Nauk SSSR 125, No. 1 (1959), 485-587; English transl., Amer. Math. Soc. Transl. (2) 47 (1965), 43-46.

46. Lidskii, V. B.: *Summability of series in terms of the principal vectors of nonselfadjoint operators*, Trudy Moskov. Mat. Obshch. 11 (1962), 3-35; English transl., Amer. Math. Soc. Transl. (2) 40 (1964), 193-228.

47. Lifshits, I. M.: *On a problem of perturbation theory related to quantum statistics*, Uspekhi Mat. Nauk 7 , No. 1 (1952), 171-180. (Russian)

48. Lifshits, I. M.: *Some problems of the dynamic theory of non-ideal crystal lattices*, Nuovo Cimento, Suppl. al vol. 3, Ser. X (1956), 716-734.

49. Lifshits, I. M.: Uchen. Zapiski Khar'kov. Gosud. Univ. 27 (1949). (Russian)

50. Lippman, B. and Schwinger, J.: *Variational principles for scattering processes. I.* Phys. Rev. 79 (1950), 469-480.

51. Møller, C.: *General properties of the characteristic matrix in the theory of elementary particles*, Kgl. Danske Videsk. Selskab Mat.-Fys. Medd. 23, No. 1 (1945), 3-48.

52. Plesner, A. I. and Rokhlin, V. A.: *Spectral theory of linear operators. II.* Uspekhi Mat. Nauk 1, No. 1 (1946), 71-191; English transl., Amer. Math. Soc. Transl. (2) 62 (1967), 29-175.

53. Povsner, A. Yu.: *On the expansion of arbitrary functions in terms of eigenfunctions of the operator* $-\Delta u + cu$, Mat. Sb. 32 (74) (1953), 109-156; English transl., Amer. Math. Soc. Transl. (2) 60 (1967), 1-49.

54. Povsner, A. Yu.: *On expansion in functions that are solutions to the scattering problem*, Dokl. Akad. Nauk SSSR 104, No. 3 (1955), 360-363. (Russian)

55. Rosenblum, M.: *Perturbation of the continuous spectrum and unitary equivalence*, Pacific J. Math. 7 (1957), 997-1010.

56. Stankevich, I. V.: *On the theory of perturbation of continuous spectrum*, Dokl. Akad. Nauk SSSR 144, No. 2 (1962), 279-282; English transl., Soviet Math. Dokl. 3 (1962), 719-722.

57. Yavriyan, V. A.: *On the spectral shift function for Sturm-Liouville operators*, Dokl. Akad. Nauk Arm SSR 38, No. 4 (1964), 193-198. (Russian)

ON NONLINEAR INTEGRAL EQUATIONS WHICH PLAY A ROLE IN THE THEORY OF WIENER-HOPF EQUATIONS. I, II*

M. G. Krein

PART I

In author's paper [6], a study is made of the theory of Wiener-Hopf integral equations

$$\chi(t) - \int_0^\infty k(t-s)\chi(s)ds = f(t) \quad (0 \leqq t < \infty) \qquad (0.1)$$

under the assumptions that the function $k(t)$ $(-\infty < t < \infty)$ belongs to $L_1(-\infty,\infty)$, $f \in E$, and the solution χ is sought in the same space E. In [6] E denotes one of the spaces belonging to an entire class, containing, in particular, all the spaces $L_p(0,\infty)$ $(1 \leqq p < \infty)$.

In [6], the author wrote (p. 170 of the English translation):

"As is known, the integro-differential equation for transfer of radiant energy under known assumptions relative to the so-called indicatrix of dispersion (see [8] or [3]) can be turned into an equation or a system of equations of Wiener-Hopf type. On the other hand, starting with certain physical arguments (the so-called "principle of invariance"), V. A. Ambartsumyan succeeded in reducing the problem of integrating the integro-differential equation of transfer of radiant energy to the solution of one or

*Translation of Nonselfadjoint Operators, Matematicheskie Issledovaniya, No. 42 (1976), 47-90, and of Spectral Properties of Operators, Matematicheskie Issledovaniya, No. 45 (1977), 67-92, Izdat. Shtiintsa, Kishinev.

more nonlinear equations. These *nonlinear* equations admit simple methods of computational solution, and furthermore, they permitted V. V. Sobolev [8] and S. Chandrasekhar [3] to obtain explicit analytic expressions for a series of characteristic functions of the problem of transfer of radiant energy. Among these are characteristic functions for solutions of the corresponding system of *linear* integral equations.

This connection between linear integral equations of Wiener-Hopf type and the corresponding nonlinear equations can be obtained by purely analytic means under certain general hypotheses concerning the kernels of the equations."

Indeed, this connection is rediscovered each time the function $k(t) \in L_1(-\infty,\infty)$ admits the representation

$$k(t) = \int_0^\infty e^{-|t|u} \, d\sigma(u) \, , \tag{0.2}$$

where $\sigma(u)$ $(0 \leqq u < \infty)$ is some function having bounded variation on every finite interval. Incidentally, in virtue of a well-known theorem of S. N. Bernshtein [2] (see also [7]), an even function $k(t)$ has a representation (0.2) with a nondecreasing function $\sigma(u)$ if and only if it is absolutely monotonic for $t < 0$, i.e., it has nonnegative derivatives of all orders for $t < 0$.

When the function $k(t)$ may be represented in the form (0.2), the theory of equation (0.2) is intimately related to the theory of nonlinear equation

$$F^{-1}(v) = 1 - \int_0^\infty \frac{F(u)\,d\sigma(u)}{u + v} \, . \tag{!}$$

To avoid repetition, we refer the reader to the relevant passages in the introduction to [6], where the relation between the general equation (!) and the well-known equation of V. A. Ambartsumyan is explained.

In the present paper we use the results of [6] in an essential way, as well as the notation introduced there. The references to paper [6] are given as follows: [6,4.21] refers to "relation (4.21) in [6]". The present work was written immediately following the paper [6] (i.e., almost twenty years ago), and was meant to be some sort of continuation of the latter. It would have probably remained unpublished had not I. A. Fel'dman expressed strong interest in it. He attentively read the manuscript, made a number of critical observations, and helped the author immensely in preparing the paper for publication.

I consider my pleasant duty to express my heartfelt gratitude to I. A. Fel'dman for the great work he did.

§ 1. THE FUNDAMENTAL SOLUTION OF EQUATION (!)

If the function $k(t)$ admits a representation (0.2) with a nondecreasing function $\sigma(u)$, then $k(t) \in L_1(-\infty,\infty)$ precisely when

$$\left(\int_0^\infty k(x)dx =\right) \quad \int_0^\infty \frac{d\sigma(t)}{t} < \infty .$$

If, however, the function $\sigma(u)$ (in general, complex) appearing in representation (0.2) has bounded variation in every finite interval, then $k(t) \in L_1(-\infty,\infty)$ as soon as

$$(0 <) \quad \int_0^\infty \frac{|d\sigma(t)|}{t} < \infty . \tag{1.1}$$

From now on, we shall assume that this condition is always fulfilled.

It follows from (0.2) that the Fourier transform

$$K(\lambda) = \int_{-\infty}^\infty k(t)e^{i\lambda t}dt \quad \left(=2\int_0^\infty k(t)\cos \lambda t\, dt\right)$$

has the form

$$K(\lambda) = \int_0^\infty \left[\frac{1}{t - i\lambda} + \frac{1}{t + i\lambda}\right]d\sigma(t) = 2\int_0^\infty \frac{td\sigma(t)}{t^2 + \lambda^2} .$$

The condition that $1 - K(\lambda) \neq 0 \quad (-\infty < \lambda < \infty)$, i.e., the condition

$$1 - 2\int_0^\infty \frac{td\sigma(t)}{t^2 + \lambda^2} \neq 0 \quad (-\infty < \lambda < \infty) \tag{1.2}$$

plays a fundamental role in the analysis of equation (0.1) (see [6]). If this condition is fulfilled, then according to Theorem 2.1 in [6], the function $1 - K(\lambda)$ admits a canonical factorization. Since $1 - K(\lambda)$ is also even, we obtain the following factorization

$$1 - 2\int_0^\infty \frac{t\,d\sigma(t)}{t^2 + \lambda^2} = G_+^{-1}(\lambda)G_+^{-1}(-\lambda) \qquad (-\infty < \lambda < \infty) . \qquad (1.3)$$

Here $G_+^{-1}(\lambda)$ and $G_+^{-1}(-\lambda)$ are holomorphic in the interior of and continuous in the closed upper half plane Π_+. Multiplying both sides of equality (1.3) by $G_+(\lambda)$, we see that the function $G_+^{-1}(-\lambda)$ is holomorphic in the half plane Π_+ with a cut along the imaginary semiaxis $[0,i\infty)$. Thus, the function $G_+^{-1}(\lambda)$ is holomorphic in the entire complex plane with a cut along the negative imaginary semiaxis $[0,-i\infty)$.

Consider the function

$$F(\lambda) = G_+^{-1}(\lambda) - 1 + \int_0^\infty \frac{G_+(it)}{t - i\lambda}\,d\sigma(t) \qquad (\lambda \notin [0,-i\infty)), \qquad (1.4)$$

which is holomorphic in the complex plane with a cut along the negative imaginary semiaxis $[0,-i\infty)$. It turns out that the interior points of this cut are regular points of $F(\lambda)$ too. Indeed, introducing the expression for $G_+^{-1}(\lambda)$ given by (1.3) into (1.4), we find that

$$\begin{aligned} F(\lambda) &= G_+(-\lambda) + iG_+(-\lambda)\int_0^\infty \frac{d\sigma(t)}{\lambda - it} + \\ &\quad + i\int_0^\infty \frac{G_+(it) - G_+(-\lambda)}{it + \lambda}\,d\sigma(t) - 1 , \end{aligned} \qquad (1.5)$$

whence it is easy to see that the points of the ray $(0,-i\infty)$ are regular for $F(\lambda)$.

It is known [6,2.3] that in the upper half plane Π_+ one has $\lim_{|\lambda|\to\infty} (1 - G_+^{-1}(\lambda)) = 0$ uniformly $(\lambda \in \Pi_+)$. Since for any $\lambda \in \Pi_+$ and $N > 0$

$$\left|\int_0^\infty \frac{G_+(it)}{t - i\lambda}\,d\sigma(t)\right| \leq c\int_0^\infty \frac{|d\sigma(t)|}{|t - i\lambda|} < c\left\{\int_0^N \frac{|d\sigma(t)|}{|t - i\lambda|} + \right.$$

$$\left. + \int_N^\infty \frac{|d\sigma(t)|}{|t|}\right\} ,$$

we see that $F(\lambda) \longrightarrow 0$ as $|\lambda| \longrightarrow \infty$ uniformly in the upper half plane Π_+. If we can show that the same happens in the lower half plane, then we have shown at the same time that

$$F(\lambda) \equiv 0 . \tag{1.6}$$

Replacing λ by $-\lambda$ in (1.5), it is not hard to verify that $F(\lambda)$ has the desired behavior in Π_- provided that the function

$$\Phi(\lambda) = \int_0^\infty \frac{G_+(\lambda) - G_+(it)}{\lambda - it} \, d\sigma(t)$$

tends uniformly to zero in Π_+ as $|\lambda| \longrightarrow \infty$.

In writing $\Phi(\lambda)$ as a sum of two integrals: $\Phi(\lambda) = \int_0^N + \int_N^\infty$, the first integral tends uniformly to zero in Π_+ as $|\lambda| \longrightarrow \infty$. Consequently, we need to show that the second integral tends uniformly to zero in Π_+ as $N \longrightarrow \infty$.

We take advantage of the fact that $G_+(\lambda)$ has the form [6,2.16]; whence

$$\frac{G_+(\lambda) - G_+(it)}{\lambda - it} = \int_0^\infty \frac{e^{i\lambda s} - e^{-ts}}{\lambda - it} \gamma(s) ds ,$$

where $\gamma(s) \in L_1(0,\infty)$.

For $\lambda \in \Pi_+$, one has the estimates

$$\left| \frac{e^{i\lambda s} - e^{-ts}}{\lambda - it} \right| \leqq e^{-ts} \left| \int_0^s e^{(i\lambda + t)u} du \right| \leqq$$

$$\leqq e^{-ts} \int_0^s e^{tu} du < \frac{1}{t} \quad (s,t > 0; \lambda \in \Pi_+) .$$

It follows that

$$\left| \frac{G_+(\lambda) - G_+(it)}{\lambda - it} \right| \leqq \frac{1}{t} \int_0^\infty |\gamma(s)| ds = \frac{C}{t} ,$$

and hence

$$\left| \int_N^\infty \frac{G_+(\lambda) - G_+(it)}{\lambda - it} \, d\sigma(t) \right| \leqq C \int_N^\infty \frac{|d\sigma(t)|}{|t|} \longrightarrow 0$$

as $N \longrightarrow \infty$.

Therefore, (1.6) is established, i.e.,

$$G_+^{-1}(\lambda) = 1 - \int_0^\infty \frac{G_+(it)}{t - i\lambda}\, d\sigma(t) \qquad (\lambda \in \text{Ext}\ (0,-i\infty)). \qquad (1.7)$$

Thus, under the assumption (1.2), equation (!) has a solution $F(s) = G_+(is)$ $(0 \leqq s < \infty)$, which we shall call the *fundamental* solution.

Remark. The previous result can be generalized to the situation when the function $k(t)$ has the following form

$$k(t) = \int_{\Pi_+} e^{i|t|z}\, d\omega(z) \qquad (-\infty < t < \infty), \qquad (1.8)$$

where the integral is understood in the sense of Stieltjes-Radon. Therefore, the $\omega(\Delta)$ in this integral is an additive function defined on the set of all open, semiopen, and closed rectangles Δ with sides parallel to the coordinate axes and contained in the upper half plane Π_+, and ω is of bounded variation on each such rectangle Δ.

Assuming that the integral (1.8) converges absolutely for any $t \in (-\infty,\infty)$, and, in addition, that

$$\int_{\Pi_+} \frac{|d\omega(z)|}{|\lambda + z|} < \infty \qquad (-\infty < \lambda < \infty),$$

one has easily that

$$K(\lambda) = 2\int_{\Pi_+} \frac{z\, d\omega(z)}{\lambda^2 - z^2} \qquad (-\lambda < \lambda < \infty).$$

The arguments that led to relation (1.7) can be readily generalized and permit us to assert that if the condition $1 - K(\lambda) \neq 0$ $(-\infty < \lambda < \infty)$ is fulfilled, then one has

$$G_+^{-1}(\lambda) = 1 + i\int_{\Pi_+} \frac{G_+(z)\, d\omega(z)}{\lambda + z} \qquad (\lambda \in \Pi_+).$$

§ 2. THE SIMPLEST PROPERTIES OF THE SOLUTIONS TO EQUATION (!)

1. It is convenient to begin our discussion with a slightly more general equation, rather than with equation (!),

namely

$$\frac{1}{F(s)} = c - \int_0^\infty \frac{F(t)d\sigma(t)}{t + s} \qquad (0 \leqq s < \infty) , \tag{2.1}$$

where c is an arbitrary complex constant.

Let us emphasize that we call a σ-measurable function $F(t)$ $(0 \leqq t < \infty)$ a *solution* of equation (2.1) if

$$\int_0^\infty \frac{|F(t)|\,|d\sigma(t)|}{t} < \infty \tag{2.2}$$

and $F(s)$ satisfies equation (2.1).

Condition (2.2) is obviously equivalent to the absolute convergence of the integral on the right-hand side of (2.1) for all $s \geqq 0$.

Since the latter integral has an analytic continuation to a function holomorphic in the domain $\mathrm{Ext}\,(-\infty,0]$, the function $F(s)$ also has an analytic continuation to a meromorphic function $F(z)$ in the same domain:

$$\frac{1}{F(z)} = c - \int_0^\infty \frac{F(t)d\sigma(t)}{t + z} \qquad (z \in \mathrm{Ext}\,(-\infty,0]) .$$

If $F(z)$ has no poles on the semiaxis $[0,\infty)$ (on the open semiaxis $(0,\infty)$), then we shall say that the solution $F(s)$ $(0 \leqq s < \infty)$ is *regular* (respectively, *regular on the semiaxis* $(0,\infty)$).

Because (2.1) implies that $\lim\limits_{s\to\infty} F(s) = c^{-1}$, when $c \neq 0$ a regular solution F is a bounded solution of equation (2.1), and conversely.

THEOREM 2.1 *If* $F(s)$ *is a solution of equation* (2.1), *then*

$$\frac{1}{F(z)F(-z)} = c^2 - 2\int_0^\infty \frac{t\,d\sigma(t)}{t^2 - z^2} \qquad (z^2 \notin [0,\infty)) . \tag{2.3}$$

PROOF. Indeed, a repeated use of relation (2.1) gives

$$[c - F^{-1}(z)][c - F^{-1}(-z)] = \int_0^\infty\int_0^\infty \frac{F(t)F(s)d\sigma(t)d\sigma(s)}{(t + z)(s - z)} =$$

$$= \int_0^\infty\int_0^\infty \left[\frac{1}{t + z} + \frac{1}{s - z}\right] \frac{F(t)F(s)d\sigma(t)d\sigma(s)}{t + s} =$$

$$= \int_0^\infty \left(\int_0^\infty \frac{F(t)d\sigma(t)}{t+s}\right)\frac{F(s)d\sigma(s)}{s-z} + \int_0^\infty \left(\int_0^\infty \frac{F(s)d\sigma(s)}{s+t}\right)\frac{F(t)d\sigma(t)}{t+z}$$

$$= \int_0^\infty [c - F^{-1}(s)]\frac{F(s)d\sigma(s)}{s-z} + \int_0^\infty [c - F^{-1}(t)]\frac{F(t)d\sigma(t)}{t+z} =$$

$$= c[c - F^{-1}(z)] + c[c - F^{-1}(-z)] - 2\int_0^\infty \frac{t\,d\sigma(t)}{t^2 - z^2},$$

whence (2.3) follows at once.

This theorem is a straightforward generalization of a result of Chandrasekhar and Crum (see [3], Ch. V, § 40).

Returning to the equation

$$\frac{1}{F(s)} = 1 - \int_0^\infty \frac{F(t)d\sigma(t)}{t+s} \quad (0 \leq s < \infty), \tag{!}$$

we then have for it the following conclusion.

THEOREM 2.2 *If* F *is a solution of equation* (!), *then*

$$1 - \int_0^\infty \frac{F(t)d\sigma(t)}{t} = F^{-1}(0) = \pm(1 - 2s_1)^{1/2}, \tag{2.4}$$

where

$$s_1 = \int_0^\infty \frac{d\sigma(u)}{u}.$$

PROOF. By Theorem 2.1,

$$\frac{1}{F(z)F(-z)} = 1 - 2\int_0^\infty \frac{u\,d\sigma(u)}{u^2 - z^2} \quad (z^2 \notin [0,\infty)).$$

Setting $z = iy$ $(y > 0)$ here and subsequently letting y go to zero, we obtain

$$\frac{1}{F^2(0)} = 1 - 2\int_0^\infty \frac{d\sigma(u)}{u}.$$

[For a full justification of this limit, see § 4, no. 1.] This equality together with equation (!) for $s = 0$ yield (2.4).

COROLLARY 2.1. *Let* $1 - 2s_1 \neq 0$. *Then*

$$\int_0^\infty \frac{F(t)d\sigma(t)}{t} = 1.$$

2. If the integrals $s_k = \int_0^\infty \frac{d\sigma(u)}{u^k}$ $(k = 1,2,3)$ are absolutely convergent, then the integrals

$$\alpha_k = \int_0^\infty \frac{F(u)\,d\sigma(u)}{u^k} \qquad (k = 1,2,3)$$

will be absolutely convergent for any regular solution F of equation (!). The same statement may be made for the irregular solutions F of equation (!) if, for example, $\sigma(u) \equiv \text{const}$ in a neighborhood of the point 0, as well as for a number of other cases.

THEOREM 2.3. *Suppose that for some solution* F *to equation* (!) *the integrals* s_k *and* α_k $(k = 1,2,3)$ *are absolutely convergent. Then*

$$s_3 = (1 - \alpha_1)\alpha_3 + \frac{\alpha_2^2}{2} . \tag{2.5}$$

PROOF. Let us recast relation (!) for F in the form

$$1 = F(s) - F(s) \int_0^\infty \frac{F(t)\,d\sigma(t)}{t + s} .$$

Multiplying this by $s^{-3}d\sigma(s)$ and integrating from 0 to ∞, we get

$$s_3 = \alpha_3 - \int_0^\infty \int_0^\infty \frac{F(s)F(t)\,d\sigma(s)\,d\sigma(t)}{s^3(t + s)} .$$

Adding this equality term-by-term to the similar equality resulting from interchanging the roles of s and t, we obtain

$$s_3 = \alpha_3 - \frac{1}{2} \int_0^\infty \int_0^\infty \left(\frac{1}{s^3} + \frac{1}{t^3}\right) \frac{F(s)F(t)\,d\sigma(s)\,d\sigma(t)}{t + s} =$$

$$= \alpha_3 - \frac{1}{2} \int_0^\infty \int_0^\infty \left(\frac{1}{st^3} - \frac{1}{s^2t^2} + \frac{1}{s^3t}\right) F(s)F(t)\,d\sigma(s)\,d\sigma(t) =$$

$$= \alpha_3 - \frac{1}{2}\,(2\alpha_1\alpha_3 - \alpha_2^2) = (1 - \alpha_1)\alpha_3 + \frac{\alpha_2^2}{2} ,$$

as claimed.

COROLLARY 2.2. *If* $1 - 2s_1 = 0$, *then*

$$\int_0^\infty \frac{F(u)\,d\sigma(u)}{u^2} = \pm\sqrt{2s_3} . \tag{2.6}$$

Indeed, if $1 - 2s_1 = 0$, then, according to (2.4),

$\alpha_1 = 1$, and thus $2s_3 = \alpha_2^2$, which is just (2.6).

It is obvious that Theorem 2.3 can be generalized, leading to expressions for the moments s_{2n-1} in terms of the moments $\alpha_1, \alpha_2, \ldots, \alpha_{2n-1}$, under the assumption that the integrals giving these moments are absolutely convergent.

§ 3. NONLINEAR EQUATIONS RELATED TO EQUATION (!)

1. Let F be a solution of equation (!). Then, according to Theorem 2.2,

$$F^{-1}(s) = 1 - \int_0^\infty \frac{F(t)\,d\sigma(t)}{t} - \int_0^\infty \left(\frac{1}{s+t} - \frac{1}{t}\right) F(t)\,d\sigma(t) =$$

$$= \pm(1 - 2s_1)^{1/2} + s\int_0^\infty \frac{F(t)\,d\sigma(t)}{t(s+t)} .$$

THEOREM 3.1. *Each solution* $F(s)$ $(0 \leq s < \infty)$ *of the equation*

$$\frac{1}{F(s)} = c - \int_0^\infty \frac{F(t)\,d\sigma(t)}{t+s} \qquad (0 \leq s < \infty) , \tag{3.1}$$

where $c^2 = 1$, *is a solution of the equation*

$$\frac{1}{F(s)} = d + s\int_0^\infty \frac{F(t)\,d\sigma(t)}{t(t+s)} \qquad (0 \leq s < \infty) , \tag{3.2}$$

where

$$d^2 = 1 - 2s_1 , \tag{3.3}$$

and conversely. [*When* $s = 0$, *equality* (3.2) *must be understood as* $F^{-1}(0) = d$.]

PROOF: The first part of the theorem for $c = 1$ is already proved. Now let F be a solution of equation (3.1) for $c = -1$; then $F_1 = -F$ will be a solution of the same equation for $c = 1$, and so it will satisfy equation (3.2) for some $d = d_1$ such that (3.3) holds. It follows that $F = -F_1$ will be a solution of (3.2) for $d = -d_1$.

Conversely, let $F(s)$ $(0 \leq s < \infty)$ be a solution of equation (3.2), where d satisfies condition (3.3). Then, in any case, F satisfies equation (2.1) for

$$c = d + \int_0^\infty \frac{F(t)d\sigma(t)}{t} .$$

Relation (2.1) implies, as we know, the relation (2.3), which in turn yields

$$F^{-2}(0) = c^2 - 2s_1 . \tag{3.4}$$

On the other hand, by setting $s = 0$ in (2.1), one obtains

$$F^{-1}(0) = d, \quad F^{-2}(0) = d^2 = 1 - 2s_1 . \tag{3.5}$$

Comparing (3.4) and (3.5), we get $c^2 = 1$, which completes the proof of the theorem.

2. When $1 - 2s_1 = 0$, equation (3.2) acquires the simple form

$$\frac{1}{F(s)} = s \int_0^\infty \frac{F(t)d\sigma(t)}{t(t+s)} . \tag{3.6}$$

Obviously, if $F_1(s)$ is some solution of equation (3.6), then $F_2(s) = -F_1(s)$ will be a solution of equation (3.6) too. On the other hand, if the corresponding constant c in (3.1) is equal to 1 for the solution $F = F_1$, then the constant corresponding to the solution $F = F_2 = -F_1$ is $c = -1$.

If $1 - 2s_1 \neq 0$, then it may happen that equation (3.1) has for some value of c, for example for $c = 1$, two distinct bounded solutions F_1 and F_2, to which will correspond two distinct values of d $(d_1 = -d_2)$ in equation (3.2) (see § 4, no. 4).

Notice also that equation (3.2) is of the same type as equation (2.1) for $d \neq 0$. Indeed, upon setting $\hat{F}(s) = F(s^{-1})$ $(0 \leqq s < \infty,\ F(\infty) = c)$ and

$$\hat{\sigma}(t) = -\int_{t^{-1}}^\infty \frac{d\sigma(u)}{u^2} \qquad (0 \leqq t < \infty) ,$$

equation (3.2) becomes

$$\frac{1}{\hat{F}(s)} = d - \int_0^\infty \frac{\hat{F}(t)d\hat{\sigma}(t)}{t+s} \qquad (0 \leqq s < \infty) .$$

At the same time

$$\int_0^\infty \frac{d\hat{\sigma}(t)}{t} = \int_0^\infty \frac{d\sigma(u)}{u} .$$

3. The following proposition plays an important role in our future discussion.

THEOREM 3.2. *Every solution* $F(z)$ *of equation* (!) *satisfies the relation*

$$\int_0^\infty \frac{F(z) - F(u)}{z - u}\, d\sigma(u) + F(z)\Big[1 - \int_0^\infty \frac{d\sigma(u)}{u + z}\Big] = 1 \qquad (3.7)$$

at any of its regular points $z \in \operatorname{Ext}(-\infty,0]$.

PROOF. According to relation (2.3), for any nonreal z

$$\frac{1}{F(z)} = F(-z)\Big[1 - \int_0^\infty \frac{d\sigma(u)}{u - z} - \int_0^\infty \frac{d\sigma(u)}{u + z}\Big] .$$

Inserting this expression for $F^{-1}(z)$ into (!) and regrouping the terms, we obtain

$$\int_0^\infty \frac{F(-z) - F(u)}{u + z}\, d\sigma(u) + F(-z)\Big[1 - \int_0^\infty \frac{d\sigma(u)}{u - z}\Big] = 1 . \qquad (3.8)$$

Regrouping the terms is legitimate provided that $F(-z) \neq \infty$. Replacing z by $-z$ in (3.8), we obtain (3.7) which is thus established for nonreal values of z such that $F(z) \neq \infty$. Since the left-hand side of (3.7) is also regular at each positive point z for which the function $F(z)$ is regular, equality (3.7) holds at these points too.

The theorem is proved.

THEOREM 3.3. *Let* $\zeta \in \operatorname{Ext}(-\infty,0]$ *be a regular point of some solution* F *to equation* (!), *and let* $Q_\zeta(z) = F(z)F(\zeta)/(z+\zeta)$. *Then*

$$\int_0^\infty \frac{Q_\zeta(z) - Q_\zeta(u)}{z - u}\, d\sigma(u) + Q_\zeta(z)\Big[1 - \int_0^\infty \frac{d\sigma(u)}{u + z}\Big] = \frac{1}{z+\zeta} . \qquad (3.9)$$

where $z \in \operatorname{Ext}(-\infty,0]$ *is any other regular point of* F *satisfying the condition* $z + \zeta \neq 0$.

PROOF. Multiplying all the terms of relation (3.7) by $F(\zeta)/(\zeta + z)$ we easily get that

$$\int_0^\infty [Q_\zeta(u) - \frac{F(u)F(\zeta)}{z+\zeta}] \frac{d\sigma(u)}{z-u} + Q_\zeta(z)[1 - \int_0^\infty \frac{d\sigma(u)}{u+\zeta}] =$$

$$= \frac{F(\zeta)}{\zeta + z} . \tag{3.10}$$

On the other hand, assuming first that z is not real, we shall have

$$\frac{F(\zeta)}{z+\zeta} \int_0^\infty \frac{F(u)d\sigma(u)}{z-u} = F(\zeta) \int_0^\infty [\frac{1}{z-u} - \frac{1}{z+\zeta}] \frac{F(u)d\sigma(u)}{\zeta+u} =$$

$$= \int_0^\infty \frac{Q_\zeta(u)d\sigma(u)}{z-u} - \frac{F(\zeta)}{z+\zeta} \int_0^\infty \frac{F(u)d\sigma(u)}{\zeta+u} =$$

$$= \int_0^\infty \frac{Q_\zeta(u)d\sigma(u)}{z-u} + \frac{F(\zeta)}{z+\zeta} [\frac{1}{F(\zeta)} - 1] .$$

Combining (3.10) and (3.11), we get (3.9).

This completes the proof of the theorem because the validity of the relation (3.9) for nonreal z which satisfy the conditions of the theorem implies that (3.9) holds for all z satisfying the same conditions.

§ 4. ADDITIONAL PROPERTIES OF THE SOLUTIONS TO EQUATION (!)

1. We require several properties of the function $w(z)$ defined by the formula

$$w(z) = \int_{-\infty}^\infty \frac{d\tau(t)}{t-z} + \text{const} \qquad (0 < \int_{-\infty}^\infty \frac{|d\tau(t)|}{1+|t|} < \infty) , \tag{4.1}$$

for all z for which

$$\int_{-\infty}^\infty \frac{|d\tau(t)|}{|t-z|} < \infty . \tag{4.2}$$

Let $\mathcal{D}_w$ denote the set of all points z satisfying (4.2). We shall show that at each real point $a \in \mathcal{D}_w$

$$\lim_{z\to a} w(z) = w(a) \tag{4.3}$$

as soon as one takes the limit inside the sector

$$\delta < |\arg(z - a)| < \pi - \delta, \tag{4.4}$$

where δ $(0 < \delta < \pi/2)$ is arbitrary. In fact, if z lies in sector (4.4), then $|z - t| \geq |z - a|\sin\delta$, and $|z - t| \geq |t - a|\sin\delta$. It follows that for any $\varepsilon > 0$

$$\begin{aligned} |w(z) - w(a)| &= \left|(z - a)\int_{-\infty}^{\infty} \frac{d\tau(t)}{(t - a)(t - z)}\right| \leq \\ &\leq \frac{1}{\sin\delta}\left[\int_{a-\varepsilon}^{a+\varepsilon} \frac{|d\tau(t)|}{|t - a|} + |z - a|\left(\int_{a+\varepsilon}^{\infty} + \int_{-\infty}^{a-\varepsilon}\right) \frac{|d\tau(t)|}{|t - a|^2}\right] . \end{aligned} \tag{4.5}$$

This proves (4.3) because the first integral in (4.5) can be made arbitrarily small by choosing ε, while the second tends, for fixed $\varepsilon > 0$, to zero as $z \longrightarrow a$.

If the condition

$$\int_{-\infty}^{\infty} \frac{|d\tau(t)|}{|t - a|^{k+1}} < \infty$$

is satisfied for $a \in \mathcal{D}_w$ and some natural k, then we shall call

$$w^{(k)}(a) = k!\int_{-\infty}^{\infty} \frac{d\tau(t)}{(t - a)^{k+1}}$$

the k-*th derivative* of the function $w(z)$ at the point a. If $\operatorname{Im} a \neq 0$, or a belongs to an interval on which the function $\tau(t)$ is constant, then a will be a regular point of the function $w(z)$, and our definition of the k-th derivative at the point a and the usual one coincide.

If the derivative $w^{(k)}(a)$ $(k > 1)$ exists in the above sense at a real point $a \in \mathcal{D}_w$, then the following relation is valid in the sector (4.4):

$$\lim_{z \to a} \frac{w^{(k-1)}(z) - w^{(k-1)}(a)}{z - a} = w^{(k)}(a) ,$$

and the proof is similar to that of (4.3).

Let us agree to say that the point $a \in \mathcal{D}_w$ is a *zero of multiplicity* $\geq k$ $(k > 0)$ of the function $w(z)$ if the derivative $w^{(k-1)}(a)$ exists at the point a and $w(a) = w'(a) = \dots = w^{(k-1)}(a) = 0$.

2. Recall that the function $F_0(s) = G_+(is)$ is a solution (the fundamental one) of the equation (see § 1)

$$F^{-1}(s) = 1 - \int_0^\infty \frac{F(t)d\sigma(t)}{t + s} \qquad (0 \leq s < \infty) \ . \tag{!}$$

As a straightforward consequence, we have the first assertion of the next theorem.

THEOREM 4.1. *If the function* σ *satisfies conditions* (1.1) *and* (1.2), *then the fundamental solution* $F_0(s)$ *of equation* (!) *is given by*

$$\ln F_0(s) = -\frac{s}{\pi} \int_0^\infty \ln w(i\lambda) \frac{d\lambda}{\lambda^2 + s^2} , \tag{4.6}$$

where

$$w(z) = 1 - 2 \int_0^\infty \frac{td\sigma(t)}{t^2 - z^2} \ .$$

Any other solution F *of equation* (!) *can be obtained from the formula*

$$F(s) = F_0(s) \prod_{j=1}^{n} \left(\frac{s + a_j}{s - a_j}\right)^{k_j} , \tag{4.7}$$

where a_j $(j = 1,2,\ldots,n)$ *are distinct zeros of the function* $w(z)$ *in the right half plane* $\mathrm{Re}\, z > 0$, *taken in arbitrary quantity, and* k_j *are arbitrary natural numbers no larger than the multiplicities of the corresponding zeros* a_j $(j = 1,2,\ldots,n)$.

If σ *is a nondecreasing function, then in formula* (4.7) *one can take only* $a_j > 0$, $k_j = 1$ $(j = 1,2,\ldots,n)$, *and* $F_0(s)$ *is the only positive, and also the only bounded, solution to equation* (!).

We emphasize here that by a solution to equation (!) we always mean a solution for which the integral on the right-hand side in (!) is absolutely convergent, i.e.,

$$\int_0^\infty \frac{|F(t)|\ |d\sigma(t)|}{t} < \infty \ . \tag{4.8}$$

Notice also that since

$$w(z) = 1 - \int_0^\infty \left[\frac{1}{t - z} + \frac{1}{t + z}\right] d\sigma(t) \ ,$$

$w(z)$ belongs to the class of functions introduced in no. 1.

PROOF. In virtue of the general formula [6,7.19], and because in our case $1 - K(\lambda) = w(i\lambda)$, formula (4.6) holds for the solution $F_0(s) = G_+(is)$ of equation (!).

Let us prove the second part of the theorem. Take any solution $F(t)$ $(0 \leqq t < \infty)$ to (!), and recall that the formula

$$F^{-1}(z) = 1 - \int_0^\infty \frac{F(t)\,d\sigma(t)}{t + z} \qquad (z \notin (-\infty, 0])$$

defines an analytic continuation of $F(t)$ to a meromorphic function in the complex plane with a cut along the ray $(-\infty,0]$. At the same time, according to Theorem 2.1

$$F^{-1}(z)F^{-1}(-z) = 1 - 2\int_0^\infty \frac{t\,d\sigma(t)}{t^2 - z^2} \qquad (\operatorname{Im} z \neq 0) . \tag{4.9}$$

Consider the function

$$\Phi(\lambda) = F^{-1}(-i\lambda) = 1 - \int_0^\infty \frac{F(t)\,d\sigma(t)}{t - i\lambda} \tag{4.10}$$

holomorphic in the complex plane with a cut along the ray $[0,-i\infty)$.

Taking into account (4.8), one can consider that $\Phi(\lambda)$ is defined for $\lambda = 0$ too; namely,

$$\Phi(0) = 1 - \int_0^\infty \frac{F(t)\,d\sigma(t)}{t} . \tag{4.11}$$

It is easy to see that $\Phi(\lambda) \longrightarrow 1$ as $|\lambda| \longrightarrow \infty$, uniformly in the half plane Π_+. Setting $\Phi(\infty) = 1$, we claim that the function $\Phi(\lambda)$ is continuous in the closed half plane Π_+. To verify this, it remains to show that $\Phi(\lambda)$ is continuous in Π_+ at the point $\lambda = 0$. But this is a consequence of the fact that for any $\varepsilon > 0$

$$|\Phi(\lambda) - \Phi(0)| = \left|\lambda\int_0^\infty \frac{F(t)\,d\sigma(t)}{(t - i\lambda)t}\right| \leqq \int_0^\varepsilon \frac{|F(t)|\,|d\sigma(t)|}{t} + |\lambda|\int_\varepsilon^\infty \frac{|F(t)|\,|d\sigma(t)|}{t^2} \tag{4.12}$$

because $|t - i\lambda| \geqq t$, $|t - i\lambda| \geqq |\lambda|$ for $t > 0$ and $\operatorname{Im}\lambda \geqq 0$. Indeed, for $\varepsilon > 0$ small enough and $|\lambda| \leqq \delta_\varepsilon$, the right-hand side of (4.12) will be arbitrarily small.

Using (4.10) and (1.3), one can rewrite relation (4.9) as

$$\Phi(\lambda)\Phi(-\lambda) = 1 - K(\lambda) = G_+^{-1}(\lambda)G_+^{-1}(-\lambda) . \tag{4.13}$$

Let ia_j $(j = 1,2,...n)$ be all the distinct zeros of the function $\Phi(\lambda)$ in the upper half plane, and let k_j $(j = 1, 2,...,n)$ be their multiplicities. Consider the function

$$\Phi_1(\lambda) = \Phi(\lambda) \prod_{j=1}^{n} \left(\frac{\lambda + ia_j}{\lambda - ia_j}\right)^{k_j} .$$

Equality (4.13) remains valid if one replaces $\Phi(\lambda)$ by $\Phi_1(\lambda)$, and since $\Phi_1(\lambda)$ has no zeros in Π_+, the resulting equality gives a canonical factorization of $1 - K(\lambda)$. By Theorem 2.1 of [6], such a factorization is unique; thus $\Phi_1(\lambda) = G_+^{-1}(\lambda)$. Therefore, one has

$$\Phi(\lambda) = G_+^{-1}(\lambda) \prod_{j=1}^{n} \left(\frac{\lambda - ia_j}{\lambda + ia_j}\right)^{k_j} . \tag{4.14}$$

Since $F(s) = \Phi^{-1}(is)$ $(0 \leqq s < \infty)$, according to the definition (4.10) of $\Phi(\lambda)$, it follows from (4.14) that $F(s) = F_0(s)R(s)$ $(0 \leqq s < \infty)$, where

$$R(s) = \prod_{j=1}^{n} \left(\frac{s + a_j}{s - a_j}\right)^{k_j} \qquad (\mathrm{Re}\, a_j > 0;\ j = 1,2,...,n) .$$

Moreover, one has

$$\int_0^\infty \frac{|F_0(t)R(t)|}{t + s} |d\sigma(t)| < \infty ,$$

because F is a solution to equation (!).

Now consider the integral

$$\int_0^\infty \frac{F(t)d\sigma(t)}{t + s} = \int_0^\infty \frac{F_0(t)R(t)}{t + s} d\sigma(t) \qquad (0 \leqq s < \infty) . \tag{4.15}$$

In order to get to the crux of the problem, first take the case $k_1 = k_2 = ... = k_n = 1$. Under this assumption, the expansion of $R(t)$ into partial fractions has the form

$$R(t) = 1 + \sum_{j=1}^{n} \frac{r_j}{t - a_j} \qquad (r_j \neq 0; j = 1,2,...,n) , \tag{4.16}$$

whence

$$\frac{R(t)}{t+s} = \frac{R(-s)}{t+s} + \sum_{j=1}^{n} \frac{r_j}{(a_j+s)(t-a_j)} . \tag{4.17}$$

Define a function $g(z)$ by means of the equality

$$g(z) = 1 - \int_0^\infty \frac{F_0(t)\,d\sigma(t)}{t+z}$$

for all z for which the integral converges absolutely. Then obviously $g(z) = G_+^{-1}(iz)$ for $z \notin (-\infty,0)$. Since $g(z)$ can be also represented in the form

$$g(z) = 1 + \int_{-\infty}^{0} \frac{d\tau(t)}{t-z} ,$$

where

$$\tau(-t) = - \int_0^t F_0(s)\,d\sigma(s) \qquad (0 \leqq t < \infty),$$

it follows that the results of no. 1 are applicable to $g(z)$.

Substituting the expansion (4.17) into (4.15), one gets

$$\int_0^\infty \frac{F(t)\,d\sigma(t)}{t+s} = R(-s) \int_0^\infty \frac{F_0(t)\,d\sigma(t)}{t+s} +$$

$$+ \sum_{j=1}^{n} \frac{r_j}{a_j+s} \int_0^\infty \frac{F_0(t)\,d\sigma(t)}{t-a_j} =$$

$$= R(-s)[1 - F_0^{-1}(s)] + \sum_{j=1}^{n} \frac{r_j}{a_j+s} [1 - g(-a_j)].$$

Furthermore, taking into account that $R(-s) = R^{-1}(s)$, and that, by (4.16), $R(-s) + \sum_{j=1}^{n} r_j/(a_j+s) = 1$, one finds that

$$\int_0^\infty \frac{F(t)\,d\sigma(t)}{t+s} = 1 - F^{-1}(s) - \sum_{j=1}^{n} \frac{r_j}{a_j+s}\, g(-a_j) . \tag{4.18}$$

Comparing this with (!), we conclude that $g(-a_j) = 0$ $(j = 1,2,\ldots,n)$.

In the general case, when $k_j \geqq 1$ $(j = 1,2,\ldots,n)$, similar considerations lead, following a more complicated computation, to the relation

$$\int_0^\infty \frac{F(t)\,d\sigma(t)}{t+s} = 1 - F^{-1}(s) + \sum_{j=1}^{n} \sum_{\ell=0}^{k_j-1} P_{j\ell}\left(\frac{1}{a_j+s}\right) g^{(\ell)}(-a_j) , \tag{4.19}$$

which replaces (4.18). Here the $P_{j\ell}(z)$ are certain polynomials in z having precisely the degrees $k_j - \ell$ ($\ell = 0,1,\ldots,k_j-1$; $j = 1,2,\ldots,n$). Comparing with (!) yields now $g^{(\ell)}(-a_j) = 0$ ($\ell = 0,1,\ldots,k_j-1$; $j = 1,2,\ldots,n$). In other words, each of the numbers $-a_j$ is a zero of multiplicity no less than k_j ($j = 1, 2,\ldots,n$) of the function $g(z)$. The converse is also obvious, i.e., given that the numbers $-a_j$ ($j = 1,2,\ldots,n$) satisfy this condition, the function $F(s)$ defined by (4.7) is a solution of equation (!).

In order to complete the proof of the second assertion of the theorem, we have only to show that if some point $-a$ ($\mathrm{Re}\, a > 0$) is a zero of multiplicity $\geqq k$ of the function $g(z)$, i.e., if

$$g^{(\ell)}(-a) = 0 \qquad (\ell = 0,1,\ldots,k-1)\ , \tag{4.20}$$

then a is a zero of multiplicity $\geqq k$ of the function $w(z)$, i.e.,

$$w^{(\ell)}(a) = 0 \qquad (\ell = 0,1,\ldots,k-1)\ , \tag{4.21}$$

and conversely, (provided that $\mathrm{Re}\, a > 0$).

To do this, take $\lambda = iz$ in (1.3), which gives

$$g(z)g(-z) = 1 - 2\int_0^\infty \frac{t\,d\sigma(t)}{t^2 - z^2}\ . \tag{4.22}$$

Since $g(z) = G_+^{-1}(iz)$, the function $g(z)$ does not vanish for $\mathrm{Re}\, z \geqq 0$. Moreover, it is holomorphic for $z \notin (-\infty,0]$. Therefore, if $g(-a) = 0$, then $w(a) = 0$ and $g(a) \neq 0$, and a is a regular point of $g(z)$. Taking into account, in addition, that the continuity and positivity of the function $F_0(s)$ together with the fact that $F_0(s) \longrightarrow 1$ as $s \longrightarrow \infty$ imply that the integrals

$$\int_0^\infty \frac{|d\sigma(t)|}{|t-a|^k}\ ,\quad \int_0^\infty \frac{|F_0(t)d\sigma(t)|}{|t-a|^k}\ ,\quad \int_0^\infty \frac{t^k|d\sigma(t)|}{|t^2-a^2|^k}$$

converge or diverge simultaneously, we conclude that for $\mathrm{Re}\, a > 0$ the existence of the derivative $g^{(k)}(-a)$ is equivalent to the existence of the derivative $w^{(k)}(a)$. Using also (4.22), it is not hard to show that as soon as the derivatives $w^{(k)}(a)$ and $g^{(k)}(-a)$ $(\mathrm{Re}\, a > 0)$ exist, one has

$$w^{(k)}(a) = \sum_{p=0}^{k} (-1)^p \binom{k}{p} g^{(p)}(-a) g^{(k-p)}(a) .$$

The discussion above makes it plain already that conditions (4.20) and (4.21) are equivalent.

To conclude the proof of the theorem it remains to consider the case of a nondecreasing function σ. Then

$$\mathrm{Im}\, w(z) = -2\, \mathrm{Im}(z^2) \int_0^\infty \frac{t d\sigma(t)}{|t^2 - z^2|} \neq 0 \quad \text{for} \quad \mathrm{Im}(z^2) \neq 0 ,$$

and so $w(z)$ can vanish only for $\mathrm{Im}(z^2) = 0$. Since by assumption the function $w(i\lambda)$ does not vanish for positive λ, it follows that the zeros of $w(z)$ in the half plane $\mathrm{Re}\, z > 0$ can lie only on the positive axis. In other words, in this case $a_j > 0$ $(j = 1,2,\ldots,n)$. Now all $k_j = 1$ $(j = 1,2,\ldots,n)$: assuming that some $k_j > 1$, we would have

$$w'(a_j) = -4a_j \int_0^\infty \frac{t d\sigma(t)}{(t^2 - a^2)^2} = 0 ,$$

which is impossible.

Finally, we still have to remark that if σ is a nondecreasing function, then, according to (4.6) and (4.7) (with $k_j = 1$), the function $F_0(s)$ is the unique positive, as well as the unique bounded, solution of equation (!).

The theorem is proved.

<u>3. Dropping condition (1.2).</u> Let us show that condition (1.2) is not essential for the existence of a solution F to equation (!). This condition states that $w(i\lambda) \neq 0$ $(-\infty < \lambda < \infty)$, but instead we shall require only that $w(0) \neq 0$, i.e., that

$$1 - 2 \int_0^\infty \frac{d\sigma(t)}{t} \neq 0 . \tag{4.23}$$

Under this assumption, we shall prove the existence of solutions to equation (!) and obtain a general formula for them. Subsequently, we shall consider the case $w(0) = 0$ too.

We shall need the following

LEMMA 4.1. *Let* a_j $(j = 1,2,\dots,n)$ *be the zeros of the function*

$$w(z) = c + \int_{-\infty}^{\infty} \frac{d\tau(t)}{t - z} \qquad \left(\int_{-\infty}^{\infty} \frac{|d\tau(t)|}{1 + |t|} < \infty \right) ,$$

with respective multiplicities $\geq k_j$ $(j = 1,2,\dots,n)$, *let* b_ℓ $(\ell = 1,2,\dots,N;\ N = k_1 + k_2 + \dots + k_n)$ *be arbitrary points of the complex plane, and set*

$$R(z) = \prod_{\ell=1}^{N} (z - b_\ell) / \prod_{j=1}^{n} (z - a_j)^{k_j} .$$

Then

$$R(z)w(z) = c + \int_{-\infty}^{\infty} \frac{R(t)\,d\tau(t)}{t - z} .$$

PROOF. First we look at the case when all $k_j = 1$ $(j = 1,2,\dots,n)$. In this situation $R(z)$ can be represented as

$$R(z) = 1 + \sum_{j=1}^{n} \frac{r_j}{z - a_j} ,$$

whence

$$\frac{R(t)}{t - z} = \frac{R(z)}{t - z} + \sum_{j=1}^{n} \frac{r_j}{(a_j - z)(t - a_j)} .$$

Therefore,

$$c + \int_{-\infty}^{\infty} \frac{R(t)\,d\tau(t)}{t - z} = c + R(z) \int_{-\infty}^{\infty} \frac{d\tau(t)}{t - z} + \sum_{j=1}^{n} \frac{r_j}{a_j - z} \int_{-\infty}^{\infty} \frac{d\tau(t)}{t - a_j} . \tag{4.24}$$

Since $w(a_j) = 0$, i.e.,

$$\int_{-\infty}^{\infty} \frac{d\tau(t)}{t - a_j} = - c \qquad (j = 1,2,\dots,n) ,$$

the right-hand side in (4.24) is equal to

$$c\left(1 - \sum_{j=1}^{n} \frac{r_j}{a_j - z}\right) + R(z) \int_{-\infty}^{\infty} \frac{d\tau(t)}{t - z} = R(z)w(z)$$

and this proves the lemma for the case under consideration.

In the general case $k_j \geq 1$ $(j = 1,2,\dots,n)$, the function $R(z)$ may be always represented in the form

$R(z) = R_1(z)R_2(z) \ldots R_k(z)$ $(k = \max(k_1,k_2,\ldots,k_n))$, where $R_p(z)$ $(p = 1,2,\ldots,k)$ is a rational function of the type considered above, with simple poles at a_j. Applying the result in the case of simple poles successively, we get

$$R_p(z)\left(c + \int_{-\infty}^{\infty} \frac{d\tau_{p-1}(t)}{t - z}\right) = c + \int_{-\infty}^{\infty} \frac{R_p(t)d\tau_{p-1}(t)}{t - z} =$$

$$= c + \int_{-\infty}^{\infty} \frac{d\tau_p(t)}{t - z} \qquad (p = 1,2,\ldots,k) ,$$

where $\tau_0(t) = \tau(t)$, and $\tau_p(t) = \int_0^t R_p(s)d\tau_{p-1}(s)$ $(p = 1,2,\ldots,k)$. It follows that

$$R(z)w(z) = c + \int_{-\infty}^{\infty} \frac{d\tau_k(t)}{t - z} = c + \int_{-\infty}^{\infty} \frac{R(t)d\tau(t)}{t - z} ,$$

as required.

Remark 4.1. Sometimes we shall write the function $w(z)$ as

$$w(z) = c + \int_{-\infty}^{\infty} \frac{d\tau(t)}{t + z} .$$

Then we obviously have

$$R(z)w(z) = c + \int_{-\infty}^{\infty} \frac{R(-t)d\tau(t)}{t + z} .$$

Now let the function $w(z)$ satisfy condition (4.23), but not necessarily condition (1.2). Denote by $i\gamma_j$ $(\gamma_j > 0,$ $j = 1,2,\ldots,m)$ all its distinct zeros on the positive imaginary axis, and let ν_j $(j = 1,2,\ldots,m)$ be the corresponding multiplicities. Select arbitrary points b_j $(j = 1,2,\ldots,m)$ such that $\mathrm{Re}\, b_j > 0$, $\mathrm{Im}\, b_j \neq 0$, $w(b_j) \neq 0$ $(j = 1,2,\ldots,m)$, and consider the functions

$$R(z) = \prod_{j=1}^{m} \left(\frac{z^2 - b_j^2}{z^2 + \gamma_j^2}\right)^{\nu_j} \quad \text{and} \quad \sigma^*(t) = \int_0^t R(s)d\sigma(s)$$

$(0 \leq t < \infty)$. Then, in virtue of Lemma 4.1,

$$w^*(z) = R(z)w(z) = R(z)\left[1 - \int_0^{\infty}\left(\frac{1}{t - z} + \frac{1}{t + z}\right) d\sigma(t)\right] =$$

$$= 1 - \int_0^{\infty}\left[\frac{R(t)}{t - z} + \frac{R(-t)}{t + z}\right] d\sigma(t) = 1 - 2\int_0^{\infty} \frac{tR(t)d\sigma(t)}{t^2 - z^2} ,$$

whence

$$w^*(z) = 1 - 2\int_0^\infty \frac{t\,d\sigma^*(t)}{t^2 - z^2} .$$

Since the function $w^*(z)$ does not vanish on the imaginary axis, Theorem 4.1 applies to the equation

$$\frac{1}{F^*(s)} = 1 - \int_0^\infty \frac{F^*(t)\,d\sigma^*(t)}{t + s} \qquad (0 \leqq s < \infty) . \tag{!*}$$

Denote by $F_0^*(s)$ its fundamental solution, given by the formula

$$\ln F_0^*(s) = -\frac{s}{\pi}\int_0^\infty \ln w^*(i\lambda)\,\frac{d\lambda}{\lambda^2 + s^2} \qquad (0 \leqq s < \infty). \tag{1.6*}$$

We show that by using $F_0^*(s)$, one can obtain all the solutions of equation (!). Let F be one of these solutions. Thus

$$\frac{1}{F(z)} = 1 - \int_0^\infty \frac{F(t)\,d\sigma(t)}{t + z} \qquad (z \notin (-\infty,0))$$

and

$$\frac{1}{F(z)F(-z)} = 1 - 2\int_0^\infty \frac{t\,d\sigma(t)}{t^2 - z^2} .$$

From the last relation it follows that $i\gamma_j$ and $-i\gamma_j$ are poles of $F(z)$ of certain multiplicities ℓ_j and ℓ_j' satisfying $\ell_j + \ell_j' = \nu_j$ $(j = 1,2,\ldots,m)$. Now construct the function

$$S(z) = S(z;\ell_1,\ldots,\ell_m) = \prod_{j=1}^{m} \frac{(z + b_j)^{\nu_j}}{(z - i\gamma_j)^{\ell_j}(z + i\gamma_j)^{\ell_j'}} . \tag{4.25}$$

Then $R(z) = S(z)S(-z)$. By virtue of Remark 4.1, we shall have

$$\frac{S(z)}{F(z)} = 1 - \int_0^\infty \frac{S(-t)F(t)\,d\sigma(t)}{t + z} = 1 - \int_0^\infty \frac{S^{-1}(t)F(t)\,d\sigma^*(t)}{t + z} .$$

Therefore, if F is a solution of equation (!), then $F^*(t) = S^{-1}(t)F(t)$ will be a solution of equation (!*). It is easy to see that the converse is also true, i.e., given a solution $F^*(t)$ of (!*), the function $F(t) = S(t)F^*(t)$ will be a solution of (!) whenever $S(t)$ is a function of the form (4.25) with exponents ℓ_j satisfying the conditions

$$0 \leq \ell_j \leq \nu_j \quad (j = 1,2,\ldots,m). \tag{4.26}$$

The resulting function $F(z)$ has poles of multiplicities ℓ_j and $\ell_j' = \nu_j - \ell_j$ at the points $i\gamma_j$ and $-i\gamma_j$, respectively $(j = 1,2,\ldots,m)$. This is a consequence of the fact that the function $w^*(z) = [F^*(z)F^*(-z)]^{-1}$ (in contrast to $w(z)$) does not vanish at any of the points $\pm i\gamma_j$ $(j = 1,2,\ldots,m)$.

On the other hand, Theorem 4.1 asserts that the general solution F^* to equation (!*) is given by

$$F^*(t) = \prod_{j=1}^{n} \left(\frac{t + a_j}{t - a_j}\right)^{k_j} F_0^*(t) \quad (0 \leq t < \infty) ,$$

where $F_0^*(t)$ is defined by equality (1.6*), and a_j $(j = 1,2,\ldots,m)$ are arbitrarily taken zeros of the function $w^*(z)$ - and so of the function $w(z)$ - in the half plane $\operatorname{Re} z > 0$, of respective multiplicities $\geq k_j$.

Thus, the general solution of equation (!) is obtained from the formula

$$F(t) = \prod_{j=1}^{n} \left(\frac{t + a_j}{t - a_j}\right)^{k_j} \cdot S(t;\ell_1,\ldots,\ell_m)F_0^*(t) , \tag{4.27}$$

where $S(z)$ is a function of the type (4.25) with exponents ℓ_j satisfying condition (4.6). Obviously, the total number of distinct functions $S(t;\ell_1,\ldots,\ell_m)$ is $N = (1+\nu_1)(1+\nu_2)\ldots(1+\nu_m)$. Set

$$F(z;\ell_1,\ldots,\ell_m) = S(z;\ell_1,\ldots,\ell_m)F_0^*(z) . \tag{4.28}$$

Obviously, the solutions $F(t;\ell_1,\ldots,\ell_m)$ of equation (!) do not depend upon the choice of the auxilliary points b_j $(j = 1,2,\ldots,m)$. Recalling that $F_0^*(z)$ is holomorphic and does not vanish in the half plane $\operatorname{Re} z > 0$, we reach the following conclusion.

1°. *The functions* $F(t;\ell_1,\ldots,\ell_m)$ *are the unique solutions of equation* (!) *such that their analytic continuations* $F(z)$ *are holomorphic (and do not vanish) in the right half plane* $\operatorname{Re} z > 0$.

Among these N solutions, the solution $F(t;\ell_1,\ldots,\ell_m)$ is individually characterized by the fact that its analytic continuation $F(z;\ell_1,\ldots,\ell_m)$ has at the points $i\gamma_j$ $(j = 1,2,\ldots,m)$ poles of respective multiplicities ℓ_j.

Setting

$$F_0(t) = F(t;0,\ldots,0) , \tag{4.29}$$

we have

$$F(z;\ell_1,\ldots,\ell_m) = \prod_{j=1}^{m} \left(\frac{z + i\gamma_j}{z - i\gamma_j}\right)^{\ell_j} F_0(z) . \tag{4.30}$$

This shows that all the solutions $F(t;\ell_1,\ldots,\ell_m)$ $(0 \leqq t < \infty)$ have equal absolute values.

Using (4.30), one can recast formula (4.27) for the general solution in the more symmetric form

$$F(t) = \prod_{j=1}^{n} \left(\frac{t + a_j}{t - a_j}\right)^{k_j} \cdot \prod_{p=1}^{m} \left(\frac{t + i\gamma_p}{t - i\gamma_p}\right)^{\ell_p} \cdot F_0(t) , \tag{4.31}$$

where $0 \leqq \ell_p \leqq \nu_p$ $(p = 1,2,\ldots,m)$, a_j $(j = 1,2,\ldots,n)$ are arbitrarily taken zeros of the function $w(z)$ in the right half plane $\operatorname{Re} z > 0$, and k_j $(j = 1,2,\ldots,n)$ are nonnegative integers no larger than the multiplicities of the corresponding zeros a_j.

Using the fact that the solution $F_0(t)$ does not depend upon the choice of the auxilliary points b_j $(j = 1,2,\ldots,m)$, it is not hard to show that instead of formula (1.6) one has

$$\ln F_0(t) = -\frac{t}{\pi}\int_0^{\infty} \widehat{\ln}\, w(i\lambda)\, \frac{d\lambda}{\lambda^2 + t^2} \quad (0 \leqq t < \infty) , \tag{4.32}$$

where $\widehat{\ln}\, w(i\lambda)$ $(-\infty < \lambda < \infty)$ denotes that branch of the logarithm of $1 - K(\lambda) = w(i\lambda)$ which tends to 0 as $\lambda \longrightarrow \infty$, is continuous at the points λ different from the zeros γ_j $(j = 1,2,\ldots,m)$ of the function $1 - K(\lambda)$, and whose argument is augumented by $\pi\nu_j$ as λ passes increasing through the point γ_j $(j = 1,2,\ldots,m)$. [The last rule can be stated alternatively as follows: to define the argument of $\widehat{\ln}\, w(i\lambda)$ $(0 \leqq \lambda < \infty)$ one goes around each zero γ_j $(j = 1,2,\ldots,m)$ on a small upper semicircle.]

To obtain (4.32), set $b_j = i\gamma_j + \varepsilon$ $(j = 1,2,\ldots,m)$, where $\varepsilon > 0$. Then we have

$$F_0^*(t) = F_0(t) \prod_{j=1}^{m} \left(\frac{t + i\gamma_j + \varepsilon}{t + i\gamma_j} \right)^{-\nu_j}$$

and

$$w^*(i\lambda) = w(i\lambda) \prod_{j=1}^{m} \left(\frac{\lambda^2 - (\gamma_j - i\varepsilon)^2}{\lambda^2 - \gamma_j^2} \right)^{\nu_j} .$$

Inserting these expressions for $F_0^*(t)$ and $w^*(i\lambda)$ into (1.6*) and subsequently passing to the limit $\varepsilon \longrightarrow 0$, we easily get (4.32).

4. If $\sigma(t)$ $(0 \leqq t < \infty)$ is a nondecreasing function then, as we have already noticed in no. 2, the zeros a_j of the function $w(z)$ in the half plane $\operatorname{Re} z > 0$ can lie only on the positive axis, in which case one must have $k_j = 1$ $(j = 1,2,\ldots, m)$. Moreover, under this assumption on $\sigma(t)$, it is easy to see that $w(i\lambda)$ is a strictly increasing function of λ^2 and so it can have at most two simple zeros $\pm\gamma$. Formula (4.31) becomes simpler:

$$F(t) = \prod_{j=1}^{n} \frac{t + a_j}{t - a_j} \left(\frac{t + i\gamma}{t - i\gamma} \right)^{\ell} F_0(t) ,$$

where $\ell = 0,1$, and a_j $(j = 1,2,\ldots,n)$ are arbitrary, distinct, positive zeros of the function $w(z)$.

THEOREM 4.2. *Let* $\sigma(t)$ $(0 \leqq t < \infty)$ *be a nondecreasing function satisfying condition* (1.1). *Then depending upon whether*

$$w(0) = 1 - 2 \int_0^\infty \frac{d\sigma(t)}{t}$$

is positive, negative, or equal to zero, equation (!) *has respectively only one bounded solution* $F_0(t)$ $(0 \leqq t < \infty)$, *exactly two distinct bounded solutions (which are complex conjugate), and - in the third case - no bounded solutions.*

PROOF. If σ is nondecreasing, then condition (1.2) is fulfilled if and only if $w(0) > 0$, and hence the corresponding assertion of our theorem is included in Theorem 4.1.

Suppose $w(0) < 0$. Then, since $w(\pm\infty) = 1$, the function $w(z)$ has zeros on the imaginary axis, and we already know that it has exactly two: $i\gamma$ and $-i\gamma$ $(\gamma > 0)$. By formula (4.31), equation (!) has exactly two bounded solutions

$$F_\ell(t) = \left(\frac{t + i\gamma}{t - i\gamma}\right)^\ell F_0(t) \qquad (0 \leq t < \infty;\ \ell = 0,1) . \qquad (4.33)$$

Since for σ real $\bar{F}$ is a solution of (!) as soon as F is, we conclude that $F_1(t) = \overline{F_0(t)}$.

Finally, let $w(0) = 0$, and assume that equation (!) has a bounded solution $F(t)$. Then we would have $F^{-1}(z)F^{-1}(-z) = w(z)$ $(\mathrm{Im}\ z \neq 0)$, and the function $F(z)$ will be continuous at the point $z = 0$ in the half plane $\mathrm{Re}\ z \geq 0$. Therefore, by passing to the limit $z \longrightarrow 0$ along the imaginary axis in the last relation, we obtain $F^{-2}(0) = w(0) = 0$, which is impossible.

Remark 4.2. Notice that, according to formula (4.31), the bounded solutions of equation (!) in the cases $w(0) > 0$ and $w(0) < 0$ are simultaneously the only solutions of (!) which are bounded in any interval (a,∞) $(a > 0)$.

If, however, $w(0) = 0$, then, as we will show in § 7, equation (!) has one and only one solution $F(t)$ that is bounded in each interval (a,∞) $(a > 0)$, although it has no bounded solutions.

§ 5. AUXILLIARY LEMMAS FROM ANALYTIC FUNCTION THEORY

Let $f(z)$ $(\not\equiv 0)$ be a function holomorphic in the interior of the upper half plane Π_+. As is known (see, for example, [1,5]), f will have the property that $\mathrm{Im}\ f(z) > 0$ $(\mathrm{Im}\ z > 0)$ if and only if it admits the representation

$$f(z) = \alpha + \beta z + \int_{-\infty}^{\infty} \left[\frac{1}{t - z} - \frac{t}{1 + t^2}\right] d\tau(t) \qquad (5.1)$$

$(\mathrm{Im}\ z > 0)$, where α is a real number, $\beta \geq 0$, and $\tau(t)$ $(\tau(0) = 0,\ \tau(-0) = 0)$ is some nondecreasing function satisfying

$$\int_{-\infty}^{\infty} \frac{d\tau(t)}{1 + t^2} < \infty .$$

If the function $\tau(t)$ is normalized for any t ($\neq 0$) in the sense that $\tau(t) = [\tau(t+0) - \tau(t-0)]/2$, then one has the Stieltjes inversion formula

$$\tau(t_2) - \tau(t_1) = \lim_{\varepsilon\downarrow 0} \frac{1}{\pi} \int_{t_1}^{t_2} \operatorname{Im} f(t + i\varepsilon)dt \tag{5.2}$$

$(t_1, t_2 \neq 0)$. In particular, if the limit

$$\lim_{\varepsilon\downarrow 0} \operatorname{Im} f(s + i\varepsilon) , \tag{5.3}$$

which we denote by $\operatorname{Im} f(s)$, exists for any s in the closed interval $[a,b]$, then we obtain

$$\tau(t) - \tau(a) = \frac{1}{\pi} \int_0^t \operatorname{Im} f(s)\, ds \quad (t \in [a,b]) . \tag{5.4}$$

Representation (5.1) together with the inversion formula (5.2) allow us to establish the following proposition (see [5], Theorem 1.5.1).

LEMMA 5.1. *In order that a function* $f(z)$ *holomorphic in the domain* $\operatorname{Ext}(0,\infty)$ *admit a representation*

$$f(z) = a + \int_0^\infty \frac{d\tau(t)}{t - z} , \tag{5.5}$$

where $a \geqq 0$ *and* $\tau(t)$ $(0 \leqq t < \infty)$ *is a nondecreasing function satisfying the condition*

$$\int_0^\infty \frac{d\tau(t)}{1 + t} < \infty, \tag{5.6}$$

it is neceessary and sufficient that: 1) $f(-x) \geqq 0$ *for* $0 < x < \infty$, *and* 2) $\operatorname{Im} f(z) \geqq 0$ *for* $\operatorname{Im} z > 0$.

Remark. Taking $z = x$ $(x < 0)$ in (5.5) and letting $x \longrightarrow -\infty$, we find that in (5.5) $a = f(-\infty)$ $(= \lim_{x\to-\infty} f(x))$.

2. One has

LEMMA 5.2. *Let* $\sigma(t)$ $(0 \leqq t < \infty)$ *be a nondecreasing function such that*

$$\int_0^\infty \frac{d\sigma(t)}{t} < \infty , \tag{5.7}$$

and let $w(z) = w_\sigma(z)$, *i.e.*,

$$w(z) = 1 - 2\int_0^\infty \frac{t\,d\sigma(t)}{t^2 - z^2} \qquad (z^2 \notin (0,\infty)) . \tag{5.8}$$

If

$$w(0) = 1 - 2\int_0^\infty \frac{d\sigma(t)}{t} > 0 , \tag{5.9}$$

then

$$w^{-1}(z) = 1 + 2\int_0^\infty \frac{t\,d\sigma_*(t)}{t^2 - z^2} , \tag{5.10}$$

where $\sigma_*(t)$ $(0 \leq t < \infty)$ *is a nondecreasing function satisfying the same condition* (5.7), *i.e.,*

$$\int_0^\infty \frac{d\sigma_*(t)}{t} < \infty .$$

Conversely, if some function $w^{-1}(z)$ $(z^2 \notin (0,\infty))$ *admits a representation* (5.10) *with the properties indicated above, then* $w(z)$ *admits a representation* (5.8) *and* $w(0) > 0$.

PROOF. Set

$$g(\zeta) = 1 - 2\int_0^\infty \frac{t\,d\sigma(t)}{t^2 - \zeta} = w(\sqrt{\zeta})$$

and notice that

$$\operatorname{Im} g(\zeta) = -2 \operatorname{Im}\zeta \int_0^\infty \frac{t\,d\sigma(t)}{|t^2 - \zeta^2|} < 0 \qquad (\operatorname{Im}\zeta > 0) ,$$

whence $\operatorname{Im} g^{-1}(\zeta) > 0$ for $\operatorname{Im}\zeta > 0$.

Since $g(-x)$ is an increasing function of x for $0 < x < \infty$ and $g(0-) = w(0)$, we have $g(-x) > 0$ for $0 < x < \infty$ as soon as $w(0) > 0$.

It follows that when $w(0) > 0$ we can claim that the function $f(\zeta) = g^{-1}(\zeta)$ satisfies conditions 1) and 2) of Lemma 5.1, and so

$$g^{-1}(\zeta) = 1 + \int_0^\infty \frac{d\tau(u)}{u - \zeta} \qquad (\zeta \notin (0,\infty)) , \tag{5.11}$$

where $\tau(t)$ $(0 \leq t < \infty)$ is a nondecreasing function satisfying condition (5.6). The first term on the right-hand side equals 1 because $g(-\infty) = 1$. Putting in (5.1) $\zeta = -x$ $(x > 0)$, and letting $x \longrightarrow 0$, we obtain

$$1 + \int_0^\infty \frac{d\tau(u)}{u} = g^{-1}(0-) = w^{-1}(0) .$$

Defining $\sigma_*(t)$ as

$$\sigma_*(t) = \frac{1}{2}\int_0^{t^2} \frac{d\tau(u)}{\sqrt{u}} = \frac{1}{2}\int_0^t \frac{d\tau(v^2)}{v} , \tag{5.12}$$

we shall have

$$\int_0^\infty \frac{d\sigma_*(t)}{t} = \frac{1}{2}\int_0^\infty \frac{d\tau(u)}{u} < \infty .$$

On the other hand, representation (5.11) may be written also as

$$g^{-1}(\zeta) = 1 + \int_0^\infty \frac{d\tau(v^2)}{v^2 - \zeta} = 1 + 2\int_0^\infty \frac{t d\sigma_*(t)}{t^2 - \zeta} \quad (\zeta \notin (0,\infty)).$$

Taking $\zeta = z^2$ here and observing that $g(z^2) = w(z)$, we obtain (5.10). Conversely, suppose that (5.10) holds for some function $w^{-1}(z)$ $(\operatorname{Im} z \neq 0)$. Then $f(\zeta) = w^{-1}(\sqrt{\zeta})$ will satisfy $\operatorname{Im} f(\zeta) > 0$ for $\operatorname{Im} \zeta > 0$, and $f(-x) > 1$ for $x > 0$. Consequently, for $g(\zeta) = w(\sqrt{\zeta}) = f^{-1}(\zeta)$ we have $\operatorname{Im}(1 - g(\zeta)) > 0$ when $\operatorname{Im} \zeta > 0$, and $1 - g(-x) > 0$ when $x > 0$.

Therefore, by Lemma 5.1, one can find a nondecreasing function $\omega(t)$ $(0 \leqq t < \infty)$ such that

$$1 - g(\zeta) = \int_0^\infty \frac{d\omega(t)}{t - \zeta} \quad (\zeta \notin (0,\infty)) . \tag{5.13}$$

Letting $\zeta \longrightarrow 0$ on the negative axis, we get

$$\int_0^\infty \frac{d\omega(t)}{t} = 1 - w(0) < \infty .$$

Furthermore, upon setting

$$\sigma(t) = \frac{1}{2}\int_0^t \frac{d\omega(v^2)}{v} \quad (0 \leqq t < \infty)$$

and passing from the function ω to the function σ in (5.5), we obtain, following the substitution $\zeta = z^2$, the desired representation (5.8). The lemma is proved.

LEMMA 5.3. *Let* $\sigma(t)$ *and* $w(z)$ *be as in the previous lemma, but suppose now that* $w(0) < 0$. *Then*

$$w^{-1}(z) = 1 - \frac{\rho}{z^2 + \gamma^2} + 2\int_0^\infty \frac{td\sigma_*(t)}{t^2 - z^2} \quad (\text{Im } z \neq 0). \quad (5.14)$$

where $\sigma_*(t)$ $(0 \leqq t < \infty)$ *is a nondecreasing function satisfying condition* (5.7), γ *is the positive root of the equation*

$$1 - 2\int_0^\infty \frac{td\sigma(t)}{t^2 + \lambda^2} = 0 , \quad (5.15)$$

and

$$\rho^{-1} = 2\int_0^\infty \frac{td\sigma(t)}{(t^2 + \gamma^2)^2} . \quad (5.16)$$

PROOF. Set, as above, $g(\zeta) = w(\sqrt{\zeta})$. Since $w(0) < 0$, we have $g(0-) < 0$. Taking into account that $g(-\infty) = 1$, we conclude that the decreasing function $g(x)$ $(-\infty < x < 0)$ has some zero $x = -\gamma^2 < 0$ $(\gamma > 0)$, and we shall have $g(x) > 0$ for $x < -\gamma^2$. Obviously, γ is the unique (positive) root of equation (5.15).

Now the function $f(\zeta) = g^{-1}(\zeta - \gamma^2)$ already fulfills conditions 1) and 2) of Lemma 5.1, and so

$$g^{-1}(\zeta - \gamma^2) = 1 + \int_0^\infty \frac{d\tau_1(u)}{u - \zeta} \quad (\zeta \notin (0,\infty)) , \quad (5.17)$$

where $\tau_1(u)$ $(0 \leqq u < \infty)$ is some nondecreasing function satisfying condition (5.6).

Since $g^{-1}(\zeta - \gamma^2)$ is regular and real (negative) for $0 < \zeta < \gamma^2$, we see from formula (5.4) that $\tau_1(u) = \text{const}$ for $0 < u < \gamma^2$. This implies that representation (5.17) becomes

$$g^{-1}(\zeta - \gamma^2) = 1 - \frac{\rho}{\zeta} + \int_{\gamma^2-0}^\infty \frac{d\tau_1(u)}{u - \zeta} , \quad (5.18)$$

where $\rho = \tau_1(0+) - \tau_1(0)$.

Putting $\zeta = \gamma^2 + z^2$ in (5.18), we get

$$w^{-1}(z) = 1 - \frac{\rho}{\gamma^2 + z^2} + \int_0^\infty \frac{d\tau(u)}{u - z^2} \quad (5.19)$$

with $\tau(u) = \tau_1(u + \gamma^2) - \tau_1(\gamma^2 - 0)$ $(0 \leqq u < \infty)$.

Letting $z \longrightarrow 0$ along the imaginary axis in (5.19), we get (5.7). Subsequently, inserting the function $\sigma_*(t)$ given by (5.12) into (5.19), we can rewrite (5.19) in the form (5.14).

By (5.19), the residue of the function $w^{-1}(z)$ at the point $z = i\gamma$ is equal to $i\rho/2\gamma$. Since, on the other hand, this residue is $1/w'(i\gamma)$, one obtains $\rho = 2\gamma/iw'(i\gamma)$, whence (5.16).

The lemma is proved. Let us remark that it could be stated in a more complete way (similar to Lemma 5.2), by adding the converse assertion.

3. Now consider the case where

$$w(0) = 1 - 2\int_0^\infty \frac{d\sigma(t)}{t} = 0 \; . \tag{5.20}$$

As above, (5.11) holds, i.e.,

$$w^{-1}(z) = 1 + \int_0^\infty \frac{d\tau(u)}{u - z^2} \qquad (z^2 \notin (0,\infty)) \; ,$$

where the function τ satisfies condition (5.6). However, due to (5.11) and (5.20), we shall have now that

$$\int_0^\infty \frac{d\tau(u)}{u} = \infty,$$

and transformation (5.12) is then not possible.

Set

$$s_k = \int_0^\infty \frac{d\sigma(u)}{u^k} \qquad (k = 1,2,\ldots) \; .$$

As we shall see below, when s_3 exists and is finite, we will have $\rho = \tau(0+) - \tau(0) > 0$. If, in addition, $s_5 < \infty$, then

$$\int_{0+}^\infty \frac{d\tau(u)}{u} < \infty.$$

LEMMA 5.4. *Let* $w(0) = 0$ *and* $s_5 < \infty$. *Then the representation*

$$w^{-1}(z) = 1 - \frac{1}{2s_3 z^2} + 2\int_0^\infty \frac{t\,d\sigma_*(t)}{t^2 - z^2} \; , \tag{5.21}$$

holds, where $\sigma_*(t)$ $(0 \leq t < \infty)$ *is a nondecreasing function, and*

$$\int_0^\infty \frac{d\sigma_*(t)}{t} = \frac{1}{4}\,\frac{s_5}{s_3^2} - \frac{1}{2} \; . \tag{5.22}$$

PROOF. By (5. 20) ,

$$w(z) = 2\int_0^\infty \frac{d\sigma(t)}{t} - 2\int_0^\infty \frac{td\sigma(t)}{t^2 - z^2} = -2z^2\int_0^\infty \frac{d\sigma(t)}{t(t^2 - z^2)} \quad (5.23)$$

and so, picking any $\gamma > 0$, we have

$$w_\gamma(z) = \frac{z^2 + \gamma^2}{z^2} w(z) = w(z) + \frac{\gamma^2}{z^2} w(z) = 1 - 2\int_0^\infty \frac{td\sigma(t)}{t^2 - z^2}$$
$$- 2\gamma^2 \int_0^\infty \frac{d\sigma(t)}{t(t^2 - z^2)} = 1 - 2\int_0^\infty \frac{(t^2 + \gamma^2)d\sigma(t)}{t(t^2 - z^2)} . \quad (5.24)$$

Alternatively,

$$w_\gamma(z) = 1 - 2\int_0^\infty \frac{td\sigma_\gamma(t)}{t^2 - z^2} \qquad (z^2 \notin (0,\infty)) ,$$

where

$$\sigma_\gamma(t) = \int_0^t \frac{s^2 + \gamma^2}{s^2} d\sigma(s) \qquad (0 \leqq t < \infty) . \quad (5.25)$$

From (5.25) and (5.7) we get

$$\int_0^\infty \frac{d\sigma_\gamma(t)}{t} = \int_0^\infty \frac{(t^2 + \gamma^2)}{t^3} d\sigma(t) < \infty.$$

Since $w_\gamma(i\gamma) = 0$ by construction, it follows that $w_\gamma(0) < 0$; hence, the second assertion of Lemma 5.3 applies to $w_\gamma(z)$, i.e.,

$$w_\gamma^{-1}(z) = -\frac{\rho_\gamma}{\gamma^2 + z^2} + 1 + 2\int_0^\infty \frac{td\omega_\gamma(t)}{t^2 - z^2} , \quad (5.26)$$

where

$$\rho_\gamma^{-1} = 2\int_0^\infty \frac{td\sigma_\gamma(t)}{(t^2 + \gamma^2)^2} = 2\int_0^\infty \frac{d\sigma(t)}{t(t^2 + \gamma^2)}$$

and $\omega_\gamma(t)$ $(0 \leqq t < \infty)$ is some nondecreasing function satisfying the condition

$$\int_0^\infty \frac{d\omega_\gamma(t)}{t} < \infty .$$

Now we take advantage of the fact that, in virtue of (5.23) and (5.24),

$$\lim_{y\to 0} w^{-1}(iy) = \lim_{y\to 0} \frac{-y^2}{(\gamma^2 - y^2)w(iy)} = -\frac{\rho}{\gamma^2} ,$$

where

$$\rho^{-1} = 2\int_0^\infty \frac{d\sigma(t)}{t^3} = 2s_3 . \tag{5.27}$$

Therefore, putting $z = iy$ $(y > 0)$ in (5.26) and letting $y \longrightarrow 0$, we obtain

$$-\frac{\rho}{\gamma^2} = 1 - \frac{\rho_\gamma}{\gamma^2} + 2\int_0^\infty \frac{d\omega_\gamma(t)}{t} . \tag{5.28}$$

Substracting this equality from (5.26) term by term, one gets

$$w_\gamma^{-1}(z) = \frac{\rho_\gamma - \rho}{\gamma^2} - \frac{\rho_\gamma}{\gamma^2 + z^2} + 2z^2\int_0^\infty \frac{d\omega_\gamma(t)}{t(t^2 - z^2)} .$$

Multiplying both sides of this equality by $1 + \frac{\gamma^2}{z^2}$ yields

$$w^{-1}(z) = (1 + \frac{\gamma^2}{z^2})\frac{\rho_\gamma - \rho}{\gamma^2} - \frac{\rho_\gamma}{z^2} + 2\int_0^\infty \frac{(z^2 + \gamma^2)d\omega_\gamma(t)}{t^2 - z^2} ,$$

and since

$$\frac{z^2 + \gamma^2}{t(t^2 - z^2)} = \frac{t^2 + \gamma^2}{t(t^2 - z^2)} - \frac{1}{t} ,$$

we then obtain, using (5.28) once more,

$$w^{-1}(z) = 1 - \frac{\rho}{z^2} + 2\int_0^\infty \frac{t^2 + \gamma^2}{t(t^2 - z^2)} d\omega_\gamma(t) \tag{5.29}$$

$(\operatorname{Im} z \neq 0)$.

Now notice that by virtue of (5.23) and (5.27)

$$w^{-1}(z) + \frac{\rho}{z^2} = \frac{2}{w(z)}\left[\int_0^\infty \frac{d\sigma(t)}{t^3} - \int_0^\infty \frac{d\sigma(t)}{t(t^2 - z^2)}\right] =$$

$$= -\frac{2\rho z^2}{w(z)}\int_0^\infty \frac{d\sigma(t)}{t^3(t^2 - z^2)} = \rho\int_0^\infty \frac{d\sigma(t)}{t^3(t^2 - z^2)} \Big/ \int_0^\infty \frac{d\sigma(t)}{t(t^2 - z^2)} .$$

Consequently,

$$\lim_{y\to 0}\left[w^{-1}(iy) - \frac{\rho}{y^2}\right] = \rho\int_0^\infty \frac{d\sigma(t)}{t^5} \Big/ \int_0^\infty \frac{d\sigma(t)}{t^3} = \frac{s_5}{2s_3^2}$$

provided $s_5 < \infty$. Calculating the same limit by starting with relation (5.29), yields

$$\frac{1}{2}\frac{s_5}{s_3^2} = 1 + 2\int_0^\infty \frac{t^2+\gamma^2}{t^3}\, d\omega_\gamma(t) .$$

The finiteness of the last integral allows us to set

$$\sigma_*(t) = \int_0^t \frac{s^2+\gamma^2}{s^2}\, d\omega_\gamma(s) \qquad (0 \leqq t < \infty) .$$

It is clear that by passing from the function ω_γ to the function σ_* in representation (5.29), we reduce it to representation (5.21). Also

$$\int_0^\infty \frac{d\sigma_*(t)}{t} = \int_0^\infty \frac{t^2+\gamma^2}{t^3}\, d\omega_\gamma(t) = \frac{1}{4}\frac{s_5}{s_3^2} - \frac{1}{2} ,$$

which completes the proof of the lemma.

4. These lemmas have an interesting application to the theory of integral equations of the form

$$g(t) - \int_{-\infty}^{\infty} k(t-s)g(s)ds = f(t) \qquad (-\infty < t < \infty), \tag{5.30}$$

where $k(t) \in L_1(-\infty,\infty)$ is an even function that is absolutely monotonic for $t \in (0,\infty)$, i.e.,

$$k(t) = \int_0^\infty e^{-|t|u}\, d\sigma(u) \qquad (-\infty < t < \infty)$$

with a nondecreasing function $\sigma(u)$ $(0 \leqq u < \infty)$ satisfying (5.7).

As is known, equation (5.30) will have a unique solution for any given $f \in L_1$ if and only if $1 - K(\lambda) = w(i\lambda) \neq 0$ $(-\infty < \lambda < \infty)$, or, which is the same, if and only if $w(0) = \min\limits_{-\infty<\lambda<\infty} w(i\lambda) > 0$.

On the other hand, Lemma 5.2 guarantees that as soon as $w(0) > 0$, one can form the kernel

$$k_*(t) = \int_0^\infty e^{-|t|u}\, d\sigma_*(t) \qquad (-\infty < t < \infty) ,$$

for which

$$1 + K_*(\lambda) = 1 + \int_{-\infty}^{\infty} e^{i\lambda t} k_*(t)dt = 1 + 2\int_{-\infty}^{\infty} \frac{t d\sigma_*(t)}{t^2+\lambda^2} = w^{-1}(i\lambda) ,$$

and so $(1 - K(\lambda))(1 + K_*(\lambda)) = 1$, whence $K_*(\lambda) = K(\lambda) + K(\lambda)K_*(\lambda)$. This relation gives

$$k_*(t) = k(t) + \int_{-\infty}^{\infty} k(t - s)k_*(s)ds \quad (-\infty < t < \infty). \quad (5.31)$$

Thus, for any given $f \in L_1(-\infty,\infty)$, the unique solution $g \in L_1(-\infty,\infty)$ to (5.30) is given by the formula

$$g(t) = f(t) + \int_{-\infty}^{\infty} k_*(t - s)f(s)ds \quad (-\infty < t < \infty) . \quad (5.32)$$

Therefore, for the symmetric kernel $k(t - s)$ $(-\infty < t,s < \infty)$ generated by an absolutely monotonic function $k(t) \in L_1(0,\infty)$, the resolvent kernel $k_*(t - s)$ (if it exists) is generated by an absolutely monotonic function $k_*(t) \in L_1(0,\infty)$ too. Conversely, let the function $k_*(t) \in L_1(-\infty,\infty)$ be absolutely monotonic for $t > 0$. Then equation (5.31) always has an even solution $k(t)$, absolutely monotonic for $t > 0$. Moreover, the solution $f \in L_1(-\infty,\infty)$ to equation (5.32) is given, for any $g \in L_1(-\infty,\infty)$, by formula (5.30).

[Equations (5.30) and (5.32) are taken in the space $L_1(-\infty,\infty)$ $(g,f \in L_1)$. All the statements concerning these equations remain valid, under the same assumptions on $k(t)$ and $k_*(t)$, when $L_1(-\infty,\infty)$ is replaced, for example, by the space $L_p(-\infty,\infty)$ $(1 < p \leq \infty)$, and also by a number of other spaces (cf. [6], § 6).]

§ 6. A MORE DETAILED ANALYSIS OF THE CASE $w(0) < 0$

1. In this case, as we know from § 4, (cf. (4.33)),

$$\begin{aligned}(t + i\gamma)F_0(t) &= (t - i\gamma)F_1(t) = (t - i\gamma)\overline{F_0(t)} = \\ &= Q_0(t) .\end{aligned} \quad (6.1)$$

Therefore, the bounded solutions of equation (!) may be written in the form $F_{0,1}(t) = Q_0(t)/(t \pm i\gamma)$ $(0 \leq t < \infty)$, where $Q_0(t)$ is some real function.

It follows from the construction that $-i\gamma$ is a first order pole of the function $F_0(z)$, and this point is the unique singularity of $F_0(z)$ in the domain $\mathrm{Ext}(-\infty,0]$. Thus $\Omega_0(z)$ is a holomorphic function in $\mathrm{Ext}(-\infty,0]$.

Using the following equality with $z = -i\gamma$,

$$\frac{1}{F_0(z)} = 1 - \int_0^\infty \frac{F_0(t)\,d\sigma(t)}{t+z} \qquad (z \notin (-\infty,0)) , \tag{6.2}$$

one obtains

$$0 = 1 - \int_0^\infty \frac{F_0(t)\,d\sigma(t)}{t - i\gamma} . \tag{6.3}$$

Substracting (6.3) from (6.2) term by term, we get

$$\frac{1}{F_0(z)} = \int_0^\infty \frac{z+i\gamma}{z-i\gamma} \frac{F_0(t)\,d\sigma(t)}{t+z} \qquad (z \notin (-\infty,0)) . \tag{6.4}$$

Dividing the terms of (6.4) by $z + i\gamma$ and introducing the function $\Omega_0(z)$ instead of $F_0(z)$, we find that

$$\frac{1}{\Omega_0(z)} = \int_0^\infty \frac{\Omega_0(t)}{t+z} \frac{d\sigma(t)}{\gamma^2 + t^2} \qquad (z \notin (-\infty,0)) . \tag{6.5}$$

In virtue of (6.1), $\Omega_0(t)/t \longrightarrow 1$ as $t \longrightarrow \infty$. On the other hand, (6.1) shows that $\Omega_0(z)$ does not vanish for $z \notin (-\infty,0)$. Hence $\Omega_0(t) > 0$ $(0 \leqq t < \infty)$.

The methods used previously imply that $\Omega_0(t)$ is the unique positive solution of the equation

$$\frac{1}{\Omega(s)} = \int_0^\infty \frac{\Omega(t)}{t+s} \frac{d\sigma(t)}{t^2+\gamma^2} \qquad (0 \leqq s < \infty) . \tag{6.6}$$

Indeed, according to Theorem 2.1 and equality (5.15) with $\lambda = \gamma$, one has

$$\frac{1}{\Omega(z)\Omega(-z)} = \int_0^\infty \frac{2t}{t^2 - z^2} \frac{d\sigma(t)}{t^2+\gamma^2} =$$

$$= -\frac{1}{z^2+\gamma^2}\left(1 - \int_0^\infty \frac{2t\,d\sigma(t)}{t^2-z^2}\right) , \tag{6.7}$$

where $\Omega(z)$ is the analytic continuation of $\Omega(t)$ in the domain $\mathrm{Ext}(-\infty,0]$.

If $Q(s) > 0 \quad (0 \leqq s < \infty)$, then

$$c = \lim_{s\to\infty} sQ^{-1}(s) = \lim_{s\to\infty} \int_0^\infty \frac{Q(t)s}{t+s} \frac{d\sigma(t)}{t^2+\gamma^2} =$$

$$= \int_0^\infty \frac{Q(t)d\sigma(t)}{t^2+\gamma^2} > 0 . \tag{6.8}$$

It follows that $Q(t) = O(t)$ when both $t \longrightarrow \infty$ and $c < \infty$. Now let us show that $c = 1$. Since one can replace s by $iy \quad (-\infty < y < \infty)$ in (6.6), one can easily see that $\lim_{y\to\pm\infty} iyQ^{-1}(iy) = c$. On the other hand, taking $z = iy \quad (|y| > \gamma)$ in (6.7), and multiplying the resulting equation by y^2, and then letting $y \longrightarrow \infty$, we get $c = 1$.

Set $F(z) = Q(z)/(z + i\gamma)$. Then $F(z)$ satisfies not only relation (6.4), but also relation (6.3) - and this is a result of (6.8) - and the equality $c = 1$. Adding (6.3) and (6.4), we see that $F(z)$ is a bounded solution of equation (!) having a unique pole at $z = -i\gamma$ in $\mathrm{Ext}(-\infty,0]$. By Theorem 4.2, $F(z)$ coincides with $F_0(z)$, and so $Q(z)$ coincides with $Q_0(z)$.

At the same time, we actually proved the following statement.

THEOREM 6.1. *Let* $\omega(t) \quad (0 \leqq t < \infty)$ *be a nondecreasing function such that*

$$(J =) \int_0^\infty t\,d\omega(t) < \infty, \qquad \int_0^\infty \frac{d\omega(t)}{t} < \infty.$$

Then the equation

$$Q^{-1}(s) = \int_0^\infty \frac{Q(t)d\omega(t)}{t+s} \qquad (0 \leqq s < \infty) \tag{6.9}$$

has a unique, positive solution $Q_0(t) \quad (0 \leqq t < \infty)$ *with the property that* $Q_0(t)/t \longrightarrow 1/\sqrt{2J}$ *as* $t \longrightarrow \infty$.

In fact, with no loss of generality, one may assume that $2J = 1$: indeed, take $\omega_1(t) = \omega(t)/2J$ and $Q_1(t) = \sqrt{2J}\,Q(t)$.

Now take any $\gamma > 0$ and put

$$\sigma(t) = \int_0^t (\gamma^2 + s^2)d\omega(s) \qquad (0 \leqq t < \infty) .$$

Then $\sigma(t)$ will satisfy condition (5.7), the corresponding function $w_\sigma(z)$ will vanish at the points $z = \pm i\gamma$, and equations (6.9) and (6.6) will be equivalent.

2. Set $\Phi_0(z) = Q_0(z)/(z^2 + \gamma^2)$. According to (6.5), we shall have

$$\frac{1}{Q_0(z)} = \int_0^\infty \frac{\Phi_0(t)\,d\sigma(t)}{t + z} \qquad (z \notin (-\infty,0)) .$$

Inserting the expression for $Q_0^{-1}(z)$ furnished by (6.7) into the above yields

$$\int_0^\infty \frac{\Phi_0(t)\,d\sigma(t)}{t - z} = - \frac{Q_0(-z)}{z^2 + \gamma^2} \left(1 - 2 \int_0^\infty \frac{t\,d\sigma(t)}{t^2 - z^2}\right)$$

$(z \notin (-\infty,0))$. When z is not real, the substitution of $-z$ for z is allowed here and gives

$$\int_0^\infty \frac{\Phi_0(t)\,d\sigma(t)}{t - z} = - \Phi_0(z)\left(1 - 2 \int_0^\infty \frac{t\,d\sigma(t)}{t^2 - z^2}\right) .$$

Consequently,

$$\int_0^\infty \frac{\Phi_0(t) - \Phi_0(z)}{t - z}\,d\sigma(t) = -\left(1 - \int_0^\infty \frac{d\sigma(t)}{t + z}\right)\Phi_0(z) \tag{6.10}$$

$(z \notin (-\infty,0))$. We should explain that although the last equality was obtained under the assumption that $\operatorname{Im} z \neq 0$, it is valid for the points $z = s > 0$ too, because these are regular points for both the right and left sides of this equality.

Since the function $w(z)$ has exactly two nonreal zeros $\pm i\gamma$, Lemma 5.3 implies

$$w^{-1}(z) = 1 - \frac{\rho}{z^2 + \gamma^2} + 2 \int_0^\infty \frac{t\,d\sigma_*(t)}{t^2 - z^2} ,$$

where σ_* is a nondecreasing function and

$$\rho^{-1} = 2 \int_0^\infty \frac{t\,d\sigma(t)}{(t^2 + \gamma^2)^2} > 0 .$$

Taking into account that $w^{-1}(z) = F_0(z)F_0(-z)$ and that $F_0(z)$ has in the complex plane with a cut along the ray $(-\infty,0)$, a unique pole $z = -i\gamma$, the same method which led to formula (1.6) gives

$$F_0(z) = 1 + \frac{\rho}{2i\gamma}\cdot\frac{F_0^{-1}(i\gamma)}{z + i\gamma} + \int_0^\infty \frac{d\sigma_*(t)}{F_0(t)(t + z)} . \tag{6.11}$$

Whence we obtain

$$\frac{F_0(z) - F_0(i\gamma)}{z - i\gamma} = \frac{\rho}{4\gamma^2 F_0(i\gamma)}\cdot\frac{1}{z + i\gamma} - \int_0^\infty \frac{d\sigma_*(t)}{F_0(t)(t+i\gamma)(t+z)} .$$

Therefore,

$$\Phi_0(z) = \frac{F_0(z)}{z - i\gamma} =$$

$$= \frac{F_0(i\gamma)}{z - i\gamma} + \frac{\rho}{4\gamma^2 F_0(i\gamma)(z+i\gamma)} - \int_0^\infty \frac{d\sigma_*(t)}{Q_0(t)(t+z)} . \tag{6.12}$$

Since the function on the left side is real (positive) for $z = t > 0$, its residues at the conjugate poles $z = \pm i\gamma$ must be conjugate, i.e., $\overline{F_0(i\gamma)} = \rho/4\gamma^2 F_0(i\gamma)$, whence $A = 2|F_0(i\gamma)| = \sqrt{\rho}/\gamma$, or

$$A = \frac{1}{\sqrt{2}\gamma}\left[\int_0^\infty \frac{t\,d\sigma(t)}{(t^2 + \gamma^2)^2}\right]^{-1/2} .$$

Let $\alpha - \frac{\pi}{2}$ be the argument of $F_0(i\gamma)$, so that $F_0(i\gamma) = -iAe^{i\alpha}/2$. To find α, take $z = i\gamma$ in (6.5). Recalling that, according to (6.1), $Q_0(i\gamma) = 2i\gamma F_0(i\gamma) = A\gamma e^{i\alpha}$, the last substitution leads to

$$\frac{e^{-i\alpha}}{A\gamma} = \int_0^\infty \frac{Q_0(t)}{t + i\gamma}\,\frac{d\sigma(t)}{t^2 + \gamma^2} ;$$

whence

$$\frac{\cos\alpha}{A\gamma} = \int_0^\infty \frac{tQ_0(t)d\sigma(t)}{(t^2 + \gamma^2)^2} > 0, \quad \frac{\sin\alpha}{A\gamma} = \int_0^\infty \frac{Q_0(t)d\sigma(t)}{(t^2 + \gamma^2)^2} > 0 .$$

Thus, $0 < \alpha < \pi/2$.

Relation (6.12) may be recasted in the form

$$\Phi_0(z) = \frac{iA}{2}\left(\frac{e^{i\alpha}}{-z + i\gamma} + \frac{e^{-i\alpha}}{z + i\gamma}\right) - \int_0^\infty \frac{d\sigma_*(t)}{Q_0(t)(t + z)} .$$

Setting

$$\phi_0(t) = A\sin(\gamma t + \alpha) - \int_0^\infty \frac{e^{-tu}d\sigma_*(u)}{Q_0(u)} , \tag{6.13}$$

we easily find that

$$\int_0^\infty \phi_0(t)e^{-tz}dt = \Phi_0(z) \qquad (\mathrm{Re}\ z > 0).$$

On the other hand, if we put

$$\Gamma(t) = \gamma A e^{-i(\gamma t+\alpha)} + \int_0^\infty e^{-tu}\frac{d\sigma_*(u)}{F_0(u)} \qquad (0 \leqq t < \infty)\ , \quad (6.14)$$

and use equalities (6.11) and $\rho/2i\gamma F_0(i\gamma) = \gamma A e^{-i\alpha}$, we obtain

$$F_0(z) = 1 + \int_0^\infty e^{-zt}\Gamma(t)dt \qquad (\mathrm{Re}\ z > 0)\ . \qquad (6.15)$$

By virtue of (6.13) and (6.14), the function ϕ_0 is defined in terms of $\Gamma(t)$ via the differential system

$$\frac{d\phi}{dt} - i\gamma\phi = \Gamma(t) \qquad (0 \leqq t < \infty), \qquad \phi(0) = 1\ .$$

Taking advantage of the results of § 15, no. 1 in [6], it is natural to assume that the following proposition holds true.

THEOREM 6.2. *Up to a factor, the function*

$$\phi_0(t) = A\sin(\gamma t + \alpha) - \int_0^\infty \frac{e^{-ts}d\sigma_*(s)}{\Phi_0(s)(s^2 + \gamma^2)}$$

is the unique bounded solution of the homogeneous integral equation

$$\phi(t) - \int_0^\infty k(t - s)\phi(s)ds = 0\ . \qquad (6.16)$$

If there exists $H > 0$ such that $\sigma(t) = \mathrm{const}$ for $0 \leqq t < H$, the we would have $k(t) = o(e^{-ht})$ as $t \longrightarrow \infty$, for any $0 < h < H$. Now it would follow from the results of § 15, no. 5 in [6] that $\phi(t) = A\sin(\gamma t + \alpha) + o(e^{-ht})$.

We shall prove the theorem under the previous assumptions concerning the nondecreasing function σ.

PROOF. By (0.2), we have for all $s > 0$ and z with $\mathrm{Re}\ z > 0$

$$\int_0^\infty e^{-zt}k(t-s)dt = \int_0^s + \int_s^\infty = e^{-zs}[\int_0^s e^{zt}k(t)dt +$$

$$+ \int_0^\infty e^{-zt}k(t)dt] = -\int_0^\infty \frac{e^{-sz} - e^{-su}}{z - u}d\sigma(u) + \int_0^\infty \frac{e^{-sz}}{z + u}d\sigma(u).$$

Therefore, given some bounded solution $\phi(t)$ $(0 \leq t < \infty)$ of equation (6.16), we multiply the left-hand side of this equation term by term by e^{-zt} $(\mathrm{Re}\, z > 0)$, and integrate with respect to t from 0 to ∞, to get

$$\int_0^\infty \frac{\Phi(z) - \Phi(u)}{z - u}\, d\sigma(u) + \left(1 - \int_0^\infty \frac{d\sigma(u)}{z + u}\right)\Phi(z) = 0 \qquad (6.17)$$

$(\mathrm{Re}\, z > 0)$, where

$$\Phi(z) = \int_0^\infty e^{-zt}\phi(t)\,dt \quad (\mathrm{Re}\, z > 0) .$$

The converse is obviously true too, i.e., if $\phi(t)$ $(0 \leq t < \infty)$ is some bounded function whose Laplace transform $\Phi(z)$ $(\mathrm{Re}\, z > 0)$ satisfies equation (6.17), then ϕ is a solution of equation (6.16).

According to (6.10), the function $\Phi_0(z)$ satisfies equation (6.17), and so $\phi_0(t)$ $(0 \leq t < \infty)$ is actually a solution of (6.16).

To complete the proof, it remains to show that if the Laplace transform $\Phi(z)$ of some bounded function $\phi(t)$ $(0 \leq t < < \infty)$ satisfies equation (6.17), then $\Phi(z) = c\Phi_0(z)$, where $c = \mathrm{const}$.

First, note the following properties of $\Phi(z)$:

a) $\Phi(z)$ is holomorphic in the half plane $\mathrm{Re}\, z > 0$, and $z\Phi(z)$ is bounded in any sector $L_\theta = \{z: |\arg z| \leq \theta\}$ $(0 < \theta < \pi/2)$;

b) $\int_0^\infty |\Phi(t)|\,d\sigma(t) < \infty$.

Indeed,

$$|\Phi(z)| \leq \int_0^\infty e^{-s\,\mathrm{Re}\, z}\, |\phi(s)|\,ds < \frac{M}{\mathrm{Re}\, z} \quad (\mathrm{Re}\, z > 0) .$$

This immediately implies property a), as well as the inequality $|\Phi(t)| < M/t$ $(t > 0)$. Taking into consideration (5.7), we get b).

According to (4.9), equality (6.17) can be rewritten as

$$F_0(-z) \int_0^\infty \frac{\Phi(u)\,d\sigma(u)}{u - z} = -\frac{\Phi(z)}{F_0(z)} \quad (\mathrm{Re}\, z > 0,\ \mathrm{Im}\, z \neq 0). \quad (6.18)$$

Pick some number θ $(0 < \theta < \pi/2)$. The left-hand side of equality (6.18) is holomorphic in the complex plane, except possibly the ray $[0,\infty)$ and the point $z = i\gamma$ at which it has a pole of order at most one. Therefore, equality (6.18) defines a function $G(z)$ holomorphic in the entire complex plane, except possibly the points $z = \infty, 0, i\gamma$. Let us show that, in fact, $G(z)$ is holomorphic at $z = \infty, 0$ too.

By virtue of property a) and the boundedness of $|F_0^{-1}(z)|$ $(z \in L_\theta)$, function $zG(z)$ is bounded in L_θ. On the other hand, property b) implies that $|G(z)| \longrightarrow$ for $z \longrightarrow \infty$, $z \notin L_\theta$. Hence the function $G(z)$ is holomorphic at infinity, and $G(\infty) = 0$.

Property b) implies also that

$$\lim_{\substack{z\to 0\\ z\notin L_\theta}} z\int_0^\infty \frac{\Phi(u)\,d\sigma(u)}{u - z} = 0 ,$$

and so $\lim\limits_{z\to 0} zG(z) = 0$ $(z \notin L_\theta)$. Since, as we already have noticed, $zG(z)$ is bounded in L_θ, $G(z)$ is holomorphic at the point $z = 0$ too.

Therefore, $G(z) \equiv c/(z - i\gamma)$, i.e., $\Phi(z) = -c\Phi_0(z)$, and the theorem is proved.

Since F_0 is a solution of equation (!), Theorem 3.2 yields

$$\int_0^\infty \frac{F_0(z) - F_0(u)}{z - u}\,d\sigma(u) + F_0(z)\left[1 - \int_0^\infty \frac{d\sigma(u)}{u + z}\right] = 1. \quad (6.19)$$

It is easy to conclude that this relation for F_0 leads to the following relation for the function $\Gamma(t)$, which is related to F_0 via the equality (6.15):

$$\Gamma(t) - \int_0^\infty k(t - s)\Gamma(s)\,ds = k(t) \quad (0 \leq t < \infty) . \quad (6.20)$$

Indeed, taking the Laplace transform of both sides of (6.20), this relation becomes (6.19) with $\operatorname{Re} z > 0$.

Notice that, according to (6.14), the function $\Gamma(t)$ is the sum of a bounded function and of a function from $L_1(0,\infty)$. It is not difficult to show that any solution of the equation

$$\tilde{\Gamma}(t) - \int_0^\infty k(t-s)\tilde{\Gamma}(s)ds = k(t) \quad (0 \leqq t < \infty) , \qquad (6.21)$$

which can be represented as the sum of a bounded function and of a function from $L_1(0,\infty)$, is given by $\tilde{\Gamma}(t) = \Gamma(t) + c\phi_0(t)$. In particular, due to the equality $\phi_0' - i\gamma\phi_0 = \Gamma$, the real function

$$\phi_0'(t) = \gamma A \cos(\gamma t + \alpha) + \int_0^\infty \frac{ue^{-tu}d\sigma_*(u)}{\Phi_0(u)(u^2+\gamma^2)} ,$$

is also a solution of equation (6.21). Clearly, here the integral on the right is an absolutely monotonic function belonging to $L_1(0,\infty)$. Our assertion that $\phi_0'(t)$ is a solution to equation (6.21) is also a straightforward consequence of the fact that $\phi_0(t)$ is a solution to equation (6.16). We need only integrate the integral in (6.16) by parts and subsequently differentiate the resulting equality term by term.

PART II

Below we present the last two sections (§§ 7 and 8) of our investigation. Naturally, we shall keep the same notations as in the first part of the paper. As in § 6 , the main object of our study is the homogeneous integral equation

$$\phi(t) - \int_0^\infty k(|t-s|)\phi(s)ds = 0 \quad (0 < t < \infty) , \qquad (7.0)$$

where $k(t)$ $(0 < t < \infty)$ is an absolutely monotonic function.

In § 8 equation (7.0) is investigated under the extra assumption that $k(t)$ decays exponentially. Equation (7.0) has also been studied under identical assumptions by E. Hopf ([4], § 15). It seems to us that by enlisting nonlinear integral equations and the theory of certain classes of analytic functions, one is able to make a more complete study of this equation.

§ 7. THE CASE $w(0) = 0$

1. The results obtained in this situation, i.e., under the assumption that

$$1 - 2\int_0^\infty \frac{d\sigma(t)}{t} = 0 \tag{7.1}$$

look more complete and may be proved more easily, if one also assumes that

$$(s_5 =) \int_0^\infty \frac{d\sigma(t)}{t^5} < \infty . \tag{7.2}$$

In applications, this condition is usually fulfilled.

THEOREM 7.1. *If the nondecreasing function* $\sigma(t)$ $(0 \leqq t < \infty)$ *satisfies conditions* (7.1) *and* (7.2), *then equation* (!) *has a unique positive solution* $F_0(t)$ $(0 \leqq t < \infty)$, *given by the formula*

$$\ln F_0(t) = -\frac{t}{\pi}\int_0^\infty \ln w(i\lambda)\frac{d\lambda}{\lambda^2 + t^2} \qquad (0 \leqq t < \infty), \tag{7.3}$$

where $\ln w(i\lambda) > 0$ $(0 < \lambda < \infty)$.

Simultaneously, this solution is the unique solution that is regular for $t > 0$. *Moreover, it satisfies*

$$\lim_{t \to 0} tF_0(t) = (2s_3)^{-1/2} . \tag{7.4}$$

PROOF: By (7.1),

$$w(z) = 2\int_0^\infty \frac{d\sigma(t)}{t} - 2\int_0^\infty \frac{t\,d\sigma(t)}{t^2 - z^2} = -2z^2\int_0^\infty \frac{d\sigma(t)}{t(t^2 - z^2)} . \tag{7.5}$$

Consequently, for any $\gamma > 0$

$$w_\gamma(z) = \frac{z^2 + \gamma^2}{z^2} w(z) = w(z) - 2\gamma^2\int_0^\infty \frac{d\sigma(t)}{t(t^2 - z^2)} =$$

$$= 1 - 2\int_0^\infty \frac{t\,d\sigma(t)}{t^2 - z^2} - 2\gamma^2\int_0^\infty \frac{d\sigma(t)}{t(t^2 - z^2)} =$$

$$= 1 - 2\int_0^\infty \frac{(t^2 + \gamma^2)\,d\sigma(t)}{t(t^2 - z^2)} .$$

Therefore, upon setting

$$\sigma_\gamma(t) = \int_0^t \frac{(s^2+\gamma^2)d\sigma(s)}{s^2} \qquad (0 \leqq t < \infty) , \tag{7.6}$$

we have

$$w_\gamma(z) = 1 - 2\int_0^\infty \frac{td\sigma_\gamma(t)}{t^2 - z^2} \qquad (\mathrm{Im}\, z \neq 0) ,$$

and

$$\int_0^\infty \frac{d\sigma_\gamma(t)}{t} = \int_0^\infty \frac{t^2+\gamma^2}{t^3}\, d\sigma(t) < \infty .$$

Since $w_\gamma(\pm i\gamma) = 0$, we see that $w_\gamma(0) < 0$, and according to the results of § 6, given the function $\sigma_\gamma(t)$, there exists a unique function $F_\gamma(z)$ having a unique singularity in the domain $\mathrm{Ext}(-\infty,0]$, namely a pole of first order at the point $z = -i\gamma$, and such that

$$\frac{1}{F_\gamma(z)} = 1 - \int_0^\infty \frac{F_\gamma(t)d\sigma_\gamma(t)}{t+z} .$$

Set

$$F_0(z) = \frac{z+i\gamma}{z} F_\gamma(z) = \frac{1}{z} Q(z) \qquad (z \in \mathrm{Ext}(-\infty,0]) , \tag{7.7}$$

where $Q(t)$ $(0 \leqq t < \infty)$ is, recalling the results of § 6 , a positive function satisfying the equation (see (6.5))

$$Q^{-1}(z) = \int_0^\infty \frac{Q(t)d\sigma_\gamma(t)}{(t+z)(t^2+\gamma^2)} = \int_0^\infty \frac{Q(t)d\sigma(t)}{t^2(t+z)} . \tag{7.8}$$

Remark 4.1 and (7.7) show that

$$F_0^{-1}(z) = 1 - \int_0^\infty \frac{t}{t-i\gamma}\cdot\frac{F_\gamma(t)d\sigma_\gamma(t)}{t+z} =$$

$$= 1 - \int_0^\infty \frac{t+i\gamma}{t}\cdot\frac{F_\gamma(t)d\sigma(t)}{t+z} ,$$

i.e.,

$$F_0^{-1}(z) = 1 - \int_0^\infty \frac{F_0(t)d\sigma(t)}{t+z} \qquad (z \in \mathrm{Ext}(-\infty,0]) .$$

Thus, we have constructed a positive solution $F_0(t)$ to equation (!) which is bounded on each interval (a,∞), $a > 0$, and satisfies

$$\lim_{s \to 0} sF(s) = Q(0) .$$

This last relation is equivalent to condition (7.4). Indeed, it follows from (7.5) and Theorem 2.1 that

$$\frac{1}{Q(z)Q(-z)} = 2 \int_0^\infty \frac{d\sigma(t)}{t(t^2 - z^2)} , \tag{7.9}$$

whence

$$Q^{-2}(0) = 2 \int_0^\infty \frac{d\sigma(t)}{t^3} .$$

According to the general formula (4.32), we shall have

$$\ln F_\gamma(t) = -\frac{t}{\pi} \int_0^\infty \hat{\ln}[\frac{(\lambda^2 - \gamma^2)w(i\lambda)}{\lambda^2}] \frac{d\lambda}{\lambda^2 + t^2} \quad (0 < t < \infty).$$

Here the logarithm $\hat{\ln}[\ldots]$ is negative for $\lambda^2 > \gamma^2$ by its definition. Since γ is an arbitrary positive number here, one can let $\gamma \longrightarrow 0$, which gives (7.3).

Now let $F(t)$ $(0 < t < \infty)$, be an arbitrary solution of equation (!). Then its analytic continuation $F(z)$ to the domain $\mathrm{Ext}(-\infty,0]$ satisfies the equation

$$F^{-1}(z)F^{-1}(-z) = w(z) \quad (\mathrm{Im}\, z \neq 0) . \tag{7.10}$$

Since σ is nondecreasing the function $w(z)$ has no real zeros, and hence the function $F(z)$ has no real poles in the domain $\mathrm{Ext}(-\infty,0]$.

Taking into account that, according to (7.1), $w(z) \longrightarrow 0$ as $z \longrightarrow 0$ along the imaginary axis, then by letting $z \longrightarrow 0$ following any sequence along the imaginary axis, we shall have $F^{-1}(z) \longrightarrow 0$ too. Passing to the limit with respect to this sequence in (!), we get

$$1 - \int_0^\infty \frac{F(t)d\sigma(t)}{t} = 0 .$$

Therefore, setting $\hat{F}(z) = zF(z)/(z + i\gamma)$, Lemma 4.1 yields

$$\hat{F}^{-1}(z) = 1 - \int_0^\infty \frac{(t - i\gamma)F(t)d\sigma(t)}{t(t + z)} = 1 - \int_0^\infty \frac{\hat{F}(t)d\sigma_\gamma(t)}{t + z} ,$$

where $\sigma_\gamma(t)$ is defined, as above, by equality (7.6).

If the function $F(t)$ is bounded on each interval (a,∞), $a > 0$, then $F(z)$ is holomorphic in the domain $\operatorname{Ext}(-\infty,0]$, and so is the function $\hat{F}(z)$, except at the point $z = -i\gamma$, where it has a first order pole. But then $\hat{F}(z) = F_\gamma(z)$; whence $\hat{F}(z) \equiv F_0(z)$.

The same can be said when the function $F(t)$ $(0 < t < \infty)$ is nonnegative. Indeed, in this case, according to (7.10) and equation (!), $F(t)$ monotonically decreases from $+\infty$ to 1 as t varies from 0 to ∞, and so the function $F(t)$ $(0 \leqq t < \infty)$ has no poles. The theorem is proved. [Notice that condition (7.2) was not used in the proof: the weaker condition $s_3 < \infty$ turned out to be sufficient.]

2. By Lemma 5.4,

$$\frac{Q(z)Q(-z)}{z^2} = -w^{-1}(z) = -1 + \frac{1}{2s_3 z^2} - 2\int_0^\infty \frac{t\,d\sigma_*(t)}{t^2 - z^2},$$

where $\sigma_*(t)$ $(0 \leqq t < \infty)$ is a nondecreasing function such that

$$1 + 2\int_0^\infty \frac{d\sigma_*(t)}{t} = \frac{s_5}{2s_3^2}.$$

The same method which gave relation (6.11) easily leads to

$$F_0(z) = 1 + \frac{1}{\sqrt{2s_3}\,z} + \int_0^\infty \frac{d\sigma_*(t)}{(t+z)F_0(t)}, \tag{7.11}$$

whence

$$F_0(z) = 1 + \int_0^\infty e^{-zt}\Gamma(t)\,dt,$$

where
$$\Gamma(t) = \frac{1}{\sqrt{2s_3}} + \int_0^\infty \frac{e^{-ut}d\sigma_*(u)}{F_0(u)}.$$

Thus $F_0(t)$ and $\Gamma(t)$ are absolutely monotonic functions. Putting

$$\phi_0(t) = 1 + \int_0^t \Gamma(s)\,ds \quad (0 \leqq t < \infty)$$

or, equivalently,

$$\phi_0(t) = 1 + \frac{1}{\sqrt{2s_3}} t + r(0) - r(t) , \tag{7.13}$$

where

$$r(t) = \int_0^\infty \frac{e^{-ut} d\sigma_*(u)}{uF_0(u)} \qquad (0 \leqq t < \infty) , \tag{7.14}$$

we shall have

$$\Phi_0(z) = \int_0^\infty e^{-zt}\phi_0(t)dt = \frac{F_0(z)}{z} = \frac{Q(z)}{z^2} \qquad (\text{Re } z > 0).$$

Now we remark that, by virtue of (7.8),

$$Q^{-1}(z) = \int_0^\infty \frac{\Phi_0(t)d\sigma(t)}{t + z} \qquad (z \in \text{Ext}(-\infty,0]) .$$

Inserting this expression for $Q^{-1}(z)$ into (7.9), we obtain

$$\int_0^\infty \frac{\Phi_0(s)d\sigma(s)}{s + z} = 2Q(-z) \int_0^\infty \frac{d\sigma(s)}{s(s^2 - z^2)} \qquad (\text{Im } z \neq 0) .$$

Replacing here z by $-z$ $(\text{Im } z \neq 0)$, we have

$$\int_0^\infty \frac{\Phi_0(s)d\sigma(s)}{s - z} = 2z^2\Phi_0(z) \int_0^\infty \frac{d\sigma(s)}{s(s^2 - z^2)} =$$

$$= 2\Phi_0(z) \int_0^\infty \{\frac{s}{s^2 - z^2} - \frac{1}{s}\}d\sigma(s) =$$

$$= \Phi_0(z)\{-1 + \int_0^\infty (\frac{1}{s - z} + \frac{1}{s + z})d\sigma(s)\} .$$

It follows that for $\text{Im } z \neq 0$

$$\int_0^\infty \frac{\Phi_0(s) - \Phi_0(z)}{s - z} d\sigma(s) = \Phi_0(z)(-1 + \int_0^\infty \frac{d\sigma(s)}{s + z}) . \tag{7.15}$$

Now letting $z \longrightarrow t$, $t > 0$, we see that the function $\Phi_0(t)$ $(0 < t < \infty)$ satisfies equation (7.15) too, i.e.,

$$\int_0^\infty \frac{\Phi_0(t) - \Phi_0(s)}{t - s} d\sigma(s) = \Phi_0(t)(1 - \int_0^\infty \frac{d\sigma(s)}{s + t}) .$$

Now notice that whenever condition (7.2) is satisfied, one has

$$\int_0^\infty k(t)t^4 dt = \int_0^\infty t^4 \int_0^\infty e^{-tu} d\sigma(u)\, dt = 4! \int_0^\infty \frac{d\sigma(u)}{u^5} < \infty,$$

and so certainly

$$\int_0^\infty k(t)tdt < \infty.$$

Recalling the results of § 15 from paper [6], it follows naturally that one has

THEOREM 7.2. *Suppose that conditions* (7.1) *and* (7.2) *are fulfilled. Then the function* $\phi_0(t)$ *is, up to a scalar factor, the unique solution of the homogeneous equation* (7.0) *which satisfies the condition*

$$|\phi(t)| = o(t^2) \quad \textit{as} \quad t \longrightarrow \infty . \tag{7.16}$$

The proof is entirely analogous to that of Theorem 6.2. If $\phi(t)$ is some solution of equation (7.0) satisfying (7.16), then, setting

$$\Phi(z) = \int_0^\infty e^{-tz}\phi(t)dt \qquad (\text{Re } z > 0)$$

and following the same arguments used in the proof of Theorem 6.1, we convince ourselves that

$$\int_0^\infty \frac{\Phi(z) - \Phi(u)}{z - u}\, d\sigma(u) + \left(1 - \int_0^\infty \frac{d\sigma(u)}{u + z}\right)\Phi(z) = 0 \tag{7.17}$$

(Re $z > 0$).

Moreover, the converse is also true, i.e., if $\phi(t)$ $(0 \leqq t < \infty)$ is some function satisfying condition (7.16) and whose transform Φ satisfies equation (7.17), then ϕ is a solution to equation (7.0). Since Φ_0 satisfies equation (7.17), ϕ_0 is a solution of the integral equation (7.0).

Furthermore, following a line of argument similar to that in the proof of Theorem 6.1, one shows that the transform $\Phi(z)$ of a function $\phi(t)$ satisfying condition (7.16) will be a solution of equation (7.17) if and only if $\Phi = c\Phi_0$. This completes the proof of the theorem.

3. Let us list some more properties of the absolutely monotonic function $r(t)$ figuring in formula (7.13). According to (7.14),

$$r(0) = \int_0^\infty \frac{d\sigma_*(u)}{uF_0(u)} < \int_0^\infty \frac{d\sigma_*(u)}{u} = \frac{1}{2}\left(\frac{s_5}{2s_3^2} - 1\right) .$$

Therefore, in formula (7.13),

$$r(t) < r(0) < \frac{1}{2}\left(\frac{s_5}{2s_3^2} - 1\right) \quad (t > 0) . \tag{7.18}$$

Moreover, $r(0)$ can be explicitly expressed in terms of the function $w(z)$. Indeed, applying (7.11),

$$1 - r(0) = 1 + \int_0^\infty \frac{d\sigma_*(t)}{Q(t)} = \lim_{s\downarrow 0} [F_0(s) - \frac{1}{s}(2s_3)^{-1/2}] =$$

$$= \lim_{s\downarrow 0} \frac{Q(s) - (2s_3)^{-1/2}}{s} .$$

Further, since by (7.4) $Q(0) = (2s_3)^{-1/2}$, we get

$$1 + r(0) = \lim_{s\downarrow 0} \frac{Q(s) - Q(0)}{s} = Q'(0+) . \tag{7.19}$$

As

$$\ln s = \frac{2s}{\pi} \int_0^\infty \frac{\ln \lambda \, d\lambda}{\lambda^2 + s^2} ,$$

it follows from (7.13) that

$$\ln Q(s) = -\frac{1}{\pi} \int_0^\infty \ln \frac{w(i\lambda)}{\lambda} \frac{s d\lambda}{\lambda^2 + s^2} \quad (s \geqq 0) ,$$

and so

$$\ln Q(s) = -\frac{1}{\pi} \int_0^\infty \ln \frac{w(i\lambda)Q^2(0)}{\lambda^2} \frac{s d\lambda}{\lambda^2 + s^2} + \ln Q(0) \quad (s \geqq 0).$$

Integrating by parts and taking into account that

$$\lim_{\lambda\to 0} \frac{w(i\lambda)Q^2(0)}{\lambda^2} = 1 ,$$

we get

$$\ln Q(s) = \frac{1}{\pi} \int_0^\infty [\frac{2}{\lambda} - \frac{iw'(i\lambda)}{w(i\lambda)}] \int_\lambda^\infty \frac{s d\mu}{\mu^2 + s^2} d\lambda + \ln Q(0) . \tag{7.20}$$

Here

$$\frac{2}{\lambda} - \frac{iw'(i\lambda)}{w(i\lambda)} = -\frac{d}{d\lambda} \ln \frac{w(i\lambda)}{\lambda^2} = -\frac{d}{d\lambda} \ln[2 \int_0^\infty \frac{d\sigma(t)}{t(t^2 + \lambda^2)}] =$$

$$= 2\lambda \int_0^\infty \frac{d\sigma(t)}{t(t^2+\lambda^2)^2} \Big/ \int_0^\infty \frac{d\sigma(t)}{t(t^2+\lambda^2)} =$$

$$= \frac{2s_5}{s_3}\lambda + o(\lambda) \quad \text{as} \quad \lambda \longrightarrow 0 \ . \tag{7.21}$$

Since $t^2 + \lambda^2 > \lambda^2$, one has

$$(0 <)\ \frac{2}{\lambda} - \frac{iw'(i\lambda)}{w(i\lambda)} < \frac{2}{\lambda} \quad \text{for} \quad \lambda > 0 \ .$$

Therefore, the functions appearing under the integral sign in (7.20) are all positive. Using also the fact that

$$\int_\lambda^\infty \frac{s\,d\mu}{\mu^2 + s^2} = \int_0^s \frac{\lambda\,dt}{t^2+\lambda^2} \ ,$$

one can recast relation (7.20) in the form

$$\ln Q(s) = \frac{1}{\pi}\int_0^s dt \int_0^\infty \left[\frac{2}{\lambda} - \frac{iw'(i\lambda)}{w(i\lambda)}\right] \frac{\lambda\,d\lambda}{t^2+\lambda^2} + \ln Q(0)$$

$(s \geqq 0)$. It follows that

$$\frac{Q'(s)}{Q(s)} = \frac{1}{\pi}\int_0^\infty \left[\frac{2}{\lambda} - \frac{iw'(i\lambda)}{w(i\lambda)}\right] \frac{\lambda\,d\lambda}{s^2+\lambda^2} \qquad (s \geqq 0) \ ,$$

and, in particular,

$$\frac{Q'(0+)}{Q(0)} = \frac{1}{\pi}\int_0^\infty \left[\frac{2}{\lambda} - \frac{iw'(i\lambda)}{w(\lambda)}\right] \frac{d\lambda}{\lambda} \ . \tag{7.22}$$

From (7.19) and (7.22) we determine the expression for $r(0)$.

If one considers the following solution instead of $\phi_0(t)$:

$$\phi_H(t) = Q^{-1}(0)\phi_0(t) = \sqrt{2s_3}\ \phi_0(t) \ , \tag{7.23}$$

then one has $\phi_H(t) = t + q(t)$, where

$$q(t) = \sqrt{2s_3}[1 + r(0) - r(t)] \qquad (\geqq \sqrt{2s_3}) \ . \tag{7.24}$$

Obviously,

$$q(\infty) = \lim_{t\to\infty} q(t) = \sqrt{2s_3}(1 + r(0)) \ , \tag{7.25}$$

and therefore, using (7.22) and (7.23),

$$q(\infty) = \frac{1}{\pi}\int_0^\infty \left[\frac{2}{\lambda} + \frac{K'(\lambda)}{1 - K(\lambda)}\right]\frac{d\lambda}{\lambda} , \tag{7.26}$$

because $w(i\lambda) = 1 - K(\lambda)$. Formula (7.26) has been obtained by E. Hopf [4, p. 85], under different assumptions on the kernel $k(|t - s|)$. [In the book [4] there is a misprint in formula (7.26), namely in the integral only $d\lambda$ appears instead of $d\lambda/\lambda$.]

Assuming that $k(t)$ has the form (0.2) and decreases exponentially (i.e., σ = const in a neighborhood of the zero point), E. Hopf proved [4] that $q(t)$ increases monotonically as $t \longrightarrow \infty$, while remaining bounded. Here we prove more, and under more general assumptions. More precisely, we show that $q(0) - q(t)$ $(0 \leqq t < \infty)$ is an absolutely monotonic function. In § 8 we shall make this statement more precise in the case where $k(t)$ satisfies Hopf's conditions.

According to (7.21), expression (7.25) may be written alternatively in the form

$$q(\infty) = \frac{2}{\pi}\int_0^\infty \left(\int_0^\infty \frac{d\sigma(t)}{t(t^2 + \lambda^2)^2} \bigg/ \int_0^\infty \frac{d\sigma(t)}{t(t^2 + \lambda^2)}\right) d\lambda \tag{7.27}$$

Since

$$\int_0^\infty \frac{d\sigma(t)}{t(t^2 + \lambda^2)} < s_3,$$

we see that

$$q(\infty) > \frac{2}{\pi s_3}\int_0^\infty\int_0^\infty \frac{d\sigma(t)}{t(t^2 + \lambda^2)}\, d\lambda = \frac{1}{2s_3}\int_0^\infty \frac{d\sigma(t)}{t^4} = \frac{s_4}{2s_3} .$$

Therefore, taking into account (7.18) and (7.25), we get the following bounds for $q(\infty)$:

$$\frac{s_4}{2s_3} < q(\infty) < \frac{\sqrt{s_3}}{\sqrt{2}}\left(1 + \frac{s_5}{2s_3^2}\right) .$$

4. The results of no. 2 allow us to obtain a formula for the resolvent of the integral equation (7.30) and to establish a number of its properties. Using (7.13) and (7.14), one has for $\Gamma(t) = \phi_0'(t)$ that

$$\Gamma(t) = (2s_3)^{-1/2} - r'(t) ,$$

where

$$- r'(t) = \int_0^\infty \frac{e^{-tu} \, d\sigma_*(u)}{F_0(u)} \qquad (0 \leqq t < \infty).$$

is an absolutely monotonic function belonging to $L_1(0,\infty)$ and satisfying the condition

$$r'(t) = O(t^{-1}) \quad \text{for} \quad t \longrightarrow 0,\infty .$$

The latter is a consequence of both the inequalities $xe^{-x} \leqq e^{-1}$ $(x \geqq 0)$, $F_0(u) \geqq 1$, and of the finiteness of the moment $\int_0^\infty u^{-1} d\sigma_*(u)$.

Reasoning as we did at the end of § 6, we can now convince ourselves that the function $\Gamma(t) = \phi_0'(t)$ is a solution of the equation

$$\gamma(t) - \int_0^\infty k(t - s)\gamma(s) = k(t) \qquad (0 < t < \infty) . \qquad (7.28)$$

However, in contrast to what happened when $w(0) < 0$, now $\Gamma(t) = \phi_0'(t)$ will be the unique bounded solution of equation (7.28).

We leave to the reader the proof of the fact that the function $\Gamma(t)$ may be obtained as a series

$$\Gamma(t) = \sum_{n=1}^{\infty} k_n(t) \qquad (0 < t < \infty) ,$$

which converges uniformly in each interval (a,∞), $a > 0$ (and, if $K(0) < 1$, i.e., $w(0) > 0$, uniformly on the entire half line). In this series $k_1(t) = k(t)$, while

$$k_n(t) = \int_0^\infty k_{n-1}(t - s)k(s)ds \qquad (n = 2,3,\ldots) .$$

Following the general rule (see 6,7.12), we set

$$\Gamma(t,s) = \Gamma(t - s) + \Gamma(s - t) + \int_0^\infty \Gamma(t - u)\Gamma(s - u)du \quad (7.29)$$

$(0 \leqq t,s < \infty)$, where we consider that $\Gamma(t) = 0$ for $t < 0$. Moreover, upon setting

$$\gamma(t,s) = -r'(t - s) - r'(s - t) + \int_0^\infty r'(t - u)r'(s - u)du,$$

and considering that $r'(t) = 0$ for $t < 0$, we shall obviously have

$$\Gamma(t,s) = \frac{1}{2\delta_3}\min(t,s) + \chi(t,s) + \gamma(t,s) \quad (0 \leqq t,s < \infty),$$

where $\chi(t,s)$ is a bounded function and $\gamma(t,s)$ generates a bounded operator in each space E (see [6], § 6, no. 2).

Let us show that $\Gamma(t,s)$ is the resolvent kernel for the equation

$$g(t) - \int_0^\infty k(t-s)g(s)ds = f(t) \quad (0 \leqq t < \infty), \tag{7.30}$$

meaning that

$$k(t-s) + \int_0^\infty k(t-u)\Gamma(u,s)du = \Gamma(t,s) \tag{7.31}$$

$(0 \leqq t,s < \infty)$.

Multiplying this equality by $e^{-\zeta s}$ $(\mathrm{Re}\,\zeta > 0)$ term by term and subsequently integrating with respect to s from 0 to ∞, we see that it is equivalent to

$$g_\zeta(t) - \int_0^\infty k(t-s)g_\zeta(s)ds = e^{-\zeta t} \quad (0 \leqq t < \infty), \tag{7.32}$$

where

$$g_\zeta(t) = e^{-\zeta t} + \int_0^\infty \Gamma(t,s)e^{-\zeta s}ds. \tag{7.33}$$

On the other hand, multiplying equation (7.32) by e^{-tz} term by term $(\mathrm{Re}\,z > 0)$, and integrating the result with respect to t from 0 to ∞, we get

$$\int_0^\infty \frac{\Phi_\zeta(z) - \Phi_\zeta(u)}{z-u}\,d\sigma(u) + \Phi_\zeta(z)\left(1 - \int_0^\infty \frac{d\sigma(u)}{u+z}\right) = \frac{1}{z+\zeta}, \tag{7.34}$$

where

$$\Phi_\zeta(z) = \int_0^\infty e^{-zt}g_\zeta(t)dt.$$

Using the relations (7.29), (7.33), and (7.12) one obtains easily

$$\Phi_\zeta(z) = F_0(z)F_0(\zeta)/(z+\zeta) \quad (\mathrm{Re}\,z > 0,\ \mathrm{Re}\,\zeta > 0).$$

By Theorem 3.3, relation (7.34) does indeed hold, and this proves (7.31) at the same time.

The existence of the resolvent kernel $\Gamma(t,s)$ for equation (7.30) does not contradict the fact that, in the case under consideration, the point $\lambda = 1$ belongs to the spectrum of the operator generated by the kernel $k(t - s)$ $(0 \leqq t,s < \infty)$, in each space E (see the definition in [6], § 6). The point is that the integral

$$\int_0^\infty \Gamma(t,s)f(s)ds \tag{7.35}$$

does not make sense for all $f \in E$ and, moreover, if it does make sense for some $f \in E$ (and all $t \geqq 0$), then it does not necessarily transform f again into an element of E. Expressing this in the language of operator theory, one may say that the integral (7.35) generates in E an unbounded, densely defined operator.

Nevertheless, the structure of the kernel $\Gamma(t,s)$ permits us, for example, to claim that whenever $f(t),\ tf(t) \in L_1(0,\infty)$ the formula

$$g(t) = f(t) + \int_0^\infty \Gamma(t,s)f(s)ds \quad (0 < t < \infty)$$

provides the unique bounded solution to equation (7.30).

§ 8. THE CASE WHEN $k(t)$ IS AN ABSOLUTELY MONOTONIC, EXPONENTIALLY DECREASING FUNCTION OF $|t|$

Let $k(t)$ be an absolutely monotonic function, and let $\sigma(u)$ $(0 \leqq u < \infty,\ \sigma(0) = 0)$ be its spectral function. If, given $k(t)$, one can find a $(a > 0)$ such that

$$k(t) = o(e^{-ht}) \quad \text{as} \quad t \longrightarrow \infty \tag{8.1}$$

for all $h < a$, then in this case, and only in it, we will have $\sigma(u) = 0$ for $0 \leqq u < a$, and representation (0.2) becomes

$$k(t) = \int_a^\infty e^{-u|t|}d\sigma(u) \quad (-\infty < t < \infty)\ . \tag{8.2}$$

Notice that for absolutely monotonic functions $k(t)$ $(t > 0)$ condition (8.1) is equivalent to

$$\int_0^\infty e^{ht} k(t)\,dt < \infty \quad \text{for} \quad h < a .$$

If one takes the largest admissible $a > 0$ in representation (8.2), i.e., if a is the first growth point of the function $\sigma(u)$, then a shall be referred to as the *exponent* of the absolutely monotonic function $k(t)$. For even functions $k(t)$ $(t > 0)$ of the form (8.2), the results of the previous sections can be sharpened.

1. For a function $k(t)$ of the type (8.2), the function $w(z) = 1 - K(iz)$ has the form

$$w(z) = 1 - 2\int_0^\infty \frac{t\,d\sigma(t)}{t^2 - z^2} ,$$

and so is holomorphic in the complex plane with the cuts $(-\infty,-a]$ and $[a,\infty)$, that is, in the domain

$$\Omega_a = \operatorname{Ext}(-\infty,-a] \cap \operatorname{Ext}[a,\infty) \tag{8.3}$$

When

$$w(0) = 1 - 2\int_0^\infty \frac{d\sigma(u)}{u} \leq 0 ,$$

the function $w(x)$ is negative for $-a < x < a$, being a decreasing function of x^2 for $0 < x^2 < a^2$. Therefore, in this case Stieltjes' inversion formula implies that one has $\sigma_*(u) = 0$ for $0 < u < a$ in the representations (5.14) and (5.21). Thus we conclude that

1) If $w(0) = 0$, then

$$w^{-1}(z) = 1 - \frac{1}{2s_3 z^2} + 2\int_0^\infty \frac{t\,d\sigma_*(t)}{t^2 - z^2} \qquad (z^2 \in \operatorname{Ext}[a,\infty)) .$$

2) If $w(0) < 0$, then

$$w^{-1}(z) = 1 - \frac{\rho}{z^2 + \gamma^2} + 2\int_0^\infty \frac{t\,d\sigma_*(t)}{t^2 - z^2} \qquad (z^2 \in \operatorname{Ext}[a,\infty)),$$

where γ is the unique positive zero of the function $w(i\lambda)$, and

$$\rho^{-1} = \int_0^\infty \frac{2t\,d\sigma(t)}{(t^2+\gamma^2)^2} .$$

Here (and from now on) it is assumed that $\sigma_*(t)$ $(a \leqq t < \infty)$ is a nondecreasing function satisfying the condition

$$\int_0^\infty \frac{d\sigma_*(t)}{t} < \infty .$$

Now let us consider the case $w(0) > 0$. Under this assumption, $g(\zeta) = w(\sqrt{\zeta})$ might have one simple zero inside the interval $(0,a^2)$ and this will happen if and only if $g(a^2-0) = w(a-0) < 0$. Thus, when $w(a-0) \geqq 0$, the function $g^{-1}(\zeta)$ is regular and positive on the interval $[0,a^2)$, and so in representation (5.10) we shall have again that $\sigma_*(u) = 0$ for $0 \leqq u < a$. If, however, $w(a-0) < 0$, then $g^{-1}(\zeta)$, being real within the interval $[0,a^2)$, will have a simple pole there. Denoting the latter by κ^2, the point $u = \kappa$ will be the unique growth point of $\sigma_*(u)$ in the interval $[0,a)$.

We are thus led to the following conclusions.

3) If $w(0) > 0$ and, in addition,

$$w(a-0) = 1 - \int_{a-0}^\infty \frac{2u\,d\sigma(u)}{u^2-a^2} \geqq 0 ,$$

then

$$w^{-1}(z) = 1 + \int_a^\infty \frac{2u\,d\sigma_*(u)}{u^2-z^2} \qquad (z \in \Omega_a) .$$

4) If $w(0) > 0$, but $w(a-0) < 0$, then

$$w^{-1}(z) = 1 + \frac{\rho}{\kappa^2-z^2} + \int_a^\infty \frac{2u\,d\sigma_*(u)}{u^2-z^2} \qquad (z \in \Omega_a) .$$

Here κ is the unique root of the equation

$$1 - \int_a^\infty \frac{2u\,d\sigma(u)}{u^2-x^2} = 0$$

inside $(0,a)$, and

$$\rho^{-1} = 2\int_a^\infty \frac{u\,d\sigma(u)}{(u^2-\kappa^2)^2} .$$

2. Recalling, on one hand, the results of [6] (see the example on pp. 257-260), and, on the other hand, Theorem 7.2, we obtain without difficulty the assertions concerning equation (7.0) which are listed below for each of the four cases considered above.

THEOREM 8.1. *Let* $1 - K(0) = 0$. *Then equation* (7.0) *has the solution*

$$\phi_0(t) = 1 + \frac{t}{\sqrt{2s_3}} + r(0) - r(t) \qquad (0 < t < \infty) ,$$

where $r(t)$ $(0 \leq t < \infty)$ *is an absolutely monotonic function having exponent* $\geq a$, *namely*

$$r(t) = \int_a^\infty \frac{e^{-ut}}{uF_0(u)}\, d\sigma_*(u) \qquad (0 \leq t < \infty) ,$$

and

$$0 < \ln F_0(u) = -\frac{1}{\pi}\int_0^\infty \ln w(i\lambda)\, \frac{u d\lambda}{\lambda^2 + u^2} \qquad (0 < u < \infty) .$$

Any other solution $\phi(t)$ *of equation* (7.0) *satisfying, for some* $h < a$, *the condition*

$$\phi(t) = o(e^{ht}) \quad \textit{as} \quad t \longrightarrow \infty$$

differs from $\phi_0(t)$ *only by a scalar factor.*

The functions ϕ_0 *and* F_0 *are related via the following additional relation:*

$$F_0(u) = u \int_0^\infty e^{-ut}\phi_0(t)dt \qquad (0 < u < \infty) .$$

Let us mention also that $F_0(t)$ is now the unique solution positive for $t > 0$, as well as the unique solution regular for $t > a$, to the equation

$$F^{-1}(t) = 1 - \int_a^\infty \frac{F(u)d\sigma(u)}{t + u} \qquad (t \geq a) .$$

The fact is that any solution F to this equation can extrapolated naturally to the entire positive axis, becoming so a solution to equation (!) (having possibly a pole in the interval $(0,a)$). Recalling the relations $F(z)F(-z) = w^{-1}(z)$ and $w(t) < 0$ $(0 < t < a)$, we see that there is no such pole, and, at the same time, we see the validity of our assertion.

As in § 7, consider the solution

$$\phi_H(t) = \sqrt{2\delta_3}\,\phi_0(t) = t + q(t) ,$$

rather than $\phi_0(t)$. Then the limit $q(\infty)$ of the monotonically increasing function $q(t)$ $(0 \leq t < \infty)$ is given by (7.26). Since

$$\int_a^\infty \frac{d\sigma(t)}{t(t^2+\lambda^2)^2} \leq \frac{1}{a^2+\lambda^2}\int_a^\infty \frac{d\sigma(t)}{t(t^2+\lambda^2)} \qquad (0 \leq \lambda < \infty) ,$$

(7.2) yields (taking into account that $\sigma(t) = 0$ for $t < a$)

$$q(\infty) \leq \frac{1}{a} . \tag{8.4}$$

This bound is interesting because it does not depend upon the choice of σ. It is precise in the sense that it is attained for a certain choice of σ, namely when σ has a unique growth point at $\lambda = a$, i.e., (see (8.2)) when

$$k(t) = ce^{-a|t|} \qquad (c > 0; \quad -\infty < t < \infty) .$$

The bound (8.4) is due to E. Hopf (see [4], Theorem VI), who obtained it in another way and under more general assumptions on the function $k \in L_1(-\infty,\infty)$, namely

$$k(t) = k(|t|) > 0 \quad (-\infty < t < \infty),$$

$$k(t) \geq a\int_t^\infty k(s)ds \qquad (0 \leq t < \infty) ,$$

and also $k(t)$ is continuous for $t > 0$, while

$$k(t) = O(\ln\frac{1}{t}), \quad t \longrightarrow 0, \quad k(t) = O(e^{-t}), \quad t \longrightarrow \infty,$$

and

$$K(0) = 1 .$$

[It is possible that E. Hopf, while considering a more general case, did not notice that the bound (5.4) is "precise", because he does not mention this fact anywhere.]

In addition to the bound (8.4), one has also the lower bound $q(\infty) \geq 1/2a$. The latter is an easy consequence of the bound $s_4/2s_3 < q(\infty)$ obtained in the previous section.

Under the assumptions of Theorem 8.1, E. Hopf also established the following relation (see [4], formula (173′)):

$$\int_0^\infty e^{at}[q(\infty) - q(t)]dt \leqq \frac{1}{a^2} - \frac{q(\infty)}{a} . \tag{8.5}$$

Here the equality is attained if and only if

$$\int_a^\infty \frac{d\sigma(t)}{t - a} = \infty . \tag{8.6}$$

Notice that the last condition is equivalent to $w(a - 0) = 1 - K(i(a - 0)) = \infty$, i.e., to the condition

$$\int_0^\infty e^{at}k(t)dt = \infty .$$

Let us show that E. Hopf's result is a simple consequence of the relations that we obtained above. Indeed, since

$$q(\infty) = \sqrt{2s_3}(1 + r(0)) ,$$

we may rewrite (8.5) in the following form after dividing each term by $\sqrt{2s_3}$:

$$\int_0^\infty e^{at}r(t)dt \leq \frac{1}{a^2\sqrt{2s_3}} - \frac{1 + r(0)}{a} . \tag{8.7}$$

To prove the last inequality, notice that, according to the formula for $r(t)$ given in Theorem 8.1,

$$\int_0^\infty e^{at}r(t)dt = \int_a^\infty \frac{d\sigma_*(u)}{(u - a)uF_0(u)} = \frac{1}{a}\int_a^\infty [\frac{1}{u - a} - \frac{1}{u}]\frac{d\sigma_*(u)}{F_0(u)} =$$

$$= \frac{1}{a} [\int_a^\infty \frac{d\sigma_*(u)}{(u - a)F_0(u)} - r(0)] .$$

On the other hand, it follows from relation (7.11), which now has become

$$F_0(z) = 1 + \frac{1}{\sqrt{2s_3}z} + \int_a^\infty \frac{d\sigma_*(u)}{(u + z)F_0(u)} ,$$

that

$$\int_{a+0}^\infty \frac{d\sigma_*(u)}{(u - a)F_0(u)} = -1 + (\sqrt{2s_3}a)^{-1} + F_0(-a + 0) ;$$

whence

$$\int_0^\infty e^{at} r(t)dt = \frac{1}{a^2\sqrt{2s_3}} - \frac{1 - r(0)}{a} + \frac{F_0(-a + 0)}{a} .$$

Since $F_0(z)F_0(-z) = w^{-1}(z)$ and $0 < -w(a - 0) \leqq \infty$, $F_0(-a + 0)$ is less or equal to zero, depending on whether $w(a - 0) > -\infty$ or $w(a - 0) = -\infty$. Consequently, relation (8.7), as well as relation (8.5) are proved, and it is clear when equality is attained in these relations.

3. From (8.7) it follows, in particular, that the expression

$$\int_0^\infty e^{-zt} r(t)dt = m_0 - m_1 z + m_2 z^2 - \dots , \tag{8.8}$$

where

$$m_k = \frac{1}{k!}\int_0^\infty t^k r(t)dt \quad (k = 0,1,\dots)$$

converges absolutely at least for $|z| \leqq a$.

Let us pause and discuss the problem of how to compute the moments $k!m_k$ $(k = 0,1,\dots)$. According to the relations from Theorem 8.1, and (8.8),

$$F_0(z) = \frac{1}{\sqrt{2s_3}z} + 1 + r(0) - m_0 z + m_1 z^2 - m_2 z^3 + \dots \tag{8.9}$$

and so, upon setting

$$\frac{F_0'(z)}{F_0(z)} = -\frac{1}{z} + \alpha_0 + \alpha_1 z + \alpha_2 z^2 + \dots , \tag{8.10}$$

we get from (8.9) and (8.10) that

$$1 + r(0) = \frac{\alpha_0}{\sqrt{2s_3}} ,$$

$$(-1)^{n+1}(n+2)m_n = \frac{\alpha_0\alpha_n + \alpha_{n+1}}{\sqrt{2s_3}} - \alpha_n m_0 + \alpha_{n-1}m_1 + \dots$$
$$+ (-1)^n \alpha_1 m_{n-1} \quad (n = 0,1,\dots) \tag{8.11}$$

In particular,

$$- 2m_0 = \frac{\alpha_0^2 + \alpha_1}{\sqrt{2s_3}} . \tag{8.12}$$

On the other hand, setting $c = 1$ in (2.3) gives

$$- \frac{F_0'(z)}{F_0(z)} + \frac{F_0'(-z)}{F_0(-z)} = \frac{w'(z)}{w(z)} ,$$

whence

$$\frac{w'(z)}{w(z)} = \frac{2}{z} - 2 \sum_{k=0}^{\infty} \alpha_{2k+1} z^{2k+1} .$$

Since

$$w(z) = 1 - 2 \sum_{k=0}^{\infty} s_{2k+1} z^{2k} \qquad (|z| < a) ,$$

and since, in our case, $2s_1 = 1$, it results from (8.10) that

$$\sum_{k=1}^{n} s_{2k+1}\alpha_{2n-2k+1} = - n s_{2n+3} \qquad (n = 0,1,\dots) . \tag{8.13}$$

In particular,

$$\alpha_1 = - s_5/s_3 ,$$

which, combined with (8.12), yields

$$m_0 = \frac{1}{2\sqrt{2s_3}} \left(\frac{s_5}{s_3} - \alpha_0^2\right) ,$$

or, alternatively,

$$\int_0^{\infty} r(t)dt = \frac{1}{2\sqrt{2s_3}} \left(\frac{s_5}{s_3} - 2s_3(1 + r(0))^2\right) .$$

The recurrence relations (8.13) allow us to compute only the numbers α_{2n+1} $(n = 0,1,\dots)$, but not the numbers α_{2n} $(n = 0,1,\dots)$. We have therefore to show how to determine the latter. Afterwards, we shall be able to determine successively the numbers m_n $(n = 0,1,\dots)$. To this end, notice that by introducing, as in § 7, the function

$$Q(z) = zF_0(z) ,$$

we get

$$\frac{Q'(z)}{Q(z)} = \alpha_0 + \alpha_1 z + \alpha_2 z^2 + \dots ,$$

and so

$$\alpha_n = \frac{1}{n!}\left(\frac{d^n}{dz^n} \frac{Q'(z)}{Q(z)} \right)_{z=0} \qquad (n = 0,1,\dots) .$$

According to § 7, no.3, we have

$$\ln Q(z) = -\frac{1}{\pi}\int_0^{\infty} \ln \frac{w(i\lambda)}{\lambda^2} \frac{z d\lambda}{\lambda^2 + z^2} = -\frac{1}{2\pi}\int_{-\infty}^{\infty} \ln \frac{w(i\lambda)}{\lambda^2} \frac{d\lambda}{i\lambda+z} =$$

$$= -\frac{1}{2\pi}\int_{L_\varepsilon}^{\infty} \ln \frac{w(i\lambda)}{\lambda^2} \frac{d\lambda}{i\lambda + z} \tag{8.14}$$

Here the path of integration L_ε $(\varepsilon > 0)$ is obtained from the real axis by replacing the segment $(-\varepsilon,\varepsilon)$ by a semicircle C_ε lying in the upper half plane and having the segment $(-\varepsilon,\varepsilon)$ as its diameter.

Now differentiate (8.14) with respect to z and then take $z = 0$ to obtain

$$\alpha_n = (-1)^{n+1} \frac{n+1}{2\pi} \int_{L_\varepsilon} \ln \frac{w(i\lambda)}{\lambda^2} \frac{d\lambda}{\lambda^{n+2}} \qquad (n = 0,1,\dots). \tag{8.15}$$

When n is odd $(n = 2k + 1)$, the last integral reduces to the integral over C_ε because $\ln \frac{w(i\lambda)}{\lambda^2}$ is an even function. After computing this integral we obtain the same expressions for α_{2k+1} in terms of the numbers s_j that may be deduced from (8.13).

Consider (8.15) for even n: $n = 2k$. Integrating (8.15) by parts $2k+1$ times, observing that the double substitutions (i.e., the terms outside the integral) give zero, and subsequently letting $\varepsilon \longrightarrow 0$, we get

$$\alpha_{2k} = -\frac{(2k+1)!}{\pi} \int_0^{\infty} \frac{d^{2k+1}}{d\lambda^{2k+1}} \ln \frac{w(i\lambda)}{\lambda^2} \frac{d\lambda}{\lambda} \qquad (k = 0,1,\dots) .$$

When $k = 0$, we recover formula (7.26) which, however, was obtained in § 7 by another method and under less restrictive assumptions concerning the function σ.

4. Now let us analyze the case $w(0) < 0$. We discussed this situation in § 6 in detail without the assumption that $\sigma(t) = 0$ for $0 \leqq t < a$. It is easy to see to which simplifications this hypothesis will lead. The next statement is a straightforward consequence of the results in § 6 (in particular, of Theorem 6.2) and of proposition 2) given at the beginning of this section.

THEOREM 8.2. *Let* $w(0) = 1 - K(0) < 0$. *Then equation* (7.0) *has the solution*

$$\phi_0(t) = A \sin(\gamma t + \alpha) - r(t)$$

Here

$$A = \frac{1}{\sqrt{2}\gamma}\left(\int_a^\infty \frac{t\,d\sigma(t)}{(t^2 + \gamma^2)^2}\right)^{-1/2},$$

$$\operatorname{tg}\alpha = \gamma\int_a^\infty \frac{\Phi_0(u)\,d\sigma(u)}{u^2 + \gamma^2} \Big/ \int_a^\infty \frac{u\Phi_0(u)\,d\sigma(u)}{u^2 + \gamma^2},$$

and the functions $\Phi_0(u)$, $\phi_0(t)$, *and* $r(t)$ *are determined from the relations*

$$\ln \Phi_0(u) = -\frac{1}{\pi}\int_0^\infty \ln \frac{w(i\lambda)}{\lambda^2 - \gamma^2}\,\frac{u\,d\lambda}{\lambda^2 + u^2} \qquad (0 \leqq u < \infty),$$

$$\Phi_0(u) = \int_0^\infty e^{-ut}\phi_0(t)\,dt$$

and

$$r(t) = \int_a^\infty \frac{e^{-ut}\,d\sigma_*(u)}{(u^2 + \gamma^2)\Phi_0(u)} \qquad (0 \leqq t < \infty).$$

Any other solution ϕ *of equation* (7.0) *that satisfies the following condition for some* $h < a$

$$\phi(t) = o(e^{ht}) \quad \text{as} \quad t \longrightarrow \infty$$

differs from ϕ_0 *only by a scalar factor.*

5. Finally, consider the case $w(0) > 0$. Under this assumption the results of § 7 in [6] show that there exists the following resolvent for equation (7.30)

$$\Gamma(t,s) = \Gamma(|t - s|) + \int_0^{\min(t,s)} \Gamma(t - u)\Gamma(s - u)du$$

$(0 \leq s,t < \infty)$, where $\Gamma(t) = \Gamma(t,0) \in L_1(0,\infty)$. In our situation, one can assert, in addition, that $\Gamma(t)$ is an absolutely monotonic function whose Laplace transform plus one,

$$F_0(u) = 1 + \int_0^\infty e^{-ut}\Gamma(t)dt ,$$

is the unique solution, positive for $u \geq 0$, to equation (!), where the lower limit in the integral can be replaced by a.

Moreover, if one has $w(a - 0) \geq 0$, i.e., assertion 3) holds true, then $F_0(u)$ is also the unique solution, positive for $u \geq a$ (as well as for $u \geq 0$), to the equation

$$F_0(u) = 1 + \int_a^\infty \frac{d\sigma_*(t)}{F_0(t)(t + u)} , \tag{8.16}$$

and $\Gamma(t)$ is expressed in terms of $F_0(u)$ via the formula

$$\Gamma(t) = \int_a^\infty \frac{e^{-tu}}{F_0(u)} d\sigma_*(u) \quad (0 \leq t < \infty) .$$

Therefore, if $w(a - 0) \geq 0$, then

$$\int_a^\infty e^{at}\Gamma(t)dt = \int_a^\infty \frac{d\sigma_*(u)}{(u - a)F_0(u)} = F_0(-a + 0) - 1 > 0 .$$

Since $w^{-1}(z) = F_0(z)F_0(-z)$, we get

$$F_0(-a + 0)F_0(a) = w^{-1}(a - 0) ; \tag{8.17}$$

whence $F_0(-a + 0) < \infty$ if and only if $w(a - 0) > 0$.

On the other hand, we see from the relation $w(i\lambda) = 1 - K(\lambda)$ that

$$w(a - 0) = 1 - \int_{-\infty}^\infty e^{-at}k(t)dt . \tag{8.18}$$

From the previous section it follows that

$$\int_0^\infty e^{ht}\Gamma(t)dt < \infty \quad (0 < h < a) ,$$

while

$$\int_0^\infty e^{at}\Gamma(t)dt < \infty$$

if and only if

$$\int_{-\infty}^\infty e^{-at}k(t)dt < 1 \ . \tag{8.19}$$

This statement may be obtained from the results of § 15 in [6], and in fact under less restrictive assumptions concerning the function $k(t)$.

According to the general propositions of the same § 15 [6], the homogeneous equation (7.0) has no solution $\phi \not\equiv 0$ satisfying the condition

$$\phi(t) = o(e^{ht}), \quad t \longrightarrow \infty$$

for any $h < a$, and the same holds true for $h = a$ too provided that (8.19) is fulfilled. As for the nonhomogeneous equation (7.30), Theorem 14.1 of [6] applies.

6. A more complicated and interesting situation arises when $w(0) > 0$ but $w(a - 0) < 0$. In this case, assertion 4) holds for the function $w^{-1}(z)$. Accordingly, representation (8.15) is replaced by

$$F_0(u) = 1 + \frac{\rho}{2\kappa F_0(\kappa)(u + \kappa)} + \int_a^\infty \frac{e^{-tu}d\sigma_*(u)}{F_0(u)} \tag{8.20}$$

($u \in \operatorname{Ext}(-\infty,-a]$), and so

$$\Gamma(t) = \frac{\rho}{2\kappa F_0(\kappa)} e^{-\kappa t} + \int_a^\infty \frac{e^{-tu}d\sigma_*(u)}{F_0(u)} \qquad (0 \leq t < \infty) \ .$$

Let us show that the absolutely monotonic function

$$R(t) = \int_a^\infty \frac{e^{-tu}d\sigma_*(u)}{F_0(u)}$$

satisfies the condition

$$\int_0^\infty e^{at}R(t)dt \leq \frac{\rho}{2\kappa F_0(\kappa)(a - \kappa)} - 1 \ , \tag{8.21}$$

where equality is attained when and only when

$$\int_{-\infty}^{\infty} e^{-at} k(t)\,dt = \infty \,. \tag{8.22}$$

Indeed, the difference between the the left and right-hand sides in (8.21) is, according to (8.20), equal to $F_0(-a+0)$. If (8.22) is fulfilled, we conclude from (8.17) and (8.18) that $F_0(-a+0) = 0$, and we have equality in (8.21). If, however, (8.22) is not satisfied, then $w(a-0) > -\infty$, and relation (8.17) implies that $F_0(-a+0) < 0$, which gives strict inequality in (8.21).

Recalling the general results of § 4 , one can claim that in the case under consideration, the function $F_0(u)$, which is the unique positive (and regular) for $u \geqq 0$ solution to equation (!), is *not* the unique, positive (and regular) for $u \geqq a$, solution to this equation. In addition to $F_0(u)$, a positive (and regular) for $u \geqq a$, solution to equation (!) is given by the function

$$F_1(u) = \frac{u+\kappa}{u-\kappa} F_0(u) \,.$$

We are interested in the function

$$\Phi_0(u) = \frac{1}{u-\kappa} F_0(u) \,,$$

which is, in a certain sense, similar to the function denoted also by Φ_0, and considered in § 7. It follows from (8.20) that

$$F_0(u) - F_0(\kappa) = \frac{\rho}{2\kappa F_0(\kappa)} \left[\frac{1}{u+\kappa} - \frac{1}{2\kappa}\right] - $$

$$- (u-\kappa) \int_a^{\infty} \frac{d\sigma_*(t)}{(t+\kappa)F_0(t)(t+u)} \,.$$

Thus, we get, after dividing by $u-\kappa$, that

$$\Phi_0(u) = \frac{F_0(u)}{u-\kappa} - \frac{\rho}{4\kappa^2 F_0(\kappa)(u+\kappa)} - \int_a^{\infty} \frac{d\sigma_*(t)}{(t+\kappa)F_0(u)(t+u)}$$

Therefore,

$$\Phi_0(u) = \int_0^{\infty} e^{-ut} \phi(t)\,dt \quad \text{for} \quad u > \kappa,$$

where

$$\phi(t) = F_0(\kappa)e^{\kappa t} - \frac{\rho}{4\kappa^2 F_0(\kappa)} e^{-\kappa t} - \int_a^\infty \frac{e^{-tu} d\sigma_*(u)}{(u+\kappa)F_0(u)} \quad (8.23)$$

Alternatively, $\phi(t)$ is the solution of the differential system

$$\phi' - \kappa\phi = \Gamma, \quad \phi(0) = 1 .$$

This is a consequence of (8.20).

Proceeding from the results of § 15 , no. 6 in [6], we reach the following conclusion.

THEOREM 8.3. *Let* $w(0) > 0$ *and* $w(a - 0) < 0$, *i.e.*

$$2\int_0^\infty k(t)dt < 1, \quad 2\int_0^\infty ch(at)\, k(t)\, dt > 1 .$$

Then equation (7.0) *has the solution*

$$\phi_0(t) = A\, sh(\kappa t + \alpha) - r(t) \quad (0 \leq t < \infty) , \quad (8.24)$$

where κ *is the unique positive root of the equation*

$$2\int_0^\infty ch(xt)\, k(t)\, dt = 1 ,$$

$$A = \frac{\rho}{2\kappa} = w'(\kappa), \quad e^\alpha = \frac{2\kappa F_0(\kappa)}{w'(\kappa)}$$

and $r(t)$ $(0 \leq t < \infty)$ *is a bounded, absolutely monotonic function having exponent* $\geq a$. *More precisely,*

$$r(t) = \int_a^\infty \frac{e^{-tu} d\sigma_*(u)}{(u+\kappa)F_0(u)} \quad (0 \leq t < \infty) ,$$

where the function $F_0(u)$ *is computed according to* (4.6), *and is itself expressed in terms of* ϕ_0 *via*

$$F_0(u) = (u - \kappa)\int_0^\infty e^{-ut}\phi_0(t)dt \quad (u \geq \kappa) .$$

Let $h < a$ *if* (8.22) *holds, and* $h = a$ *in the opposite case. Then any solution* ϕ *of equation* (7.0) *satisfying*

$$\phi(t) = o(e^{ht}) \quad as \quad t \longrightarrow \infty$$

may differ from ϕ_0 *only by a scalar factor.*

PROOF. In fact, it is easy to see that the equalities (8.23) and (8.24) give one and the same fu tion ϕ_0. The fact that ϕ_0 satisfies equation (7.0) follows from the general propositions of paper [6] (see Theorem 9.2). Moreover, the same Theorem 9.2 in [6] implies the last assertion of our theorem too.

To conclude, let us mention that the results of this section are fully applicable to the Hvol'son - Milne integral equation, in which the absolutely monotonic function k(t) is defined by

$$k(t) = \frac{\omega}{2} \int_1^\infty e^{-|t|u} \, d(\ln u) \, .$$

REFERENCES

1. Ahiezer, N. I. and Glazman, I. M.: *Theory of Linear Operators in Hilbert space*, "Nauka", Moscow, 1966; English transl., Ungar, New York, 1961.

2. Bernshtein, S. N.: *Sur les fonctions absolument monotones*, Acta Math., 52 (1928), 1-66.

3. Chandrasekhar, S.: *Radiative Transfer*, Oxford Univ. Press, New York, 1950.

4. Hopf, E.: *Mathematical Problems of Radiative Equilibrium*, Cambridge Univ. Press, London, 1933.

5. Kac, I. S. and Krein, M. G.: *R-functions - analytic functions which map the upper half plane into itself*, Supplement I added to the Russian translation of the book by F. V. Atkinson, Discrete and Continuous Boundary Value Problems, "Mir", Moscow, 1968, 629-647; English transl., Amer. Math. Soc. Transl. (2) 103 (1974), 1-18.

6 Krein, M. G.: *Integral equations on a half-line with kernel depending upon the difference of the arguments*, Uspekhi. Mat. Nauk 13, No.5 (1958), 3-120; English transl., Amer. Math. Soc. Transl. (2) 22 (1962), 163-288.

7. Krein, M. G. and Nudel'man, A.A.: *The Markov Moment Problem and Extremal Problems*, "Nauka", Moscow, 1973; English transl. Transl. Math. Monographs, vol. 50, Amer. Math. Soc., Providence, R. I., 1977.

8. Sobolev, V. V.: *Transfer of Radiant Energy in the Atmosphere of Stars and Planets*, Gosudarstv. Izdat. Tekhn. Teor. Lit., Moscow, 1956. (Russian)

ON A PAIR INTEGRAL EQUATION AND ITS TRANSPOSE*

I. C. Gohberg and M. G. Krein

In the present paper we study an integral equation of the form

$$\begin{aligned} \phi(t) - \int_{-\infty}^{\infty} k_1(t - s)\phi(s)ds &= f(t) \quad (-\infty < t < 0), \\ \phi(t) - \int_{-\infty}^{\infty} k_2(t - s)\phi(s)ds &= f(t) \quad (0 < t < \infty), \end{aligned} \tag{!}$$

where $k_1(t)$ and $k_2(t)$ are arbitrary functions belonging to the space $L = L_1(-\infty,\infty)$.

The pair integral equation (!) may be rewritten as a single equation

$$\phi(t) - \int_{-\infty}^{\infty} k_\pi(t,s)\phi(s)ds = f(t) \quad (-\infty < t < \infty), \tag{!}$$

where the kernel $k_\pi(t,s)$ is defined by the equalities

$$k_\pi(t,s) = \begin{cases} k_1(t - s) & (-\infty < t < 0,\ -\infty < s < \infty), \\ k_2(t - s) & (0 < t < \infty,\ -\infty < s < \infty). \end{cases}$$

Apparently, the integral equation (!) was considered in a sufficiently general setting for the first time by I. M. Rapoport [11], who discovered that the problem of finding the solutions of (!) may be reduced to a certain Riemann-Hilbert problem on the line. At the same time, a first method for constructing the solutions to equation (!) in a sufficiently general case was found.

*Translation of Teoreticheskaya i Prikladnaya Matematika, No. 1 (1958), 58-81.

However, aside from the fact that this method for constructing the solution turned out to be far from the simplest one, I. M. Rapoport did not succeed, proceeding from it, to make a complete investigation of the integral equation (!) (explicit conditions for the solvability of equation (!) were not obtained, the structure of its resolvent, the role of the transposed equation, and other aspects, were not clarified).

To a certain extent, this may be explained by the fact that I. M. Rapoport considered equation (!) in the space $L_2(-\infty,\infty)$ (i.e., he required that both the right-hand side $f(t)$ and the solution $\phi(t)$ belong to L_2) under the assumption that the functions $k_{1,2}(t)$ belong to L. In this general setting, he was led to a Riemann-Hilbert problem on the line which, as he himself noticed, has not been studied before (because the coefficients belonged to a rather large class of functions not satisfying, in general, Hölder's condition).

Nevertheless, by taking a different direction, one does succed in giving a complete analysis of the integral equation (!) not only in the class L_2, but also in several more classes: L_p $(p \geqq 1)$, M, C among others, and this is the goal of the present paper.

This paper is based on the methods of M. G. Krein which were developed and applied to the Wiener-Hopf integral equation

$$g(t) - \int_0^{\infty} k(t - s)g(s)ds = h(t) \quad (0 < t < \infty), \tag{0.1}$$

and are discussed in his paper [9], where a short historical review of the work on equation (0.1) is to be found too.

As could be expected, the theory of the integral equation (!) turned out to be closely related to the theory of the transposed equation

$$\psi(t) - \int_{-\infty}^{\infty} k_{\pi}(s,t)\psi(s)ds = f(t), \qquad (!^{\tau})$$

which can be writen in a more explicit way as

$$\psi(t) - \int_{-\infty}^{0} k_1(s - t)\psi(s)ds - \int_0^{\infty} k_2(s - t)\psi(s)ds = f(t). \qquad (!^{\tau})$$

Let us mention that both equations (!) and $(!^{\tau})$ in the class L_2 were studied by Yu. I. Cherskii [1,2], under the assumption that the Fourier transforms $K_1(\lambda)$ and $K_2(\lambda)$ of the functions $k_1(t)$ and $k_2(t)$ satisfy Hölder conditions on the extended real line. This author obtained a number of essential additions to I. M. Rapoport's work. However, these are all included in the more complete and more precise results of the present paper.

We should add that equations (!) and $(!^{\tau})$ have been studied also by Yu. I Cherskii and F. D. Gakhov [3,4] under special assumptions concerning the way the functions $k_1(t)$ and $k_2(t)$ decrease or increase as $t \longrightarrow \pm\infty$, and under corresponding assumptions on the behavior of the functions ϕ, ψ, and f as $t \longrightarrow \pm\infty$. Their investigations are not related to the present work.

The results of V. A. Fok [5] and M. G. Krein [9] concerning the Wiener-Hopf equation (0.1) with an "exponentially decreasing" kernel (when $e^{h|t|}k(t) \in L$ for some $h > 0$) will be generalized to the equations (!) and $(!^{\tau})$ in a future publication. Also, we intend to develop there the theory of the integral equation $(!^{\tau})$ for the particular setting encountered in problems concerning the density variation of monoenergetic neutrons in two half spaces separated by a planar boundary (see, for example, the work of G. F. Bat' and D. A. Zaretskii). In this part of our work we shall also suggest some possible improvements, following from our results, to the investigations by Yu. I. Cherskii and F. D. Gakhov.

Finally, we point out that the recent work of the authors [8] allows us to generalize fully (at least in the existence proofs) the present results to the case when the "kernels" $k_1(t)$ and $k_2(t)$ are in the equations (!) and $(!^{\tau})$ some square, n-th order matrix-functions having elements from L, while ϕ, ψ, and f are n-dimensional vector-functions.

§ 1. AUXILLIARY PROPOSITIONS

1. The investigation of the pair equation (!) is closely tied to that of the Wiener-Hopf equation (0.1). As we mentioned earlier, the latter is discussed in detail in paper [9]. Since we have to deal in the following with various results from this paper, we discuss them briefly now.

Let L denote the space of all complex-valued, measurable and summable functions $f(t)$ $(-\infty < t < \infty)$, for which the integral

$$\|f\|_L = \int_{-\infty}^{\infty} |f(t)|\, dt$$

is finite. We take this integral to be the norm in L.

We shall denote by $L_+^{(p)}$ $(p \geq 1)$ the space $L_p(0,\infty)$ of complex-valued functions $f(t)$ $(0 \leq t < \infty)$, summable to the p-th power. As it is known, the conjugate of $L_+ = L_+^{(1)}$ is the space M_+ of bounded, complex-valued functions $f(t)$ $(0 \leq t < \infty)$, equipped with the norm

$$\|f\|_M = \operatorname*{ess\,sup}_{0<t<\infty} |f(t)| .$$

Let M_+^c and M_+^u be the subspaces of M_+ consisting of all continuous and, respectively, uniformly continuous functions. Finally, let C_+ denote the subspace of M_+^c consisting of all functions for which the limit

$$f(\infty) = \lim_{t\to\infty} f(t)$$

exists, and let C_+^0 $(\subset C_+)$ be the subspace of all functions for which this limit is zero: $f(\infty) = 0$.

From now on, we write E_+ when referring to one of the spaces

$$L_+^{(p)} \quad (p \geq 1), \quad M_+, \quad M_+^c, \quad M_+^u, \quad C_+, \quad C_+^0 .$$

To simplify our discussion, let us agree that a function $f(t)$, defined on the entire real axis t $(-\infty < t < \infty)$, belongs to E_+ if $f(t) \in E_+$ on the positive semiaxis $(0 \leq t < \infty)$, while $f(t)$ vanishes identically on the negative semiaxis $(-\infty < t < 0)$.

2. We shall denote by W the ring (see [9,6], where the notation R is used) of all functions $\Phi(\lambda)$ $(-\infty < \lambda < \infty)$ of the form

$$\Phi(\lambda) = c + \int_{-\infty}^{\infty} \phi(t)e^{i\lambda t}dt ,$$

where c is a constant and $\phi(t)$ is an arbitrary function in L. It is obvious that all functions $\Phi(\lambda) \in W$ are continuous on the extended real line and that $\Phi(\pm\infty) = c$. The notation $W_{\pm}$ refers to the subrings of W consisting respectively of all functions $\Phi(\lambda)$ $(-\infty < \lambda < \infty)$ of the form

$$\Phi(\lambda) = c + \int_{0}^{\infty} \phi(\pm t)e^{\pm i\lambda t}dt \quad (\phi \in L) . \tag{1.1}$$

It is clear that if $\Phi \in W_+$ (resp. W_-), then formula (1.1) provides a definition of this function for all complex numbers in the upper (resp. lower) half plane. Following this definition, $\Phi(\lambda)$ becomes continuous in the closed upper (resp. lower) half plane, and holomorphic in the open upper (resp. lower) half plane $\mathrm{Im}\,\lambda > 0$ (resp. $\mathrm{Im}\,\lambda > 0$). Consider the operator $\mathbb{P}_+$ which projects the ring W onto the subring W_+ according to the rule

$$\mathbb{P}_+(c + \int_{-\infty}^{\infty} \phi(t)e^{i\lambda t}dt) = c + \int_{0}^{\infty} \phi(t)e^{i\lambda t}dt \quad (\phi \in L).$$

It is plain that $\mathbb{P}_+$ may be defined analytically by

$$\mathbb{P}_+(\Phi(\lambda)) = \Phi(\infty) + \frac{1}{2\pi i}\int_{-\infty}^{\infty} \frac{\Phi(\mu) - \Phi(\lambda)}{\mu - \lambda}\, d\mu \quad (\mathrm{Im}\,\lambda > 0).$$

An essential role in the study of the pair equation (!) is played by the factorization problem for functions from W. That is to say, we wish to represent a given function $G(\lambda) \in W$ as a product

$$G(\lambda) = G_+(\lambda)G_-(\lambda) \quad (-\infty < \lambda < \infty) , \tag{1.2}$$

where G_+ (resp. G_-) is a function continuous in the closed upper (resp. lower) half plane, and holomorphic in the open upper (resp. lower) half plane. A factorization (1.2) is called *canonical* if each of the factors $G_{\pm}(\lambda)$ does not vanish in the corresponding closed half plane, and *proper* if at least one of the factors $G_{\pm}(\lambda)$ does not vanish in the corresponding closed half plane.

Let the function $k(t) \in L$ have the property

$$1 - K(\lambda) \neq 0 \quad (-\infty < \lambda < \infty) , \tag{1.3}$$

where

$$K(\lambda) = \int_{-\infty}^{\infty} k(t)e^{i\lambda t}dt .$$

Then, according to Wiener's theorem (see [9,6]), the function $[1 - K(\lambda)]^{-1}$, together with $1 - K(\lambda)$, belongs to the ring W. Denote by ν the index of the function $[1 -K(\lambda)]^{-1}$, i.e.,

$$\nu = - \operatorname{ind}[1 - K(\lambda)] = - \frac{1}{2\pi} \int_{-\infty}^{\infty} d_{\lambda}\arg[1 - K(\lambda)] .$$

We are interested in the problem of factoring the function $[1 - K(\lambda)]^{-1}$, and this is discusses in detail in [9], the paper to which we have referred.

When factoring the function $[1 - K(\lambda)]^{-1}$ it is natural to normalize the factors $G_{\pm}(\lambda)$ by imposing the condition

$$G_{+}(\infty) = 1 .$$

If condition (1.3) is satisfied, then $[1 - K(\lambda)]^{-1}$ always admits a proper factorization

$$[1 - K(\lambda)]^{-1} = G_{+}(\lambda)G_{-}(\lambda) \quad (-\infty < \lambda < \infty) , \tag{1.4}$$

and the factors $G_{\pm}(\lambda)$ belong to the rings $W_{\pm}$, respectively.

When $\nu \neq 0$, the proper factorization (1.4) is not unique. If $\nu > 0$, then $G_{+}(\lambda) \neq 0$ $(\operatorname{Im} \lambda \geqq 0)$, while if $\nu < 0$, then $G_{-}(\lambda) \neq 0$ $(\operatorname{Im} \lambda \leqq 0)$.

If $\nu = 0$, then the proper factorization (1.4) is unique and becomes a canonical factorization.

3. In any space E_{+} a function $k \in L$ generates a bounded linear operator

$$\mathbb{K}g(t) = \int_{-\infty}^{\infty} k(t - s)g(s)ds \quad (0 \leqq t < \infty) ,$$

and

$$||\mathbb{K}||_{E_+} \leqq ||k||_L .$$

Provided the condition (1.3) is satisfied, we say, following [9], that the *index* of equation (0.1) is the number

$$\nu = - \operatorname{ind}[1 - K(\lambda)].$$

If $\nu = 0$, then equation (0.1) has a unique solution $g(t) \in E_+$ for any right-hand side $h(t) \in E_+$, this solution being

$$g(t) = h(t) + \int_0^{\infty} \gamma(t,s)h(s)ds \quad (0 \leqq t < \infty) . \qquad (1.5)$$

Here

$$\gamma(t,s) = \gamma_1(t - s) + \gamma_2(t - s) + \int_0^{\min(t,s)} \gamma_1(t - u)\gamma_2(s - u)du ,$$

and the functions $\gamma_j(t)$ $(-\infty < t < \infty;\ j = 1,2)$ belong to L_+ and are defined by the equalities

$$1 + \int_0^{\infty} \gamma_1(t)e^{i\lambda t}dt = \exp(-\int_0^{\infty} \ell(t)e^{i\lambda t}dt) ,$$

$$1 + \int_0^{\infty} \gamma_2(t)e^{i\lambda t}dt = \exp(-\int_{-}^{\infty} \ell(t)e^{i\lambda t}dt) ,$$

with

$$\int_{-\infty}^{\infty} \ell(t)e^{i\lambda t}dt = \ln(1 - K(\lambda)) .$$

Setting

$$G_+(\lambda) = 1 + \int_0^{\infty} \gamma_1(t)e^{i\lambda t}dt ,$$

$$G_-(\lambda) = 1 + \int_0^{\infty} \gamma_2(t)e^{-i\lambda t}dt , \qquad (1.6)$$

equality (1.4) will be obviously fulfilled, and it will represent the canonical factorization of the function $[1 - K(\lambda)]^{-1}$. The functions $\gamma_1(t)$ and $\gamma_2(t)$ may be defined by this property.

The factors $G_\pm(\lambda)$ can be derived from the formulas

$$\ln G_+(\lambda) = -\frac{1}{2\pi i}\int_{-\infty}^{\infty}\frac{\ln(1-K(\mu))}{\mu-\lambda}\,d\mu \qquad (\mathrm{Im}\ \lambda > 0)\ ,$$

$$\ln G_-(\lambda) = \frac{1}{2\pi i}\int_{-\infty}^{\infty}\frac{\ln(1-K(\mu))}{\mu-\lambda}\,d\mu \qquad (\mathrm{Im}\ \lambda < 0)\ .$$

4. Before we formulate the properties of equation (0.1) for $\nu > 0$, we introduce the notion of a D_+-*chain* of functions, taken from [9].

An ordered system of functions $\chi_0, \chi_1, \ldots, \chi_{\nu-1}$ belonging to the space L_+ is called a D_+-*chain* if:

1) all the functions χ_j $(j = 0,1,\ldots, -1)$ are absolutely continuous;

2) $\chi_{j+1}(t) = \frac{d}{dt}\chi_j(t),\quad \chi_j(0) = 0 \quad (j = 0,1,\ldots,\nu-2)$;

3) $\chi_{\nu-1}(0) = 1,\quad \frac{d}{dt}\chi_{\nu-1} \in L_+$.

Conditions 1), 2) and 3) are equivalent to the following ones:

$$\chi_j(t) = \int_0^t \chi_{j+1}(s)ds \qquad (j = 0,1,\ldots,\nu-2),$$

$$\chi_{\nu-1}(t) = 1 + \int_0^t \chi_\nu(s)ds,$$

where $\chi_\nu(t)$ is some function from L_+.

Clearly, the Laplace transforms of the functions in a D_+-chain are related via the relations

$$X_{j+1}(\lambda) = -i\lambda X_j(\lambda) \qquad (j = 0,1,\ldots,\nu-2)\ .$$

In particular, the linear independence of the functions forming a D_+-chain follows from these relations.

Notice also that the functions of a D_+-chain belong to all the spaces E_+.

If $\nu > 0$, then equation (0.1) has solutions in E_+ for any right-hand side $h(t) \in E_+$. One of these solutions can be derived from formula (1.5), where the $\gamma_j(t) \in L_+$ $(j = 1,2)$

are defined by the equalities (1.6), the $G_{\pm}(\lambda)$ appearing there being the factors of some proper factorization of the function $[1 - K(\lambda)]^{-1}$.

In this case the homogeneous equation

$$\chi(t) - \int_0^\infty k(t - s)\chi(s)ds = 0 \quad (0 < t < \infty) \tag{1.7}$$

has exactly ν linearly independent solutions in any space E_+. The set of all solutions to (1.7) does not depend upon the particular choice of the space E_+, and has a basis forming a D_+-chain.

5. When $\nu < 0$, the homogeneous equation (1.7) has in each space E_+ a unique solution - the null one. The nonhomogeneous equation (0.1) is solvable if and only if the right-hand side $h(t) \in E_+$ satisfies the condition

$$\int_0^\infty h(t)\omega(t)dt = 0 , \tag{1.8}$$

where $\omega(t) \in E_+$ is any solution to the transposed homogeneous equation

$$\omega(t) - \int_0^\infty k(s - t)\omega(s)ds = 0 .$$

Whenever condition (1.8) is satisfied, the solution of (0.1) is given by the formula (1.5), where the functions $G_{\pm}(\lambda)$ yield a proper factorization (1.4) of $[1 - K(\lambda)]^{-1}$.

From this discussion it follows, in particular, that all the values ζ for which

$$\zeta - K(\lambda) \neq 0 \quad (-\infty < \lambda < \infty), \quad \text{and} \quad \nu_\zeta = -\mathrm{ind}(\zeta - K(\lambda)) \neq 0 ,$$

belong to the spectrum of the operator $\mathbb{K}$ (these points of the spectrum are Φ-points with d-characteristic of the form $(\nu_\zeta, 0)$ or $(0, |\nu_\zeta|)$).

It turns out that in addition to the points we just mentioned, the spectrum of operator $\mathbb{K}$ consists of all the points on the curve $\zeta = K(\lambda)$ $(-\infty < \lambda < \infty)$. These are neither Φ-points nor $\Phi_{\pm}$-points of the operator $\mathbb{K}$.

[Here we adhere to the terminology that we used in our paper [7].

Let us recall the required definitions from [7]. A linear operator A acting in a Banach space B is said to be *normally solvable* if the condition $l(y) = 0$, where $l \in B^\tau$ is any solution to the homogeneous equation $A^\tau l = 0$ is necessary and sufficient for the solvability of the equation $Ax = y$. Here B^τ denotes the conjugate space of B, and A^τ - the operator conjugate (transpose) to A.

Denote by α_ζ and β_ζ the dimensions of the spaces of solutions of the equations $(A - \zeta I)x = 0$ and $(A - \zeta I)^\tau l = 0$, respectively. A number ζ is called a Φ-*point* of the operator A if the operator $A - \zeta I$ is normally solvable and the numbers α_ζ and β_ζ are finite. If the operator $A - \zeta I$ is normally solvable, but only one of the numbers α_ζ and β_ζ is finite, then ζ is called a $\Phi_\pm$-*point* (depending upon which of these numbers is finite). The pair of points $(\alpha_\zeta, \beta_\zeta)$ is called the d-*characteristic* of the operator A at the point ζ. The set of all Φ-points of the operator A is called its Φ-*set*.]

§ 2. THE PAIR INTEGRAL EQUATION AND ITS TRANSPOSE

1. From now on E will denote the space of all complex-valued functions $f(t)$ $(-\infty < t < \infty)$ such that both $f(t)$ and $f(-t)$ $(0 < t < \infty)$ belong to E_+.

Depending upon whether E is the space $L^{(p)} = L_p(-\infty,\infty)$ or a subspace of the space $M(-\infty,\infty)$, the norm in E is given by one of the equalities

$$\|f\|_p = \left(\int_{-\infty}^{\infty} |f(t)|^p dt\right)^{1/p},$$

$$\|f\|_M = \operatorname*{ess\,sup}_{-\infty<t<\infty} |f(t)| .$$

Let $k_1(t)$ and $k_2(t)$ be two functions from L. The equality

$$\mathbb{K}_\pi \phi(t) = \int_{-\infty}^{\infty} k_\pi(t,s)\phi(s)ds , \tag{2.1}$$

where

$$k_\pi(t,s) = \begin{cases} k_1(t - s) & (-\infty < t < 0;\ -\infty < s < \infty) , \\ k_2(t - s) & (0 < t < \infty;\ -\infty < s < \infty) , \end{cases} \tag{2.2}$$

defines a bounded linear operator acting in each of the spaces E, and whose norm satisfies

$$\|\mathbb{K}_\pi\|_E \leqq \|k_1\|_L + \|k_2\|_L . \qquad (2.3)$$

We should mention that in all spaces E, except $L^{(p)}$ $(p > 1)$, the convergence of the integrals

$$\int_{-\infty}^{\infty} k_j(t - s)\phi(s)ds \qquad (j = 1,2) \qquad (2.4)$$

is understood in the usual sense. In $L^{(p)}$ $(p > 1)$, integral (2.4) should be understood as the limit in the $L^{(p)}$-metric of the integral

$$\int_{-N}^{N} k_j(t - s)\phi(s)ds \qquad (j = 1,2)$$

as N tends to infinity:

$$\int_{-\infty}^{\infty} k_j(t - s)\phi(s)ds = \underset{N\to\infty}{\text{l.i.m.}} \int_{-N}^{N} k_j(t - s)\phi(s)ds .$$

The conjugate or, more precisely, the transpose of the operator $\mathbb{K}_\pi$ acting in the space L, will be the operator

$$\mathbb{K}_\pi^\tau \phi(t) = \int_{-\infty}^{\infty} k_\pi(s,t)\phi(s)ds ,$$

acting in M. In a more detailed way,

$$\mathbb{K}_\pi^\tau \phi(t) = \int_{-\infty}^{0} k_1(s - t)\phi(s)ds + \int_0^{\infty} k_2(s - t)\phi(s)ds . \qquad (2.5)$$

Operator $\mathbb{K}_\pi^\tau$ is bounded and linear not only in M, but also in any space E, and its norm satisfies the same bound (2.3).

Given the operator $\mathbb{K}_\pi^\tau$ in the space L, then its transpose will be the operator $\mathbb{K}_\pi$.

Therefore, the pair integral equation (!) and the integral equation

$$\psi(t) - \int_{-\infty}^{0} k_1(s - t)\psi(s)ds - \int_0^{\infty} k_2(s - t)\psi(s)ds = f(t) \qquad (!^\tau)$$

are each the transpose of the other.

It is suitable to study the pair equation (!) and its transpose $(!^\tau)$ in parallel with the Wiener-Hopf equation

$$g(t) - \int_0^{\infty} k(t - s)g(s)ds = h(t) \qquad (2.6)$$

$(0 < t < \infty;\ g,h \in E_+)$ having a specially chosen kernel $k(t - s)$.

Before writting down the expression for $k(t)$, let us impose to the functions $k_j(t)$ $(j = 1,2)$, besides the condition $k_j(t) \in L$ $(j = 1,2)$, the additional natural restrictions

$$1 - K_j(\lambda) = 1 - \int_{-\infty}^{\infty} k_j(t)e^{i\lambda t}dt \neq 0 \tag{2.7}$$

$(-\infty < \lambda < \infty;\ j = 1,2)$. Then, in virtue of a well-known theorem of N. Wiener [9, § 1], function $[1 - K_1(\lambda)]^{-1}[K_2(\lambda) - K_1(\lambda)]$ will be the Fourier transform of a function from L, and this is precisely the function we denote by $k(t)$, i.e.,

$$\int_{-\infty}^{\infty} k(t)e^{i\lambda t}dt = [1 - K_1(\lambda)]^{-1}[K_2(\lambda) - K_1(\lambda)] . \tag{2.8}$$

We call the number

$$\nu = -\operatorname{ind}(1 - K(\lambda)) = -\operatorname{ind}\left(\frac{1 - K_2(\lambda)}{1 - K_1(\lambda)}\right) , \tag{2.9}$$

which is the index of equation (2.6), the *index of equation* (!). We shall explain below why this is a natural definition.

The same index will be attached to the equation

$$\phi(t) - \int_{-\infty}^{0} k_1(t - s)\phi(s)ds - \int_0^{\infty} k_2(t - s)\phi(s)ds = f(t) .$$

Thus, if the pair integral equation (!) has index ν, then the transposed equation (!) will have index $\nu^{\tau} = -\nu$.

2. Let $a(t,s)$ $(-\infty < t,s < \infty)$ be some kernel inducing in space E a bounded linear operator

$$\mathbb{A}\phi(t) = \int_{-\infty}^{\infty} a(t,s)\phi(s)ds.$$

We shall say that the kernel $b(t,s)$ $(-\infty < t,s < \infty)$ is a *right resolvent* of the kernel $a(t,s)$, if $b(t,s)$ induces in E a bounded linear operator

$$\mathbb{B}\phi(t) = \int_{-\infty}^{\infty} b(t,s)\phi(s)ds ,$$

and given any function $f(t) \in E$, the function

$$g(t) = f(t) + \int_{-\infty}^{\infty} b(t,s)f(s)ds \tag{2.10}$$

is one of the solutions of the equation

$$g(t) - \int_{-\infty}^{\infty} a(t,s)g(s)ds = f(t) \; . \tag{2.11}$$

Similarly, the kernel $b(t,s)$ will be referred to as a *left resolvent* of the kernel $a(t,s)$ if $a(t,s)$ is a right resolvent of $b(t,s)$.

In other words, the kernel $b(t,s)$ will be a left resolvent for the kernel $a(t,s)$ provided that each time equation (2.11) is solvable for some $f \in E$, its solution $g \in E$ is unique and is given by formula (2.10). Notice too that if $b(t,s)$ is simultaneously a left and a right resolvent of the kernel $a(t,s)$, then it represents the full resolvent of $a(t,s)$, and so $I + \mathbb{B} = (I - \mathbb{A})^{-1}$.

§ 3. THE PAIR AND TRANSPOSED INTEGRAL EQUATIONS WITH ZERO INDEX

We begin by investigating the pair integral equation (!) in the space L under the assumption that $\nu = 0$.

First suppose that the pair equation has a solution $\phi \in L$. Consider the functions

$$b_+(t) = - f(t) + \phi(t) - \int_{-\infty}^{\infty} k_1(t - s)\phi(s)ds \quad (0 < t < \infty),$$

$$b_+(t) = 0 \quad (-\infty < t < 0);$$

$$b_-(t) = - f(t) + \phi(t) - \int_{-\infty}^{\infty} k_2(t - s)\phi(s)ds \quad (-\infty < t < 0),$$

$$b_-(t) = 0 \quad (0 < t < \infty).$$

Then $\phi(t)$ will be the solution of the following system of two equations

$$\begin{cases} \phi(t) - \displaystyle\int_{-\infty}^{\infty} k_1(t - s)\phi(s)ds = f(t) + b_+(t), \\ \phi(t) - \displaystyle\int_{-\infty}^{\infty} k_2(t - s)\phi(s)ds = f(t) + b_-(t). \end{cases} \quad (-\infty < t < \infty)$$

Taking the Fourier transform of these two equations, we get

$$(1 - K_1(\lambda))\Phi(\lambda) = F(\lambda) + B_+(\lambda) ,$$
$$(1 - K_2(\lambda))\Phi(\lambda) = F(\lambda) + B_-(\lambda) , \tag{3.1}$$

where Φ, F, and $B_\pm$ are the Fourier transforms of the functions ϕ, f, and $b_\pm$, respectively, and $B_\pm \in W_\pm$.

Therefore, the solution Φ of the system (3.1) is given by the formula

$$\Phi = \frac{F + B_+}{1 - K_1} = \frac{F + B_-}{1 - K_2} . \tag{3.2}$$

According to (2.8) and (3.2), function B_+ satisfies the equation

$$(1 - K)B_+ - B_- = KF . \tag{3.3}$$

Under our assumptions, $[1 - K(\lambda)]^{-1}$ admits a canonical factorization [9, § 2]

$$[1 - K(\lambda)]^{-1} = G_+(\lambda)G_-(\lambda) \qquad (-\infty < \lambda < \infty) .$$

It follows that equation (3.3) can be rewritten in the form

$$G_+^{-1}B_+ - G_-B_- = G_-KF .$$

Applying the projection $\mathbb{P}_+$ (see § 1, no.1) here to both sides, we get

$$G_+^{-1}B_+ = \mathbb{P}_+(G_-KF) ,$$

because $G_-B_- \in W_-$ and $B_-(\infty) = 0$. Thus

$$B_+ = G_+ \mathbb{P}_+(G_-KF) .$$

To summarize, if the pair integral equation (!) has a solution in L, then this solution is unique and its Fourier transform can be expressed in terms of the known functions as

$$\Phi = [1 - K_1]^{-1}[F + G_+ \mathbb{P}_+(G_-KF)] . \tag{3.4}$$

Conversely, let $f(t)$ be any function belonging to L, and define Φ by (3.4). Denoting the function $G_+ \mathbb{P}_+(G_-KF)$ $(\in W_+)$ by B_+, one has

$$(1 - K_1)\Phi = F + B_+ \ .$$

Considering Φ and B_+ as Fourier transforms of some functions $\phi(t) \in L$ and $b_+(t) \in L$, we find that ϕ satisfies the equation

$$\phi(t) - \int_{-\infty}^{\infty} k_1(t - s)\phi(s)ds = f(t) + b_+(t) \qquad (-\infty < t < \infty).$$

Taking into account that $b_+(t)$ vanishes for all negative t, we get

$$\phi(t) - \int_{-\infty}^{\infty} k_1(t - s)\phi(s)ds = f(t) \qquad (-\infty < t < 0) \ .$$

Furthermore, from the equality

$$B_+ = G_+ \mathbb{P}_+(G_-KF)$$

it follows that the difference

$$G_-^{-1}G_+^{-1}B_+ - KF$$

is some function from W_- which vanishes at infinity. Denoting it by B_-, we get

$$(1 - K)B_+ = KF + B_- \ ,$$

or

$$(1 - K_2)(B_+ + F) = (1 - K_1)(B_- + F) \ .$$

Consequently,

$$(1 - K_2)\Phi = F + B_- \ .$$

The last relation implies that ϕ is in L and also satisfies the second equation

$$\phi(t) - \int_{-\infty}^{\infty} k_2(t - s)\phi(s)ds = f(t) \qquad (0 \leqq t < \infty) \ .$$

Therefore, given any right-hand side $f \in L$, equation

(!) has a unique solution, whose Fourier transform is given by the formula (3.4).

It results from (3.4) that the Fourier transform of the solution to the pair equation (!) can be alternatively represented as

$$\Phi = (1 - K_1)^{-1}(F + H),$$

where $H(\lambda)$ is the Fourier transform of a function $h(t) \in L_+$ which satisfies the equation

$$h(t) - \int_0^\infty k(t - s)h(s)ds = \int_{-\infty}^\infty k(t - s)f(s)ds \quad (0 < t < \infty)$$

and vanishes for $t < 0$. Thus, the solution ϕ of the pair equation (!) for the case $\nu = 0$ can be written as

$$\phi(t) = f(t) + h(t) + \int_{-\infty}^\infty k_{-1}(t - s)f(s)ds + \int_0^\infty k_{-1}(t - s)h(s)ds$$

where $k_{-1}(t)$ is a function belonging to L and whose Fourier transform is given by

$$\int_{-\infty}^\infty k_{-1}(t)e^{i\lambda t}dt = - K_1(\lambda)[1 - K_1(\lambda)]^{-1}.$$

Denoting by $\gamma(t,s)$ the resolvent of the kernel $k(t-s)$ in E_+, we obtain the final formula

$$\phi(t) = f(t) + \int_{-\infty}^\infty \gamma_\pi(t,s)f(s)ds , \tag{3.5}$$

where

$$\gamma_\pi(t,s) = k(t - s) + k_{-1}(t - s) + \int_0^\infty k_{-1}(t-r)k(r-s)dr + \int_0^\infty \gamma(t,r)k(r-s)dr + \int_0^\infty\int_0^\infty k_{-1}(t-r)\gamma(r,u)k(u-s)dudr . \tag{3.6}$$

It is easy to show that the equalities

$$\begin{aligned} \gamma_\pi(t,s) - \int_{-\infty}^\infty k_\pi(t,r)\gamma_\pi(r,s)dr &= k_\pi(t,s), \\ \gamma_\pi(t,s) - \int_{-\infty}^\infty \gamma_\pi(t,r)k_\pi(r,s)dr &= k_\pi(t,s), \end{aligned} \tag{3.7}$$

hold for all s $(-\infty < s < \infty)$ and almost all t $(-\infty < t < \infty)$.

The kernel $\gamma_\pi(t,s)$ is a sum of kernels and composition of kernels, each of them generating a bounded linear operator not only in L, but also in any space E. Therefore, $\gamma_\pi(t,s)$ generates in any space E a bounded linear operator

$$\Gamma_\pi\phi(t) = \int_{-\infty}^{\infty} \gamma_\pi(t,s)\phi(s)ds \ .$$

Equalities (3.7) state that the operator $I + \mathbb{K}_{\_}$ has, in each space E, a bounded inverse, which is precisely the operator $I + \Gamma_\pi$:

$$(I + \mathbb{K}_{\_})^{-1} = I + \Gamma_\pi .$$

Therefore, $\gamma_\pi(t,s)$ is the resolvent of the kernel $k_\pi(t,s)$ in any space E.

We have reached the following theorem.

THEOREM 1. *Suppose that the functions* $k_j(t) \in L$ $(j=1,2)$ *satisfy condition* (2.7) *and that the index of the pair integral equation* (!) *is equal to zero* $(\nu = 0)$. *Then, given any right-hand side* $f \in E$, *equation* (!) *has one and only one solution* $\phi \in E$. *This solution is given by formula* (3.5), *and so the kernel* $\gamma_\pi(t,s)$ *defined by* (3.6) *is the resolvent of the pair kernel* $k_\pi(t,s)$.

Notice that ν always equals zero provided the functions $k_j(t)$ $(j = 1,2)$ are even, because in this case the functions $K_j(\lambda)$ $(j = 1,2)$ are even too.

The same assertion can be made whenever $k_j(t)$ $(j = 1,2)$ are Hermitian functions, i.e., $k_j(-t) = \overline{k_j(t)}$ $(j = 1,2)$.

Observing that equation $(!^\tau)$ is the transpose of the pair equation (!), and recalling the properties of the kernel $\gamma_\pi(t,s)$, the next result follows easily.

THEOREM 2. *If the conditions of Theorem 1 are fulfilled, then for any right-hand side* $f \in E$, *the transposed equation* $(!^\tau)$ *has one and only one solution* $\psi \in E$. *This solution is given by the formula*

$$\psi(t) = f(t) + \int_{-\infty}^{\infty} \gamma_\pi(s,t)f(s)ds \ . \qquad (3.8)$$

As we already mentioned, Theorem 2 is a consequence of the properties of the kernel $\gamma_\pi(t,s)$ and of Theorem 1. In what follows it is essential that Theorem 2 may be proved without appealing to Theorem 1. This can indeed be done using the same arguments as in the proof of Theorem 1, applied in the same way. We sketch this proof, restricting ourselves to the case $E = L$.

Assuming that ψ and f belong to the space L, and taking the Fourier transform of both sides in $(!^\tau)$, we see that equation $(!^\tau)$ is equivalent to the following one:

$$(1 - K_2(-\lambda))\Psi_+(\lambda) + (1 - K_1(-\lambda))\Psi_-(\lambda) = F(\lambda) , \qquad (3.9)$$

where

$$\Psi_+(\lambda) = \int_0^\infty \psi(t)e^{i\lambda t}dt \quad \text{and} \quad \Psi_-(\lambda) = \int_{-\infty}^0 \psi(t)e^{i\lambda t}dt .$$

To solve equation (3.9), we rewrite it as

$$[1 - K(-\lambda)]\Psi_+(\lambda) + \Psi_-(\lambda) = [1 - K_1(-\lambda)]^{-1}F(\lambda) . \qquad (3.10)$$

Since the assumptions of Theorem 2 imply that $[1 - K(\lambda)]^{-1}$ admits a canonical factorization

$$[1 - K(-\lambda)]^{-1} = G_+(-\lambda)G_-(-\lambda) = G_+^\tau(\lambda)G_-^\tau(\lambda) ,$$

where

$$G_+^\tau(\lambda) = G_-(-\lambda) \quad \text{and} \quad G_-^\tau(\lambda) = G_+(-\lambda) ,$$

we obtain

$$(G_+^\tau)^{-1}\Psi_+ + G_-^\tau\Psi_- = G_-^\tau[1 - K_1(-\lambda)]^{-1}F.$$

Applying the projection $\mathbb{P}_+$, we find Ψ_+. Subsequently, (3.10) shows that for any right-hand side $f \in L$, the Fourier transform $\Psi = \Psi_+ + \Psi_-$ of the solution to equation $(!^\tau)$ has the form

$$\Psi(\lambda) = [1 - K_1(-\lambda)]^{-1}F + K(-\lambda)G_+^\tau\mathbb{P}_+[G_-^\tau(1 - K_1(-\lambda))^{-1}F] .$$

Finally, the solution of equation $(!^\tau)$ is given by the formula

$$\psi(t) = f(t) + \int_{-\infty}^{\infty} \gamma_\pi^\tau(t,s)f(s)ds \quad (-\infty < t < \infty), \qquad (3.11)$$

where $\gamma_\pi^\tau(t,s) = \gamma_\pi(s,t)$, as can be easily seen.

To conclude this section, let us mention that if follows from the theorems proved above that each of the homogeneous equations

$$\phi(t) - \int_{-\infty}^{\infty} k_\pi(t,s)\phi(s)ds = 0 \quad (-\infty < t < \infty) , \qquad (+)$$

and

$$\psi(t) - \int_{-\infty}^{\infty} k_\pi(s,t)\psi(s)ds = 0 \quad (-\infty < t < \infty) \qquad (+^\tau)$$

has, in the case $\nu = 0$, a unique solution in any space E, namely, the null solution.

§ 4. THE INTEGRAL EQUATION WITH POSITIVE INDEX

1. There is very little to add to the arguments of the previous section in order to establish the following statement.

THEOREM 3. *Let the functions* $k_j(t) \in L$ $(j = 1,2)$ *satisfy condition* (2.7) *and let the index of the pair equation* (!) *be positive* $(\nu > 0)$. *Then given any right-hand side* $f \in E$, *the function* $\phi(t) \in E$ *defined by formula* (3.5) *is one of the solutions to* (!). *Therefore, the kernel* $\gamma_\pi(t,s)$ *given by* (3.6) *is a right resolvent of the pair kernel* $k_\pi(t,s)$.

PROOF. Under the assumptions of the theorem, $[1 - K(\lambda)]^{-1}$ admits a proper factorization (1.4), whose factors $G_\pm(\lambda)$ belong to the rings $W_\pm$, respectively, and such that $G_+(\lambda) \neq 0$ $(\text{Im}\ \lambda \geqq 0)$.

Given any function $f \in L$, one can easily see that the function $\Phi(\lambda) \in W$ defined by (3.4) is the Fourier transform of some solution $\phi \in L$ to the pair equation (!).

Thus, formula (3.5), where $\gamma_{1,2}(t) \in L_+$ are given by the equalities (1.6), provides a particular solution to (!). As in the case $\nu = 0$, it is not hard to persuade oneself that the last assertion remains valid when the space L is replaced by any of the spaces E.

Thus, if $\nu > 0$, the kernel $\gamma_\pi(t,s)$ is a right resolvent of the kernel $k_\pi(t,s)$.

A similar analysis of the second proof of Theorem 2 leads to the following result.

THEOREM 4. *Let the functions* $k_j(t) \in L$ $(j = 1,2)$ *satisfy condition* (2.7) *and let the index of the integral equation* $(!^\tau)$ *be positive* $(\nu^\tau = -\nu > 0)$. *Then given any right-hand side* $f \in E$, *the function* $\psi \in E$ *defined by formula* (3.11) *is one of the solutions to equation* $(!^\tau)$. *Therefore, the kernel* $\gamma_\pi^\tau(t,s) = \gamma_\pi(s,t)$ *defined by equality* (3.6) *is a right resolvent of the kernel* $k_\pi^\tau(t,s) = k_\pi(s,t)$.

2. Now we study the homogeneous pair equation (+) or, equivalently, the equations

$$\begin{cases} \phi(t) - \int_{-\infty}^{\infty} k_1(t-s)\phi(s)ds = 0 & (-\infty < t < 0), \\ \phi(t) - \int_{-\infty}^{\infty} k_2(t-s)\phi(s)ds = 0 & (0 < t < \infty), \end{cases} \tag{+}$$

in the space E.

This study is facilitated by establishing the connection between the solutions to (+) and those to the homogeneous Wiener-Hopf equation

$$\chi(t) - \int_0^{\infty} k(t-s)\chi(s)ds = 0 \quad (0 < t < \infty) . \tag{4.1}$$

Let $\chi \in E_+$ be an arbitrary solution to (4.1). Then the function

$$\phi(t) = \chi(t) + \int_{-\infty}^{\infty} k_1(t-s)\chi(s)ds \tag{4.2}$$

belongs to E and satisfies equation (+). In fact,

$$\phi(t) - \int_{-\infty}^{\infty} k_j(t-s)\phi(s)ds = \chi(t) - \int_{-\infty}^{\infty} \ell_j(t-s)\chi(s)ds$$

$(j = 1,2)$, where

$$\ell_j(t) = -k_1(t) + k_j(t) - \int_{-\infty}^{\infty} k_j(t-s)k_{-1}(s)ds \quad (j=1,2).$$

The last relations imply nothing more and nothing less than the fact that $\phi(t)$ is a solution to the homogeneous equation (+), because it is simple to check that

$$\ell_1(t) = 0 \quad \text{and} \quad \ell_2(t) = k(t) \quad (-\infty < t < \infty).$$

The converse statement may be similarly verified: if $\phi \in E$ is a solution to equation (+), then the function

$$\chi(t) = \phi(t) - \int_{-\infty}^{\infty} k_1(t-s)\phi(s)ds$$

belongs to E_+ and satisfies the homogeneous equation (4.1).

It follows from this relation between the solutions to equation (4.1) and (+) that equation (+) has the same solutions in all spaces E. When equation (+) has positive index ν, its solutions form a ν-dimensional subspace.

The situation is similar in the case of the homogeneous transposed equation

$$\psi(t) - \int_{-\infty}^{0} k_1(t-s)\psi(s)ds - \int_0^{\infty} k_2(t-s)\psi(s)ds = 0. \quad (+^{\tau})$$

Namely, let the function $\omega(t) \in E_+$ be a solution to the homogeneous equation

$$\omega(t) - \int_0^{\infty} k(s-t)\omega(s)ds = 0 . \tag{4.3}$$

Then the function $\psi(t)$ defined by

$$\psi(t) = \int_0^{\infty} k(s-t)\omega(s)ds \tag{4.4}$$

or, which is the same, by

$$\psi(t) = \begin{cases} \omega(t) & (0 < t < \infty) , \\ \int_0^{\infty} k(t-s)\omega(s)ds & (-\infty < t < 0) , \end{cases}$$

belongs to E and satisfies equation $(+^{\tau})$. Indeed, by inserting the function $\psi(t)$ defined in (4.4) into the left-hand side of equation $(+^{\tau})$, we get

$$\psi(t) - \int_{-\infty}^{0} k_1(s-t)\psi(s)ds - \int_{0}^{\infty} k_2(s-t)\psi(s)ds =$$

$$= \psi(t) - \int_{-\infty}^{\infty} k_1(s-t)\psi(s)ds -$$

$$- \int_{-\infty}^{\infty} [k_2(s-t) - k_1(s-t)]\,\omega(s)ds = \int_{-\infty}^{\infty} m(s-t)\omega(s)ds ,$$

where

$$m(t) = k(t) + k_1(t) - k_2(t) - \int_{-\infty}^{\infty} k_1(t-s)k(s)ds .$$

Taking the Fourier transform of the function $m(t)$ and recalling (2.8), we see that $m(t) = 0$.

It is not difficult to prove the converse statement too: if $\psi(t) \in E$ is a solution to equation $(+^{\tau})$, then the function $\omega(t) = \psi(t)$ $(0 < t < \infty)$ satisfies equation (4.3).

In order to make a more detailed investigation of the linear manifold of all solutions to equations (+) and $(+^{\tau})$ in E, we give the following definitions.

An ordered system of functions $\psi_0, \psi_1, \ldots, \psi_{\nu-1}$ $(\in L)$ will be called as D^{τ}-*chain* if

$$\psi_j(t) = \int_{-\infty}^{t} \psi_{j+1}(s)ds \qquad (j = 0,1,\ldots,\nu-1) ,$$

where $\psi_{\nu}(t)$ is some function from L.

The elements of a D^{τ}-chain are linearly independent, because their Fourier transforms are related by

$$\Psi_{j+1}(\lambda) = -i\lambda\Psi_j(\lambda) \qquad (j = 0,1,\ldots,\nu-2) . \tag{4.5}$$

An ordered system of functions $\phi_0, \phi_1, \ldots, \phi_{\nu-1}$ $(\in L)$ will be called a D-*chain* if:

1) the functions $\phi_0, \phi_1, \ldots, \phi_{\nu-2}$ are absolutely continuous, and so is the function $\phi_{\nu-1} - \eta$, where $\eta(t) = (1 + \text{sign } t)/2$;

2) $\phi_{j+1} = \frac{d}{dt}\phi_j$ $\quad (j = 0,1,\ldots,\nu-2)$, and

$$\frac{d}{dt}(\phi_{\nu-1} - \eta) \in L .$$

Obviously, conditions 1) and 2) could be replaced by the following ones:

$$\phi_j(t) = \int_{-\infty}^{t} \phi_{j-1}(s)ds \qquad (j = 0,1,\ldots,\nu-2),$$

and

$$\phi_{\nu-1}(t) = \eta(t) + \int_{-\infty}^{t} \phi_\nu(s)ds ,$$

where $\phi_\nu(t)$ is some function from L.

The relations (4.5) hold for the functions of a D-chain too, and this proves their linear independence.

The following connection exists between the D-chains of both types introduced above and the D_+-chains (see § 1 , no.4).

If the system of functions $\chi_0,\chi_1,\ldots,\chi_{\nu-1}$ *is a* D_+*-chain, and* $k_0(t) \in L$ *is arbitrary, then the functions*

$$\psi_j(t) = \int_0^{\infty} k_0(t - s)\chi_j(s)ds \quad (j = 0,1,\ldots,\nu-1) \tag{4.6}$$

form a D^τ*-chain, while the functions*

$$\phi_j(t) = \chi_j(t) + \int_0^{\infty} k_0(t - s)\chi_j(s)ds \quad (j=0,1,\ldots,\nu-1) \tag{4.7}$$

form a D*-chain.*

Indeed, from (4.6) it follows that

$$\int_{-\infty}^{t} \psi_{j+1}(s)ds = \int_0^{\infty} \ell(t - s)\chi_{j+1}(s)ds =$$

$$= \int_0^{\infty} k_0(t - s)\chi_j(s)ds = \psi_j(s)$$

$(j = 0,1,\ldots,\nu-2)$, where

$$\ell(t) = \int_{-\infty}^{t} k_0(s)ds .$$

Denoting by $\chi_\nu \in L_+$ the function for which the equality

$$\chi_{\nu-1} = \eta(t) + \int_0^{t} \chi_\nu(s)ds$$

holds, and by $\psi_\nu \in L$ - the function

$$\psi_\nu(t) = \int_0^\infty k_0(t - s)\chi_\nu(s)ds + k_0(t) ,$$

we get

$$\int_{-\infty}^t \psi_\nu(s)ds = \int_0^\infty \ell(t - s)\chi_\nu(s)ds + \int_{-\infty}^t k_0(s)ds =$$

$$= \int_0^\infty k_0(t - s)\chi_{\nu-1}(s)ds = \psi_{\nu-1}(t) .$$

Thus, the system of functions (4.6) is a D^τ-chain. To verify that (4.7) is a D-chain, we note that

$$\phi_j = \psi_j + \chi_j \quad (j = 0,1,\ldots,\nu-1) ,$$

whence

$$\int_{-\infty}^t \phi_{j+1}(s)ds = \int_{-\infty}^t \psi_{j+1}(s)ds + \int_{-\infty}^t \chi_{j+1}(s)ds =$$

$$= \psi_j(t) + \chi_j(t) = \phi_j(t) \quad (j = 0,1,\ldots,\nu-2),$$

and

$$\int_{-\infty}^t [\psi_\nu(s) + \chi_\nu(s)]ds = \psi_{\nu-1}(t) + \chi_{\nu-1}(s) - \eta(t) =$$

$$= \phi_{\nu-1}(t) - \eta(t) .$$

Summarizing the discussion above, and taking into account the properties of the solutions to equation (1.7) formulated earlier, we reach the following conclusions.

THEOREM 5. *Let the conditions of Theorem 3 be fulfilled. Then the homogeneous pair equation* (+) *has the same solutions in all the spaces* E. *These solutions form a* ν*-dimensional subspace having a* D-*chain as its basis.*

We add that the D-chain appearing in the theorem can be obtained from the corresponding D_+-chain of solutions to equation (4.1) via formula (4.2).

THEOREM 6. *Let the conditions of Theorem 3 be fulfilled. Then the homogeneous equation* $(+^\tau)$ *has the same solutions in all the spaces* E. *These solutions form a* ν^τ*-dimensional space having* *a D^{τ}-chain as its basis.*

The D^{τ}-chain mentioned here is obtained from the corresponding D_{+}-chain of solutions to equation (4.3) via formula (4.4).

§ 5. THE INTEGRAL EQUATIONS WITH NEGATIVE INDEX

Consider the homogeneous pair equation (+) under the assumption that its index is negative ($\nu < 0$) and, as above, the functions $k_j(t) \in L$ $(j = 1,2)$ satisfy conditions (2.7).

On the basis of the above relation between the solutions of the equation (+) and (4.1) (see § 4, no.2), we conclude that, in the case under consideration, the homogeneous equation (+) has the unique, null solution in any of the spaces E. Consequently, if the nonhomogeneous equation (!) is solvable for some right-hand side $f \in E$, then its solution is unique.

Similar considerations lead us to the same type of conclusions concerning the solutions of an equation $(+^{\tau})$ having negative index ν^{τ}.

THEOREM 7. *Let the functions* $k_j(t) \in L$ $(j = 1,2)$ *have property* (2.7), *and let the index of equation* (!) *be negative. Then for the pair equation* (!) *to be solvable in* E *it is necessary and sufficient that the function* $f \in E$ *satisfy the condition*

$$\int_{-\infty}^{\infty} f(t)\psi(t)dt = 0 \tag{5.1}$$

for all solutions $\psi(t)$ *of equation* $(+^{\tau})$. *If condition* (5.1) *is satisfied, then the solution to equation* (!) *is unique and is given by formula* (3.5). *Thus, the kernel* $\gamma_{\pi}(t,s)$ *defined by the equality* (3.6) *is a left resolvent of the pair kernel* $k_{\pi}(t,s)$.

PROOF. In the space E consider the equation transpose to (!):

$$g - \mathbb{K}_{\pi}^{\tau} g = f$$

where, as earlier,

$$\mathbb{K}_{\pi}^{\tau}g(t) = \int_{-\infty}^{\infty} k_{\pi}(s,t)g(s)ds .$$

Under the assumptions of the theorem, the index ν^{τ} $(= -\nu)$ of equation $(!^{\tau})$ is positive, and so the conditions for applying Theorem 4 are satisfied. This means that the kernel $k_{\pi}^{\tau}(t,s)$ has a right resolvent $\gamma_{\pi}^{\tau}(t,s) = \gamma_{\pi}(s,t)$.

Let us define the operator Γ^{τ} acting in the space E according to the rule

$$\Gamma^{\tau}f(t) = \int_{-\infty}^{\infty} \gamma_{\pi}(s,t)f(s)ds .$$

It is clear that the function

$$g = f + \Gamma^{\tau}f \quad (\in E)$$

satisfies the equation $(!^{\tau})$, for any right-hand side $f \in E$.

Now consider the equation

$$g - \mathbb{K}_{\pi}^{\tau}g = (I - \mathbb{K}_{\pi}^{\tau})f ,$$

whose solutions are the functions

$$g_1 = f \quad \text{and} \quad g_2 = (I + \Gamma^{\tau})(I - \mathbb{K}_{\pi}^{\tau})f .$$

The difference $g_2 - g_1$ is a solution of the homogeneous equation $(+^{\tau})$. By Theorem 6, equation $(+^{\tau})$ has exactly $|\nu|$ linearly independent solutions. Denoting these solutions by $\psi_1, \psi_2, \ldots, \psi_{|\nu|}$, we get

$$g_2 - g_1 = \sum_{j=1}^{|\nu|} c_j\psi_j ,$$

or

$$(I + \Gamma^{\tau})(I - \mathbb{K}_{\pi}^{\tau})f = f + \sum_{j=1}^{|\nu|} c_j(f)\psi_j \quad (f \in E) . \tag{5.2}$$

Let $\mathbb{C}$ denote the finite rank operator $(I + \Gamma^{\tau})(I - \mathbb{K}_{\pi}^{\tau}) - I$, and choose in the intersection of all the spaces E some arbitrary functions $\omega_j'(t)$ $(j = 1,2,\ldots,|\nu|)$ such that

$$\int_{-\infty}^{\infty} \omega_j'(t)\psi_k(t)dt = \delta_{jk} \quad (j,k = 1,2,\ldots,|\nu|) .$$

Multiplying both sides of (5.2) by ω_j' $(j = 1,2,\ldots,|\nu|)$ and integrating them from $-\infty$ to ∞, we get

$$c_j(f) = \int_{-\infty}^{\infty} (\mathbb{C}f)(t)\omega_j'(t)dt = \int_{-\infty}^{\infty} f(t)\omega_j(t)dt ,$$

where

$$\omega_j(t) = \mathbb{C}^\tau\omega_j' = [(I - \mathbb{K}_\pi)(I + \mathbb{\Gamma}) - I]\omega_j'$$

$$(j = 1,2,\ldots,|\nu|) .$$

Obviously, the functions ω_j $(j = 1,2,\ldots,|\nu|)$ belong to the intersection of all the spaces E.

Therefore, the operator $\mathbb{C}$ is given in any space E by

$$\mathbb{C}f = \sum_{j=1}^{|\nu|} \psi_j(t) \int_{-\infty}^{\infty} f(s)\omega_j(s)ds .$$

This means that the operator $\mathbb{C}^\tau$ can be written as

$$\mathbb{C}^\tau f = \sum_{j=1}^{|\nu|} \omega_j(t) \int_{-\infty}^{\infty} f(s)\psi_j(s)ds .$$

Comparing the different representations of $\mathbb{C}^\tau$, we see that the equality

$$(I - \mathbb{K}_\pi)(I + \mathbb{\Gamma})f = f + \sum_{j=1}^{|\nu|} \omega_j(t) \int_{-\infty}^{\infty} f(s)\psi_j(s)ds$$

holds for all $f \in E$. Thus, (5.1) is a sufficient condition for the solvability of equation (!). The necessity of (5.1) is plain. Moreover, the discussion above shows that the kernel $\gamma_\pi(t,s)$ is a left resolvent of the pair kernel $k_\pi(t,s)$.

Using the same method, on proves the following result.

THEOREM 8. *Let the functions* $k_j(t) \in L$ $(j = 1,2)$ *satisfy* (2.7), *and let the index of equation* $(!^\tau)$ *be negative* $(\nu^\tau = -\nu < 0)$. *Then in order that equation* $(!^\tau)$ *have a solution in* E *it is necessary and sufficient that the function* $f \in E$ *satisfy the condition*

$$\int_{-\infty}^{\infty} f(t)\phi(t)dt = 0 \tag{5.3}$$

for all solutions $\phi(t)$ *of equation* (+). *If* (5.3) *is satisfied, then the solution of equation* $(!^\tau)$ *is unique and is given by formula* (3.11). *The kernel* $\gamma_\pi^\tau(t,s) = \gamma_\pi(s,t)$ *defined by equality* (3.6) *is a left resolvent of the kernel* $k_\pi^\tau(t,s) = k_\pi(s,t)$.

§ 6. THE SPECTRUM OF THE OPERATOR $\mathbb{K}_\pi$

First of all, let us remark that the theorems already proved imply that the spectrum of the operator $\mathbb{K}_\pi$ contains all the complex numbers ζ such that

$$\zeta - K_1(\lambda) \neq 0 \ , \qquad \zeta - K_2(\lambda) \neq 0 \quad (-\infty < \lambda < \infty), \tag{6.1}$$

and

$$\nu_\zeta = - \operatorname{ind} \frac{\zeta - K_2(\lambda)}{\zeta - K_1(\lambda)} \neq 0 \ . \tag{6.2}$$

The points ζ of the spectrum enjoying properties (6.1) and (6.2) are, as we have proved, Φ-points of the operator $\mathbb{K}_\pi$. The d-characteristic of the operator $\mathbb{K}_\pi$ at these points has the form $(\nu_\zeta, 0)$ or $(0, |\nu_\zeta|)$, depending upon whether the index ν_ζ is positive or negative. So we have to look only at the points of the curves

$$\zeta = K_1(\lambda) \quad \text{and} \quad \zeta = K_2(\lambda) \qquad (-\infty < \lambda < \infty) \ .$$

Using certain arguments that we have already applied in [7, § 10], we show that not only are the points of the curves $\zeta = K_j(\lambda)$ $(j = 1,2)$ not regular points of the operator $\mathbb{K}_\pi$, but, in addition, they are neither Φ-points nor $\Phi_\pm$-points of this operator.

Consider the space E_+^{II} of two-dimensional vector-functions $f(t) = \{f_1(t), f_2(t)\}$ $(0 < t < \infty)$ having coordinates $f_j(t) \in E_+$ $(j = 1,2)$, with the norm

$$||f|| = ||f_1||_{E_+} + ||f_2||_{E_+} \ .$$

The space E_+^{II} is equivalent to the space E under an isomorphism which takes each function $f \in E$ to a vector $f \in E_+^{II}$ according to the rule

$$f_1(t) = f(-t), \quad f_2(t) = f(t) \qquad (0 < t < \infty) .$$

This equivalence allows us to consider $\mathbb{K}_\pi$ to be an operator acting in E_+^{II}. Namely,

$$f = \mathbb{K}_\pi g \quad (f,g \in E_+^{II})$$

means that

$$\begin{cases} f_1(t) = \int_0^\infty k_1(s - t)g_1(s)ds + \int_0^\infty k_1(t + s)g_2(s)ds , \\ f_2(t) = \int_0^\infty k_2(t + s)g_1(s)ds + \int_0^\infty k_1(t - s)g_2(s)ds , \end{cases}$$

or, briefly,

$$f_1 = \mathbb{K}_{11}g_1 + \mathbb{K}_{12}g_2 ,$$

$$f_2 = \mathbb{K}_{21}g_1 + \mathbb{K}_{22}g_2 .$$

Thus

$$\mathbb{K}_\pi = \begin{pmatrix} \mathbb{K}_{11} & \mathbb{K}_{12} \\ \mathbb{K}_{21} & \mathbb{K}_{22} \end{pmatrix} .$$

Consider, along with $\mathbb{K}_\pi$, the operators $\mathbb{D}$ and $\mathbb{T}$ defined by the equalities

$$\mathbb{D} = \begin{pmatrix} \mathbb{K}_{11} & 0 \\ 0 & \mathbb{K}_{22} \end{pmatrix} \quad \text{and} \quad \mathbb{T} = \begin{pmatrix} 0 & \mathbb{K}_{12} \\ \mathbb{K}_{21} & 0 \end{pmatrix} .$$

The operators $\mathbb{K}_{12}$ and $\mathbb{K}_{21}$ acting in E_+ are compact. This fact is proved in [7, § 10, no. 3] for the case $E_+ = L_+$ and can be easily generalized to any space E_+. Consequently, $\mathbb{T}$ is a compact operator.

This shows (see [7, § 4 and § 8]) that the Φ-sets, as well as the $\Phi_\pm$-sets of the operators $\mathbb{K}$ and $\mathbb{D}$ coincide. The operator $\mathbb{D}$ splits into the direct sum of the two operators $\mathbb{K}_{11}$ and $\mathbb{K}_{22}$, each of them acting in a space equivalent to E_+. It follows that the $\Phi_\pm$-set of the operator $\mathbb{D}$ is the intersection of the $\Phi_\pm$-sets of the operators $\mathbb{K}_{11}$ and $\mathbb{K}_{22}$.

As is known (see [9]), the $\Phi_{\pm}$-sets of the operator $\mathbb{K}_{jj}$ $(j = 1,2)$ coincide with its Φ-set, and this contains all the points of the complex plane, except the zero point and the points of the curve $\zeta = K_j(\lambda)$ $(j = 1,2)$.

Therefore, the $\Phi_{\pm}$-sets of the operator $\mathbb{K}_{\pi}$ coincide with its Φ-set, and the latter consists of all the complex points except the points of the curves $\zeta = K_j(\lambda)$ $(j = 1,2)$ and the zero point.

We have reached the following conclusion.

THEOREM 9. *The spectrum of the operator* $\mathbb{K}_{\pi}$ *in the space* E *consists of the closure* S_{ζ} *of the set of all points of the curves*

$$\zeta = K_1(\lambda)\ , \quad \zeta = K_2(\lambda) \qquad (-\infty < \lambda < \infty) \tag{6.3}$$

and of the open set S_0 *of all points* $\zeta \notin S$ *for which* $\nu_{\zeta} \neq 0$ *(i.e., condition* (6.2) *is satisfied). Moreover,* S_0 *is the* Φ*-set of the operator* $\mathbb{K}_{\pi}$*, while the points* $\zeta \in S_{\zeta}$ *are neither* Φ*-, nor* $\Phi_{\pm}$*-points of* $\mathbb{K}_{\pi}$.

[Obviously, to obtain the closure S_{ζ} we simply add the one point $\zeta = 0$ to the points of the curves (6.3).]

The theorem remains valid if one replaces the operator $\mathbb{K}_{\pi}$ by its transposed $\mathbb{K}_{\pi}^{\tau}$.

§ 7. A REMARK CONCERNING THE DISCRETE ANALOGUES OF EQUATIONS (!) AND (!$^{\tau}$)

A discrete analogue of the pair integral equation is the infinite system of equations

$$\begin{cases} \sum\limits_{k=-\infty}^{\infty} a_{j-k}^{(1)}\xi_k = \eta_j & (j = -1,-2,\ldots)\ , \\ \sum\limits_{k=-\infty}^{\infty} a_{j-k}^{(2)}\xi_k = \eta_j & (j = 0,1,2,\ldots)\ . \end{cases} \tag{7.1}$$

It is known that the whole theory of the integral equation (0.1) carries over [9,8] to the system of linear equations

$$\sum_{k=0}^{\infty} a_{j-k}\xi_k = \eta_j \qquad (j = 0,1,2,\ldots),$$

and, in doing so, both the formulations and the proofs of the theorems become, in a certain sense, simpler. Similarly, the theory of the systems of equations (7.1) and of the corresponding transposed systems

$$\sum_{k=-\infty}^{-1} a_{k-j}^{(1)}\xi_k + \sum_{k=0}^{\infty} a_{k-j}^{(2)}\xi_k = \eta_j \qquad (j = 0,\pm 1,\pm 2,\ldots)$$

can be developed along the same lines as the above discussion of the pair integral equation (!) and its transpose $(!^{\tau})$, and again, by doing so, the theory acquires a simpler form.

In this approach, the conditions $k_{1,2}(t) \in L$ now become

$$\sum_{k=-\infty}^{\infty} |a_k^{(p)}| < \infty \qquad (p = 1,2) .$$

Further, the role of the spaces E and E_+ is now played by the corresponding spaces of sequences (E) and (E_+) (for the latter, see [9,§ 13]. Finally, conditions (2.7) are replaced by

$$a_p(e^{i\phi}) = \sum_{k=-\infty}^{\infty} a_k^{(p)} e^{ik\phi} \neq 0 \qquad (0 \leqq \phi \leqq 2\pi;\quad p = 1,2),$$

and the index ν is taken to be the increment of the argument of the function $a_2(\zeta)/a_1(\zeta)$ as ζ traces out the unit circle in the positive direction, divided by 2π.

REFERENCES

1. Cherskii, Yu. I.: *On certain singular integral equations*, Uchenye Zapiski Kazan. Gosud. Univ., 113, No. 10 (1953), 43-55. (Russian)

2. Cherskii, Yu. I.: *Integral equations of convolution type*, Auto-synopsis of Candidate Dissertation, Tbilisi, 1956. (Russian)

3. Cherskii, Yu. I. and Gakhov, F. D.: *Singular integral equations of convolution type and a planar, Riemann - type problem*, Uchenye Zapiski Kazan. Gosud. Univ., 114, No. 8 (1954), 21-33. (Russian)

4. Cherskii, Yu. I. and Gakhov, F. D.: *Singular integral equations of convolution type*, Izv. Akad. Nauk SSSR 20, No. 1 (1956), 33-52. (Russian)

5. Fok, V. A.: *On certain integral equations of mathematical physics*, Mat. Sb. 14 (56), No. 1-2 (1944), 3-50. (Russian)

6. Gel'fand, I. M., Raikov, D. A. and Shilov (Šilov), G. E.: *Commutative normed rings*, Uspekhi Mat. Nauk 1, No. 2 (1946), 48-146; English transl., Amer. Math. Soc. Transl. (2) 5 (1957), 115-220.

7. Gohberg, I. C. and Krein, M. G.: *The basic propositions on defect numbers, root numbers and indices of linear operators*, Uspekhi Mat. Nauk 12, No. 2 (1957), 43-118; English transl., Amer. Math. Soc. Transl. (2) 13 (1960), 185-264.

8. Gohberg, I. C. and Krein, M. G.: *Systems of integral equations on a half line with kernel depending on the difference of the arguments*, Uspekhi Mat. Nauk 13, No. 2 (1958), 3-72; English transl., Amer. Math. Soc. Transl. (2) 14 (1960), 217-287.

9. Krein, M. G.: *Integral equations on a half-line with kernel depending upon the difference of the arguments*, Uspekhi Mat. Nauk 13, No. 5 (1958), 3-120; English transl., Amer. Math. Soc. Transl. (2) 22 (1962), 163-288.

10. Rapoport, I. M.: *On a class of infinite systems of algebraic equations*, Dokl. Akad. Nauk Ukrain. SSR 3 (1948), 6-10 (Russian).

11. Rapoport, I. M.: *On certain "pair" integral and integro-differential equations*, Sb. Trudov Inst. Mat. Akad. Nauk Ukrain. SSR 12 (1949), 102-118. (Russian)

NEW INEQUALITIES FOR THE CHARACTERISTIC NUMBERS OF INTEGRAL EQUATIONS WITH SMOOTH KERNELS*

I. C. Gohberg and M. G. Krein

In [4] the following proposition was established (see also [3], pp. 119-123).

Let $\mu_1 \leqq \mu_2 \leqq \ldots$ be the complete system of characteristic numbers of the integral equation

$$\varphi(t) = \mu\int_a^b G(t,s)\varphi(s)\,ds$$

with continuous, Hermitian-nonnegative kernel $G(t,s)$. If the kernel $G(t,s)$ has continuous derivatives

$$G_{pq}(t,s) = \frac{\partial^{p+q}G(t,s)}{\partial t^p \partial s^q} \qquad (p,q = 1,2,\ldots,r), \tag{0.1}$$

then

$$\sum_{n=1}^{\infty} \frac{n^{2r}}{\mu_n} < \infty. \tag{0.2}$$

This is, in a certain sense, a precise estimate. Indeed, let $\mu_1 \leqq \mu_2 \leqq \ldots$ be arbitrary positive numbers such that the series (0.2) converges. Then, for example, the real, symmetric kernel

$$G(t,s) = \frac{2}{b-a}\sum_{n=1}^{\infty}\frac{1}{\mu_n}\sin\frac{\pi(2n-1)(t-a)}{2(b-a)}\sin\frac{\pi(2n-1)(s-a)}{2(b-a)} \tag{0.3}$$

---

*Translation of Matematicheskie Issledovaniya, Vol. 5, No. 1 (1970), 22-39.

has the μ_n-s as its characteristic numbers, and its partial derivatives (0.1) exist and are continuous.

The result given above has been developed and expanded in a whole series of investigations [2,8,9].

A natural question arises: how could series (0.2) be estimated in terms of the kernel $G(t,s)$ and its derivatives ?

In the present paper we shall answer this question. For example, we shall prove that whenever $G(a,a) = 0$ (or $G(b,b) = 0$), and $r = 1$, one has

$$\sum_{n=1}^{\infty} \frac{(2n-1)^2}{\mu_n} \leqq \frac{4(b-a)^2}{\pi^2} \int_a^b G_{11}(s,s)\,ds \,. \tag{0.4}$$

Obviously, here the equality is attained for the kernel $G(t,s)$ defined by (0.3). This partial result, as well as many more general ones involving weighted and higher-dimensional integral equations, are obtained in this paper following a simple application of certain inequalities for the eigenvalues of linear operators. In the case of matrices, these inequalities have a very simple form.

Namely, let G be a Hermitian, nonnegative, n-th order matrix, and let D be a matrix of the same order. Then the following relations hold true:

$$\sum_{j=1}^{k} \lambda_j(G)\lambda_{n-j+1}(D^*D) \leqq \sum_{j=1}^{k} \lambda_j(DGD^*) \quad (k=1,\dots,n). \tag{0.5}$$

[For an arbitrary Hermitian, n-th order matrix H, one denotes by $\lambda_1(H) \geqq \lambda_2(H) \geqq \dots \geqq \lambda_n(H)$ the complete collection of its eigenvalues.]

Relations (0.5) can be easily deduced from A. R. Amir-Moéz's inequalities [1] (see also [5]), using a lemma due to H. Weyl, G. Hardy, J. Littlewood, and G. Pólya (see [3], p. 37). Relations (0.5) may be considered as the "counterinequalities" of the inequalities

$$\sum_{j=1}^{k} \lambda_j(DGD^*) \leqq \sum_{j=1}^{k} \lambda_j(G)\lambda_j(D^*D) \qquad (k = 1,2,\dots,n),$$

which are themselves particular cases of the well-known inequalities of A. Horn (see, for example, [3], p. 55).

The bound (0.4) will be obtained as a direct consequence of the infinite dimensional generalizations of the relations (0.5) which, in turn, follow easily from the infinite dimensional generalizations of the A. R. Amir-Moéz's inequalities that were derived by A. S. Markus [6], who used the lemma due to Weyl, Hardy, Littlewood, and Pólya that we already mentioned.

The operator analogues of inequalities (0.5) have various applications. In particular, they allow to connect, by means of certain relations, the critical Eulerian forces in the stability problem for a continuum (a rod or a plate) with the eigenfrequencies of the transverse oscillations of the given continuum and with those of a second one, related to it (a rod and a string, or a plate and a membrane).

The authors are grateful to A. S. Markus for his valuable comments.

§1. A THEOREM ABOUT A LOWER BOUND FOR THE EIGENVALUES OF PRODUCTS OF OPERATORS

In this section we discuss the infinite dimensional generalizations of inequalities (0.5) and their more precise versions. Let X be a positive operator acting in the separable Hilbert space H. If the operator X^{-1} exists and is compact, then we shall denote by $\mu_1(X) \leq \mu_2(X) \leq \ldots$ the complete system of eigenvalues of the operator X. The numbers $\lambda_j(X^{-1}) = \mu_j(X)^{-1}$ $(j = 1,2,\ldots)$ form the complete system of eigenvalues for the operator X^{-1}, arranged in nonincreasing order. Accordingly, given any nonnegative compact operator Y, we write $\lambda_j(Y)$ for the eigenvalues of Y, arranged in nonincreasing order.

THEOREM 1. *Let* D *be a closed operator having a domain* Δ_D *dense in* H, *and let* G *be a nonnegative compact operator. Assume that the following conditions are satisfied:*

a) the operator D^{-1} *exists and is compact;*

b) $GH \subseteq \Delta_D$;

c) *the closure of the operator* DGD* *is compact.* *Then given any positive* p, *one has*

$$\sum_{j=1}^{k} \lambda_j^p(DGD^*) \geq \sum_{j=1}^{k} \mu_j^p(D^*D)\lambda_j^p(G), \quad (k= 1,2,\ldots) . \qquad (1.1)$$

[Here the operator D*D is defined via the expression $D^*D = \{D^{-1}(D^*)^{-1}\}^{-1}$.]

The proof of this theorem is based on the following two propositions.

I. *Let* A *and* B *be two nonnegative, compact operators acting in* H. *Then*

$$\prod_{k=1}^{m} \lambda_k(AB) \leq \prod_{k=1}^{m} \lambda_k(A)\lambda_{j_k}(B) \quad , \qquad (1.2)$$

where $j_1, j_2, \ldots, j_m$ *is an arbitrary collection of distinct natural numbers.*

This result was established by A. S. Markus [6].

II. *Let* $a_1 \geq a_2 \geq \ldots \geq a_m$ *and* $b_1 \geq b_2 \geq \ldots \geq b_m$ *be two systems of real numbers which satisfy the relations*

$$\sum_{j=1}^{k} a_j \leq \sum_{j=1}^{k} b_j \qquad (k = 1,2,\ldots,m) . \qquad (1.3)$$

Then for any convex function $\Phi(x)$ $(-\infty \leq x < \infty)$ *satisfying* $\Phi(-\infty) = 0$, *the following relations hold true:*

$$\sum_{j=1}^{k} \Phi(a_j) \leq \sum_{j=1}^{k} \Phi(b_j) \qquad (k = 1,2,\ldots,m) .$$

This result is due to H. Weyl, G. Hardy, J. Littlewood and G. Pólya (see [3], p. 37).

PROOF OF THEOREM 1. Consider the two compact operators $A = DGD^*$ and $B = D^{*-1}D^{-1}$. According to relations (1.2), one has

$$\prod_{k=1}^{m} \lambda_k(A) \geq \prod_{k=1}^{m} \{\lambda_{j_k}(AB)/\lambda_{j_k}(B)\} \quad , \qquad (1.4)$$

for any collection of distinct natural numbers $j_1, j_2, \ldots, j_m$.

Now let $j_1, j_2, \ldots, j_m$ be a permutation of the numbers $1,2,\ldots,m$, selected in such a way that the following inequalities be satisfied:

$$\frac{\lambda_{j_1}(AB)}{\lambda_{j_1}(B)} \geqq \frac{\lambda_{j_2}(AB)}{\lambda_{j_2}(B)} \geqq \frac{\lambda_{j_m}(AB)}{\lambda_{j_m}(B)} \quad . \tag{1.5}$$

Next, consider the systems of numbers $a_k = \ln \lambda_k(A)$ and $b_k = \ln(\lambda_{j_k}(AB)/\lambda_{j_k}(B))$ $(k = 1,2,\ldots,m)$. From (1.4) it results that these systems satisfy (1.3). Choosing $\Phi(x) = \exp(px)$ $(p > 0)$ and applying to these systems Proposition II, we obtain

$$\sum_{k=1}^{m} \lambda_k^p(A) \geqq \sum_{k=1}^{m} [\lambda_k^p(AB)/\lambda_k^p(B)] \quad (m = 1,2,\ldots) \; . \tag{1.6}$$

It is well known that for arbitrary compact operators X and Y the nonzero eigenvalues of the operators XY and YX coincide. Consequently,

$$\lambda_k(AB) = \lambda_k(D^{-1}AD^{*-1}) = \lambda_k(G) \quad (k = 1,2,\ldots).$$

Furthermore,

$$\lambda_k(B) = \lambda_k(D^{*-1}D^{-1}) = 1/\mu_k(DD^*) \quad (k = 1,2,\ldots)$$

The last equalities show that relations (1.6) and (1.1) are identical.

The theorem is proved.

It goes without saying that Theorem 1 remains valid when D is understood to be an operator acting from the space H into another Hilbert space H_1. Then the adjoint operator D^* will take H_1 into H, and the operators D^*D and DGD^* will act in H and H_1, respectively. This remark will be used in § 7 .

Remark. Analyzing the proof of Theorem 1, one easily sees that in fact we proved the following more general fact. Suppose the conditions of Theorem 1 are satisfied, and let $\Phi(t)$ $(0 \leqq t < \infty;\ \Phi(0) = 0)$ be a function which becomes convex after the change of variables $t = \exp x$ $(-\infty < x < \infty)$. Then one has the relations

$$\sum_{j=1}^{k} \Phi(\lambda_j(DGD^*)) \geqq \sum_{j=1}^{k} \Phi(\mu_j(D^*D)\lambda_j(G)) \quad (k = 1,2,\ldots) .$$

COROLLARY. *Suppose that the conditions of Theorem 1 are satisfied, and, in addition, the operator* DGD* *has a finite trace. Then*

$$\operatorname{sp}(DGD^*) \geqq \sum_{j=1}^{\infty} \mu_j(D^*D)\lambda_j(G) . \tag{1.7}$$

This is a consequence of relations (1.1) where one takes $p = 1$ and then one passes to the limit $n \longrightarrow \infty$.

Notice that the bound (0.4) given in the introduction can be deduced from relation (1.7). The latter, as well as relation (1.1) for $p = 1$, can be proven directly without appealing to propositions I and II. This is our next task.

§2. A DIRECT PROOF OF RELATION (1.7)

The proof is based on the following lemma.

LEMMA. *Let* B *be a positive definite operator such that the operator* B^{-1} *exists and is compact. Then the relations*

$$\sum_{j=1}^{k} (Be_j, e_j) \geqq \sum_{j=1}^{k} \mu_j(B) \quad (k = 1,2,\ldots) \tag{2.1}$$

hold true for any orthonormal system of vectors $e_1, e_2, \ldots, e_k$ $(k = 1,2,\ldots)$ *belonging to the domain* Δ_B *of the operator* B.

PROOF. Denote by P_k the orthogonal projection of the Hilbert space H onto the subspace spanned by $e_1, e_2, \ldots, e_k$. The operator P_kBP_k has rank at most k. Therefore

$$\operatorname{sp}(P_kBP_k) = \sum_{j=1}^{k} (Be_j, e_j) = \sum_{j=1}^{k} \lambda_j(P_kBP_k) . \tag{2.2}$$

In the subspace P_kH one can choose a subspace N_j of dimension j with the property that

$$\lambda_{k-j+1}(P_k B P_k) = \max_{\varphi \in N_j} \frac{(B\varphi,\varphi)}{(\varphi,\varphi)} \qquad (j = 1,2,\ldots,k).$$

On the other hand,

$$\mu_j(B) \leqq \max_{\varphi \in N_j} \frac{(B\varphi,\varphi)}{(\varphi,\varphi)} ,$$

by the well-known minimax properties of the eigenvalues. Thus

$$\mu_j(B) \leqq \lambda_{k-j+1}(P_k B P_k) \qquad (j = 1,2,\ldots,k). \tag{2.3}$$

Finally, (2.2) and (2.3) imply (2.1), and the proof of the lemma is complete.

Now we verify the relations (1.1) for $p = 1$. Let $e_1, e_2, \ldots$ be the system of eigenvectors of the operator G: $Ge_j = \lambda_j e_j$. Since the operator DGD^* is compact, so is the closure T of the operator $DG^{1/2}$. Now $\lambda_j(T^*T) = \lambda_j(TT^*)$ $(j = 1,2,\ldots)$ implies that

$$\lambda_j(DGD^*) = \lambda_j(G^{1/2}D^*DG^{1/2}) \geqq \lambda_j(F_k) ,$$

and

$$\lambda_j(F_k) = 0 \quad (j = k+1,k+2,\ldots),$$

where $F_k = P_k G^{1/2} D^* D G^{1/2} P_k$, and P_k is the orthogonal projection onto the subspace spanned by $e_1, e_2, \ldots, e_k$.

Therefore,

$$\sum_{j=1}^{k} \lambda_j(DGD^*) \geqq \sum_{j=1}^{k} \lambda_j(F_k) = \operatorname{sp} F_k .$$

Taking into account that

$$\operatorname{sp} F_k = \sum_{j=1}^{k} (F_k e_j, e_j) = \sum_{j=1}^{k} \lambda_j(G)(D^*De_j, e_j) ,$$

we obtain

$$\sum_{j=1}^{k} \lambda_j(DGD^*) \geqq \sum_{j=1}^{k} \lambda_j(G)\xi_j ,$$

where $\xi_j = (D^*De_j, e_j)$. Using the Abel transformation

$$\sum_{j=1}^{k} \lambda_j(G)\xi_j = \lambda_k(G) \sum_{j=1}^{k} \xi_j + \sum_{j=1}^{k-1} (\lambda_j(G) - \lambda_{j+1}(G)) \cdot \sum_{r=1}^{j} \xi_r , \qquad (2.4)$$

we deduce from (2.1) and (2.4) the relations

$$\sum_{j=1}^{k} \lambda_j(DGD^*) \geq \lambda_k(G) \sum_{j=1}^{k} \mu_j(D^*D) +$$

$$+ \sum_{j=1}^{k-1} (\lambda_j(G) - \lambda_{j+1}(G)) \cdot \sum_{r=1}^{j} \lambda_r(D^*D) = \sum_{j=1}^{k} \mu_j(D^*D)\lambda_j(G) ,$$

and this completes the proof.

We remark also that by using this last proof and a proposition from [3] (p. 34) for the case of an n-dimensional space H, one can easily prove the following statement.

1°. *The equality*

$$\sum_{j=1}^{r} \lambda_j(DGD^*) = \sum_{j=1}^{r} \lambda_j(G)\lambda_{n-j+1}(D^*D) \quad (j = 1,2,\ldots,r)$$

is valid for some r $(1 \leq r \leq n)$ *if and only if the following conditions are satisfied*

$$D^*De_j = \lambda_{n-j+1}(D^*D)e_j, \quad Ge_j = \lambda_j(G)e_j \quad (j = 1,2,\ldots,r),$$

and

$$\lambda_j(G)\lambda_{n-j+1}(D^*D) \geq \lambda_{r+1}(DGD^*) \quad (j = 1,2,\ldots,r).$$

§3. THE CASE OF NONSELFADJOINT OPERATORS

In order to obtain relations of the form (0.4) for integral operators with non-Hermitian kernels (these relations are not given in the present paper), it is desirable to have a theorem which gives lower bounds for the s-numbers of a product of operators in terms of the s-numbers of its factors. Recall (see [3], Ch. II) that the s-numbers of a compact operator X are, by definition, the numbers $s_j(X) = [\lambda_j(X^*X)]^{1/2}$ $(j = 1,2,\ldots)$.

THEOREM 2. *Let* A *be a closed operator with a dense domain* Δ_A, *and let* B *be a compact operator. Assume that the following conditions are satisfied:*

a) the operator A^{-1} *exists and is compact;*

b) $BH \subseteq \Delta_A$ *and the operator* AB *is compact.*

Then the following relations hold true for any positive number p:

$$\sum_{j=1}^{k} s_j^p(AB) \geq \sum_{j=1}^{k} s_j^p(B)/s_j^p(A^{-1}) \qquad (k = 1,2,\ldots).$$

The proof does not differ essentially from that of Theorem 1.

§4. APPLICATIONS TO INTEGRAL EQUATIONS

We start by generalizing a statement made in the introduction. [For the sake of simplicity, we shall restrict ourselves to the case of kernels having only the derivatives $G_{pq}(t,s)$, $p,q = 0,1$. The results obtained here can be extended to the case $p,q = 0,1,2,\ldots,r$ with no difficulty.]

THEOREM 3. *Let* $G(t,s)$ $(a \leq t,s \leq b)$ *be a continuous, Hermitian-nonnegative kernel having continuous partial derivatives* $G_{pq}(t,s)$ $(p,q = 0,1)$. *If*

$$(\delta(G) =)\ G(a,a)G(b,b) - |G(a,b)|^2 = 0\ , \tag{4.1}$$

then

$$\frac{1}{(b-a)^2} \sum_{n=1}^{\infty} \frac{((-1)^{n-1}\theta + 2\pi(n-1))^2}{\mu_n} \leq \int_a^b G_{11}(s,s)\,ds, \tag{4.2}$$

where $\mu_1 \leq \mu_2 \leq \ldots$ *is the complete sequence of characteristic numbers of the integral equation*

$$\varphi(t) - \mu\int_a^b G(t,s)\varphi(s)\,ds = 0 \tag{4.3}$$

and the number θ $(0 \leq \theta \leq \pi)$ *is defined by*

$$\theta = \arccos[2\ \mathrm{Re}\ h/(1+|h|^2)], \quad \left(h = -\frac{G(a,a)}{G(b,a)} = -\frac{G(a,b)}{G(b,b)}\right)$$

whenever $G(a,b) \neq 0$, *and* θ *is taken to be* π *for* $G(a,b) = 0$.

Equality in (4.2) is attained when in the Mercer expansion

$$G(t,s) = \sum_{j=1}^{\infty} \frac{\varphi_j(t)\overline{\varphi_j(s)}}{\mu_j} \tag{4.4}$$

of the kernel G, *the system* $\{\varphi_j(t)\}_1^{\infty}$ *of orthonormal fundamental functions for equation* (4.3) *is simultaneuously the sequence of fundamental functions for the boundary value problem*

$$y'' + \nu y = 0, \quad y(a) = hy(b), \quad y'(b) = \bar{h}y'(a) , \tag{4.5}$$

when $G(a,b) \neq 0$, *and for problem*

$$y'' + \nu y = 0, \quad y(a) = y'(b) = 0 \quad (y'(a) = y(b) = 0) ,$$

when $G(a,a) = 0$ *(respectively* $G(b,b) = 0$).

PROOF. Consider the first (and more difficult) situation when $G(a,b) \neq 0$, and take $h = -G(a,a)/G(b,a) = -G(a,b)/G(b,b)$. For this choice of h,

$$G(a,a) + 2\,\mathrm{Re}[G(a,b)\bar{h}] + G(b,b)|h|^2 = 0.$$

Let (4.4) be the Mercer expansion of the kernel $G(t,s)$ relative to the orthonormal system of fundamental functions of the integral equation (4.3). Using this expansion, one can recast the last equality in the form

$$\sum_{j=1}^{\infty} \frac{|\varphi_j(a) + h\varphi_j(b)|^2}{\mu_j} = 0 ,$$

i.e., $\varphi_j(a) + h\varphi_j(b) = 0$ $(j = 1,2,\ldots)$, and so

$$G(a,s) + hG(b,s) \equiv 0 . \tag{4.6}$$

Denote by D_h the operator in $L_2(a,b)$ generated by the operation $-id/dx$ and the boundary condition

$$y(a) = hy(b) . \tag{4.7}$$

Therefore, the domain $\Delta_h = \Delta_{D_h}$ of operator D_h consists of all absolutely continuous functions $y \in L_2(a,b)$ satisfying condition (4.7) and having a derivative $y' \in L_2(a,b)$. For $y \in \Delta_h$, $D_h y = -iy'$, by definition.

It is easy to see that the adjoint D_h^* of the operator D_h will be generated (in the sense just mentioned) by the operation $-id/dx$ and the boundary condition

$$\bar{h}y(a) = y(b) \tag{4.8}$$

It follows that the spectrum of D_h will coincide with the spectrum of the boundary value problem (4.5).

Now considering the differential equation and the first boundary condition, we find that $y(x) = \mathrm{const}(\sin\sqrt{\nu}(x-a) + h\sin\sqrt{\nu}(b-x))$. Introducing this expression for $y(x)$ in the second boundary condition, we see that the spectrum $\{\nu_n\}_1^\infty$ of eigenvalues of the operator $D_h^* D_h$ is determined by the equation

$$\cos\sqrt{\nu}(b-a) = 2\,\mathrm{Re}\,h/(1+|h|^2) \qquad (\nu > 0),$$

whence

$$\nu_n^{(h)} = \frac{1}{(b-a)^2}[(-1)^{n-1}\theta + 2\pi(n-1)]^2 \quad (n = 1,2,\ldots). \tag{4.9}$$

The continuous differentiability with respect to t and s of the kernel $G(t,s)$ and identity (4.6) together imply that for the integral operator

$$(Gf)(t) = \int_a^b G(t,s)f(s)\,ds ,$$

the operator $D_h G D_h^*$ is well-defined on $\Delta_{D_h^*}$. Moreover, integrating by parts we get

$$((D_h G D_h^*)f)(x) = \int_a^b G_{11}(x,s)f(s)\,ds.$$

Therefore, the operator $D_h G D_h^*$ admits a compact closure, and

$$\mathrm{sp}(D_h G D_h^*) = \int_a^b G_{11}(s,s)\,ds.$$

At the same time, we are able to persuade ourselves that, in the case under consideration (i.e., $G(a,b) \neq 0$), relation (4.2) is the concrete expression of the general relation (1.7), taking into account our particular choice of the operators G and $D=D_h$.

The cases when $G(a,a) = 0$ or $G(b,b) = 0$ should be considered as partial (limit) cases when $h = 0$ and $h = \infty$, respectively. In the first (second) case, we shall have $G(a,s) \equiv 0$ $(G(b,s) \equiv 0)$, and in the definition of the operator D_h in terms of the operation $-id/dx$, the boundary condition $y(a) = hy(b)$ is replaced by the condition $y(a) = 0$ $(y(b) = 0)$, while the boundary value problem (4.5) is replaced by the problem

$$y'' + \nu y = 0, \quad y'(a) = y(b) = 0 \quad (y(a) = y'(b) = 0).$$

The spectrum of these symmetric problems is given by formula (4.9), where one takes $\theta = \pi/2$. From now on, the argument proceeds precisely as above, and so (4.9) is now established for all cases.

The last statement of the theorem results from the observation that if one has $\varphi_n'' = -\nu_n\varphi_n$ $(n = 1,2,\ldots)$ in the Mercer expansion (4.4), then

$$\int_a^b |\varphi_n'(x)|^2 dx = -\int_a^b \varphi_n(x)\overline{\varphi_n''(x)}dx = \nu_n \quad (n = 1,2,\ldots),$$

and so

$$\int_a^b G_{11}(s,s)ds = \sum_{n=1}^{\infty} \frac{1}{\mu_n}\int_a^b |\varphi_n'(s)|^2 ds = \sum_{n=1}^{\infty} \frac{\nu_n}{\mu_n}.$$

The theorem is proved.

Using Theorem 1 itself rather then its corollary, we obtain the following stronger result.

2°. *Under the assumptions of Theorem 3, one has* $(p > 0)$

$$(b-a)^{2p} \sum_{j=1}^{k} \left(\frac{1}{\mu_j'}\right)^p \geq \sum_{j=1}^{k} [\nu_j^{(h)}/\mu_j]^p, \quad (k = 1,2,\ldots), \tag{4.10}$$

where $\mu_1' \leq \mu_2' \leq \ldots$ *is the sequence of characteristic numbers of the integral equation*

$$\varphi(t) = \int_a^b G_{11}(t,s)\varphi(s)ds$$

and $\mu_1 \leq \mu_2 \leq \ldots$ *is the sequence of characteristic numbers of the initial integral equation* (4.3).

We remark that when at least one of the conditions $G(a,a) = 0$ or $G(b,b) = 0$ is satisfied, relations (4.10) become

$$\sum_{j=1}^{k} \left(\frac{1}{\mu_j'}\right)^p \geq \left[\frac{\pi}{4(b-a)^2}\right]^p \sum_{j=1}^{k} \left[\frac{(2j-1)^2}{\mu_j}\right]^p \quad (k = 1,2,\ldots), \qquad (4.11)$$

and if the kernel is periodic, i.e., $G(a,s) \equiv G(b,s)$, the corresponding form is

$$\sum_{j=1}^{k} \left(\frac{1}{\mu_j'}\right)^p \geq \left[\frac{4\pi^2}{4(b-a)^2}\right]^p \sum_{j=1}^{k} \left[\left(\frac{1}{\mu_{2j}}\right)^p + \left(\frac{1}{\mu_{2j-1}}\right)^p\right] j^{2p}$$

for in the first case $\theta = \pi/2$, while in the second $\theta = 0$.

If $G(t,s)$ represents the Green function for the bending of an elastic rod S with fixed extremities $x = a$ and $x = b$ which are fastened in some way (rigidly, elastically, or by hinges), then the numbers μ_j $(j = 1,2,\ldots)$ are proportional to the squares of the frequencies p_j $(j = 1,2,\ldots)$ of the natural oscillations of the rod S $(\mu_j = \rho p_j^2$, $j = 1,2,\ldots$, where ρ is the linear density of $S)$, while μ_j' are the critical contracting forces in the Euler problem for the longitudinal stability of the rod (see, for example, [7]). Since in this case $G(a,s) \equiv G(b,s) \equiv 0$, one has the relations (4.11) relating the frequencies p_j $(j = 1,2,\ldots)$ with the critical forces μ_j' $(j = 1,2,\ldots)$ of the rod S.

§5. INTEGRAL EQUATIONS HAVING A POSITIVE DEFINITE $\delta(G)$

Given a Hermitian-nonnegative kernel $G(t,s)$, assume that relation (4.1) is not satisfied, i.e. $\delta(G) = G(a,a)G(b,b) - |G(a,b)|^2 > 0$. Then one has

$$\delta = G(a,a) + 2\,\mathrm{Re}\,[G(a,b)\bar{h}] + G(b,b)|h|^2 > 0$$

for any complex h.

Consider the kernel

$$G^{(h)}(t,s) = \frac{1}{\delta}\begin{vmatrix} G(t,s) & G(a,s) + hG(b,s) \\ G(t,a) + \bar{h}G(t,b) & \delta \end{vmatrix} .$$

It is not hard to verify that the kernel $G^{(h)}$ is, together with G, Hermitian-nonnegative, and that $G^{(h)}(a,s) + hG^{(h)}(b,s) \equiv 0$, i.e., for $G^{(h)}$ condition (4.1) is already satisfied.

Relation (4.2) applied to $G^{(h)}$ gives

$$\int_a^b G_{11}(s,s)ds - \frac{1}{\delta}\int_a^b |\alpha'(s)|^2 ds \geqq \sum_{j=1}^{\infty} \frac{\nu_j^{(h)}}{\mu_j(G^{(h)})} ,$$

where $\alpha(t) = G(t,a) + \bar{h}G(t,b)$. It is known that since the kernel $G^{(h)}(t,s)$ differs from $G(t,s)$ by the degenerate kernel $-\frac{1}{\delta}\alpha(t)\overline{\alpha(s)}$, one has

$$\mu_j(G) \leqq \mu_j(G^{(h)}) \leqq \mu_{j+1}(G),$$

and so

$$\int_a^b G_{11}(s,s)ds \geqq \frac{1}{\delta}\int_a^b |\alpha'(s)|^2 ds + \sum_{j=1}^{\infty} \frac{\nu_j^{(h)}}{\mu_{j+1}} ,$$

where $\mu_j = \mu_j(G)$ $(j = 1,2,\dots)$.

Of course, in the right-hand integral one can interchange the roles of the extremities a and b without modifying h.

§6. APPLICATIONS TO INTEGRAL EQUATIONS WITH A WEIGHT

The previous discussion carries over, with certain complications, to the case of weighted integral equations

$$\varphi(t) = \int_a^b G(t,s)\varphi(s)p(s)ds , \tag{6.1}$$

where $p(s)$ $(\in L_1(a,b))$ is some nonnegative function.

Assume, for the sake of definiteness, that the kernel G satisfies the conditions of Theorem 3, condition (4.1) included, and that $G(a,b) \neq 0$. Now the definition of the operator D_h will become slightly more complicated. The domain Δ_h of D_h will be the set of all absolutely continuous functions $y(x)$ $(a \leq x \leq b)$ which satisfy (4.7) and can be represented in the form

$$y(x) = \int_a^x p(s)f(s)ds$$

where $f \in L_p^{(2)}(a,b)$, i.e., $\int_a^b p(s)|f(s)|^2 ds < \infty$.

For $y \in \Delta_h$, set $D_h y = -if$. It is easy to see that the operator D_h^*, the adjoint of D_h in the Hilbert space $L_p^{(2)}(a,b)$, is defined in the same way as D_h was, the only modification being that the boundary condition (4.7) is replaced by condition (4.8). In other words, $D_h^* = D_{1/\bar{h}}$.

The spectrum of eigenfrequencies of the operator $D_h^* D_h$ is nothing else but the spectrum of the boundary value problem

$$\frac{d}{dx}\frac{1}{p}\frac{dy}{dx} + \nu y = 0, \quad y(a) = hy(b), \quad \frac{1}{p}\frac{dy}{dx}\bigg|_{x=b} = \bar{h}\frac{1}{p}\frac{dy}{dx}\bigg|_{x=a} . \tag{6.2}$$

Performing the change of variables

$$t = \int_a^x p(s)ds, \quad y(x) = z(t),$$

the boundary value problem (6.2) becomes problem (4.5) with $a = 0$ and $b = T$, where

$$T = \int_a^b p(s)ds .$$

Therefore, the spectrum of problem (6.2) is given by the expression

$$\nu_n^{(h)} = \frac{1}{T^2}[(-1)^{n-1}\theta + 2\pi(n-1)]^2 \quad (n = 1,2,\ldots) . \tag{6.3}$$

In $L_p^{(2)}(a,b)$ consider the Hermitian operator G:

$$(Gf)(t) = \int_a^b G(t,s)f(s)p(s)ds.$$

Then one has: $((D_h G D_h^*)f,g)_p = \int_a^b (D_h G D_h^* f)(t)\overline{g(t)}p(t)dt =$

$$= \int_a^b\int_a^b G_{11}(t,s)f(s)\overline{g(t)}dsdt .$$

It is not hard to show that the operator $D_h G D_h^*$ has a compact closure with finite trace provided

$$\int\limits_{p(s)\neq 0} \frac{G_{11}(s,s)}{p(s)}\, ds < \infty, \tag{6.4}$$

and the trace $sp(D_h G D_h^*)$ is precisely this integral.

When condition (3.11) is satisfied, we obtain

$$\sum_{j=1}^{\infty} \frac{\nu_j^{(h)}}{\mu_j} \leqq \int_a^b \frac{G_{11}(s,s)}{p(s)}\, ds\ , \tag{6.5}$$

where $\mu_1 \leqq \mu_2 \leqq \ldots$ is the complete sequence of characteristic numbers of the weighted integral equation (6.1), while $\nu_1^{(h)} \leqq \leqq \nu_2^{(h)} \leqq \ldots$ are given by formula (6.3).

Relations (4.10) all remain valid in the case under consideration too. However, now $\mu_1' \leqq \mu_2' \leqq \ldots$ are the characteristic numbers of the weighted integral equation

$$p(t)\varphi(t) = \mu \int\limits_{p(s)\neq 0} G_{11}(t,s)\varphi(s)\, ds.$$

The following statement is a consequence of relation (6.5).

Suppose that the Hermitian-nonnegative kernel $G(t,s)$ $(a \leqq t,s \leqq b)$ has continuous derivatives $G_{pq}(t,s)$ $(p,q = 0,1)$, and let the weight $p(t)$ $(a \leqq t \leqq b)$ satisfy condition (6.4). Then the spectrum $\mu_1 \leqq \mu_2 \leqq \ldots$ of the weighted integral equation (6.1) satisfies the relation

$$\sum_{n=1}^{\infty} \frac{n^2}{\mu_n} \leqq \infty. \tag{6.6}$$

Notice that both relation (6.5) and its consequence (6.6) were proven under the assumption that (4.1) is satisfied. However, the arguments in §5 allow us to drop this restriction.

§7. APPLICATIONS TO HIGHER-DIMENSIONAL INTEGRAL EQUATIONS

The previous discussion is also applicable to the higher dimensional integral equation

$$\varphi(x) - \mu\int_{\Omega} G(x,y)\varphi(y)dy = 0 \qquad (x \in \Omega),$$

where $x = \{x_1,x_2,\ldots,x_m\}$ and $y = \{y_1,y_2,\ldots,y_m\}$ are points of a bounded domain Ω in the m-dimensional space.

For example, consider the situation when the positive kernel $G(x,y)$ has continuous, first-order, partial derivatives with respect to x_j and y_j, and also continuous, mixed, second-order derivatives

$$\frac{\partial^2 G(x,y)}{\partial x_j \partial y_k} \qquad (j,k = 1,2,\ldots,m) .$$

In addition, we shall assume that $G(x,y)$ vanishes whenever x or y belongs to the boundary $\partial\Omega$ of the domain Ω.

For the operator D, one takes the closed operator in $L_2(\Omega)$ defined by the expression $\operatorname{grad} f$, with the domain consisting of all smooth functions which vanish on $\partial\Omega$. Thus, the operator D will act from $L_2(\Omega)$ into the space of m-dimensional vector-functions $L_m^2(\Omega)$. Its adjoint is the operator $D^*f = \operatorname{div} f$, acting from $L_m^2(\Omega)$ into $L_2(\Omega)$.

The spectrum $\nu_1 \leqq \nu_2 \leqq \ldots$ of the operator D^*D in $L_2(\Omega)$ is precisely the spectrum of the boundary value problem

$$\Delta u - \nu u = 0, \qquad u\Big|_{\partial\Omega} = 0.$$

It is easy to check that under our assumptions DGD^* is an integral operator in the space $L_m^2(\Omega)$ and has the matrix kernel

$$G_{DD}(x,y) = \left\| \frac{\partial^2 G(x,y)}{\partial x_j \partial y_k} \right\|_{j,k=1}^{m} .$$

It results from the general relation (1.7) that

$$\sum_{n=1}^{\infty} \frac{\nu_n}{\mu_n} \leq \int_{\Omega} \left(\sum_{j=1}^{m} \left. \frac{\partial^2 G(x,y)}{\partial x_j \partial y_j} \right|_{x_j = y_j} \right) dx .$$

Taking into account the well-known asymptotic formula for the numbers ν_n, we obtain

$$\sum_{n=1}^{\infty} \frac{n^{2/m}}{\mu_n} \leq \infty .$$

Inequalities (4.10) can be extended too, in an obvious way, to the case under consideration, if one understands $\mu_1' \leq \mu_2' \leq \dots$ to be the characteristic numbers of the integral equation

$$\Phi(x) - \mu \int_{\Omega} G_{DD}(x,y) \Phi(y) dy = 0$$

in the space of m-dimensional vector-functions $\Phi(x) = \{\varphi_1(x), \varphi_2(x), \dots, \varphi_m(x)\}$ belonging to $L_m^2(\Omega)$.

When $m = 2$ and one takes G to be the Green function of a plate which leans freely upon or is fastened along the contour $\partial\Omega$, then the numbers μ_j will be proportional to the squares of the frequencies of the free, transverse, harmonic oscillations of the plate; the numbers ν_j will be proportional to the squares of the frequencies of a membrane fastened along the contour $\partial\Omega$; and, finally, the numbers μ_j' will be proportional to the critical forces in the problem of the swelling of the plate. Therefore, the inequalities of type (4.10) that we discussed will relate these three types of quantities.

Generalizing the arguments from §6 accordingly, one can broaden the results given above to higher-dimensional weighted equations and to systems of integral equations.

REFERENCES

1. Amir-Moéz, A. R.: *Extreme properties of eigenvalues of a hermitian transformation and singular values of the sum and product of linear transformations*, Duke Math. J. 23 (1956), 463-476.

2. Birman, M. Sh. and Solomyak, M. Z.: *On estimates of singular numbers of integral operators. I, II, III.* Vestnik Lenigrad. Gosud. Univ. 7 (1967), 45-53; 13 (1967), 21-28; 1 (1969), 35-48; English transl. of *III* -Vest. LSU Math. 2 (1975), 9-27.

3. Gohberg, I. C. and Krein, M. G.: *Introduction to the Theory of Linear Nonselfadjoint Operators*, "Nauka", Moscow, 1965; English transl., Transl. of Math. Monographs, vol. 18, Amer. Math. Soc., Providence, R. I. 1969.

4. Krein, M. G.: *On the characteristic numbers of differentiable symmetric kernels*, Mat. Sb. 2 (1937), 725-732. (Russian)

5. Lidskii, V. B.: *Inequalities for eigenvalues and singular numbers*, Supplement to F. R. Gantmakher's book *Theory of Matrices*, 2nd ed., "Nauka", Moscow, 1966. (Russian)

6. Markus, A. S.: *The eigenvalues and singular values of the sum and product of linear operators*, Uspekhi Mat. Nauk 19, No. 4 (1964), 93-123; English transl., Russian Math. Surveys 19, No. 4 (1964), 91-120.

7. Nudel'man, Ya. L.: *Methods for the Determination of Natural Frequencies and Critical Forces for Rod Systems*, Gostekhizdat, Moscow, 1949. (Russian)

8. Paraska, V. I.: *On asymptotics of eigenvalues and singular numbers of linear operators which increase smoothness*, Mat. Sb. 68 (110) (1965), 623-631. (Russian)

9. Sobolevskii, P. E.: *On the s-numbers of integral operators*, Uspekhi Mat. Nauk 22, No. 2 (1967), 114-115. (Russian)

A CONTRIBUTION TO THE THEORY OF S-MATRICES OF CANONICAL DIFFERENTIAL EQUATIONS WITH SUMMABLE POTENTIAL*

M. G. Krein and F. E. Melik-Adamyan

1. Let

$$J = \begin{pmatrix} 0 & I_n \\ -I_n & 0 \end{pmatrix}$$

(where I_n denotes the identity matrix of order n), and let $V(r)$ $(0 \leqq r < \infty)$ be a Hermitian matrix-function of order $2n$. Using the matrix-functions J and $V(r)$, one can write a canonical differential equation with complex parameter λ:

$$J \frac{dY}{dr} = \lambda Y + V(r)Y \tag{1}$$

for the $(2n\times n)$-matrix function $Y(r;\lambda)$ $(0 \leqq r < \infty)$.

From now on, unless we state otherwise, system (1) will be considered under the assumption that the potential $V(r)$ is summable:

$$V(r) = V^*(r) \in L^1_{2n\times 2n}(0,\infty) .$$

[$L^p_{n\times m}(a,b)$ denotes the class of $(n\times m)$-matrix functions with elements belonging to $L^p(a,b)$.]

Under this hypothesis, equation (1) has a unique solution $Y = X(r;\lambda)$ which admits the asymptotics

---

*Translation of Doklady Akademii Nauk Armyanskoi SSR, Vol. 46, No. 4 (1968), 150-155.

$$X(r;\lambda) = \begin{pmatrix} X_1(r;\lambda) \\ \\ X_2(r;\ ) \end{pmatrix} = \begin{pmatrix} I_n \\ \\ -iI_n \end{pmatrix} e^{i\lambda r} S(\lambda) + \begin{pmatrix} I_n \\ \\ iI_n \end{pmatrix} e^{-i\lambda r} + o(1) \tag{2}$$

as $r \longrightarrow \infty$ (and satisfies the boundary condition $X_2(0;\lambda) = 0$). Here $S(\lambda)$ $(-\infty < \lambda < \infty)$ is a unitary matrix-function of order n, called the *S-matrix* of equation (1).

It is known (see, for example, [1]) that following a simple substitution that preserves the summability of the potential and the S-matrix of the equation, the canonical differential equation (1) is transformed into an equation whose potential $V(r)$ is J-Hermitian: $JV = -VJ$. A potential satisfying this condition will be called *normalized*.

The existence of the solution $X(r;\lambda)$ to equation (1) having the asymptotic (2), as well as a number of structural properties of the S-matrix $S(\lambda)$, were established in [7] under the assumption that the potential $V(r)$ $(\in L^1_{2n\times 2n}(0,\infty))$ is real. In the case of a Hermitian, summable potential $V(r)$ (and even for the infinite dimensional canonical equation) the existence of the solution $X(r;\lambda)$ having the asymptotics (2) was proved in [1]. There it was shown that the matrix-function $S(\lambda)$ defined by the asymptotic relation (2), can be interpreted also as the scattering matrix in the framework of time-dependent scattering theory.

Here we generalize the results of paper [7] and establish a proposition that is, in a certain sense, definitive.

THEOREM 1. *Let* $S(\lambda)$ $(-\infty < \lambda < \infty)$ *be a unitary matrix-function of order* n: $S^*S = I_n$. *In order that* $S(\lambda)$ *be the S-matrix of a canonical differential equation* (1) *with summable potential* $V(r)$ *it is necessary and sufficient that the following two conditions hold.*

1) The matrix-function $S(\lambda)$ *admits the representation*

$$S(\lambda) = I_n - \int_{-\infty}^{\infty} \Gamma(t) e^{-i\lambda t} dt , \tag{3}$$

where $\Gamma \in L^1_{n\times n}(-\infty,\infty)$.

2) The partial indices of the matrix-function $S(\lambda)$

are all zero.

If these conditions are satisfied, then there exists a unique canonical differential equation with a normalized potential $V \in L^1_{2n\times 2n}(0,\infty)$ *and having the given matrix-function* $S(\lambda)$ *as its* S-*matrix.*

Let us explain the meaning of the second condition. As it is shown in [5], if some matrix-function $S(\lambda)$ (not necessarily unitary) satisfies condition 1) and $\det S(\lambda) \neq 0$ $(-\infty < \lambda < \infty)$, then it has a unique right factorization

$$S(\lambda) = S_-(\lambda)D(\lambda)S_+(\lambda) \qquad (-\infty < \lambda < \infty) \tag{4}$$

such that:

$$\left.\begin{array}{l} S_-(\lambda) = I_n - \int_0^\infty s_-(t)e^{-i\lambda t}dt\ , \\ \text{where } s_- \in L^1_{n\times n}(0,\infty) \text{ and } \det S_-(\lambda) \neq 0 \text{ for } \operatorname{Im}\lambda \leqq 0; \end{array}\right\} \tag{5}$$

$$\left.\begin{array}{l} S_+(\lambda) = I_n + \int_0^\infty s_+(t)e^{i\lambda t}dt\ , \\ \text{where } s_+ \in L^1_{n\times n}(0,\infty) \text{ and } \det S_+(\lambda) \neq 0 \text{ for } \operatorname{Im}\lambda \geqq 0; \end{array}\right\} \tag{6}$$

and $D(\lambda)$ is a diagonal matrix

$$D(\lambda) = \|\delta_{jk}\zeta^{\nu_j}\|\ ,$$

where $\zeta = \frac{\lambda - i}{\lambda + i}$ and $\nu_1 \geqq \nu_2 \geqq \dots \geqq \nu_n$ are integers.

The numbers ν_j $(j = 1,2,\dots,n)$ are called the *partial (right) indices* of the matrix-function $S(\lambda)$. If $S(\lambda)$ satisfies condition 1) and is unitary, then $|\det S(\lambda)| = 1$, and so $S(\lambda)$ admits the factorization (4). Condition 2) of Theorem 1 means that $D(\lambda) \equiv I_n$; whence

$$S(\lambda) = S_-(\lambda)S_+(\lambda) \qquad (-\infty < \lambda < \infty)\ . \tag{7}$$

2. In the process of proving Theorem 1 one establishes a factorization theorem of independent interest.

Before stating this, let us agree on the following

notation: given $\Gamma \in L^1_{n\times n}(0,\infty)$, we denote by $\mathbf{\Gamma}$ the operator acting in the Hilbert space $L^2_{n\times 1}(0,\infty)$ as follows:

$$(\mathbf{\Gamma} f)(t) = \int_0^\infty \Gamma(t+s)f(s)ds \qquad (f \in L^2_{n\times 1}(0,\infty)) .$$

The norm of the operator $\mathbf{\Gamma}$ in $L^2_{n\times 1}(0,\infty)$ will be denoted by $\|\mathbf{\Gamma}\|$.

[The scalar product in $L^2_{n\times 1}(a,b)$ is introduced as usual: if $f = \{f_j\}_1^n$ and $g = \{g_j\}_1^n \in L^2_{n\times 1}(a,b)$, then $(f,g) = \sum_{j=1}^n \int_a^b f_j(s)\overline{g_j(s)}ds$.]

THEOREM 2. *In order that a given matrix-function* $\Gamma_0 \in L^1_{n\times n}(0,\infty)$ *admit an extension* $\Gamma \in L^1_{n\times n}(-\infty,\infty)$ *(i.e.,* $\Gamma(t) = \Gamma_0(t)$ *for* $t \in [0,\infty)$*) for which the matrix-function* $S(\lambda)$ *defined by equality* (3) *is unitary and has zero partial indices it is necessary and sufficient that* $\|\mathbf{\Gamma}_0\| < 1$. *If this condition is fulfilled, then the required extension* Γ $(\in L^1_{n\times n}(-\infty,\infty))$ *is unique.*

Let us sketch a procedure for constructing the matrix-function $S(\lambda)$ given $\Gamma_0 \in L^1_{n\times n}(0,\infty)$. It turns out that the condition $\|\mathbf{\Gamma}_0\| < 1$ ensures the existence and uniqueness in the class $L^1_{2n\times n}(0,\infty)$ of the solution to the equation

$$\begin{pmatrix} g_-(t) \\ g_+(t) \end{pmatrix} + \int_0^\infty \begin{pmatrix} 0 & \Gamma_0(t+s) \\ \Gamma_0^*(t+s) & 0 \end{pmatrix} \begin{pmatrix} g_-(s) \\ g_+(s) \end{pmatrix} ds = \begin{pmatrix} \Gamma(s) \\ \Gamma_0^*(s) \end{pmatrix} \tag{9}$$

$$(0 \leqq t < \infty) .$$

Let

$$G_-(\lambda) = I_n - \int_0^\infty g_-(t)e^{-i\lambda t}dt \qquad (\mathrm{Im}\ \lambda \leqq 0),$$

and

$$G_+(\lambda) = I_n - \int_0^\infty g_+(t)e^{i\lambda t}dt \qquad (\mathrm{Im}\ \lambda \geqq 0). \tag{10}$$

Then $G_-(\lambda)$ (resp. $G_+(\lambda)$) is an n-th order matrix-function holomorphic in the open, and continuous in the closed upper (resp. lower) half plane. Moreover, the determinants

$\det G_{\pm}(\lambda) \neq 0$ in the corresponding (closed) half planes. It follows that the matrix-function

$$S(\lambda) = G_{-}(\lambda)G_{+}^{-1}(\lambda) \qquad (-\infty < \lambda < \infty) \tag{11}$$

is unitary and all its partial indices are zero. Subsequently, one may easily verify that $S(\lambda)$ admits the representation (3), where Γ $(\in L^{1}_{n\times n}(-\infty,\infty))$ is the required extension of the given $\Gamma_0 \in L^{1}_{n\times n}(0,\infty)$.

3. We indicate two procedures for recovering equation (1) with summable potential V from its S-matrix $S(\lambda)$. One of them (the one we begin with) is an appropriately elaborated extension to equation (1) of the procedure worked out in [2] for the case of a vector differential equation of second order. We should mention that this procedure has been already carried out for equation (1) in papers [3,4], under other assumptions concerning the potential V. The second procedure that we shall discuss is a development of the one indicated in [6,7].

Given $\Gamma \in L^{1}_{n\times n}(0,\infty)$, we let $\boldsymbol{\Gamma}_r$ $(0 \leq r < \infty)$ denote the operator

$$(\boldsymbol{\Gamma}_r f)(t) = \int_0^\infty \Gamma(r + t + s)f(s)ds \qquad (f \in L^{2}_{n\times 1}(0,\infty))$$

acting in the Hilbert space $L^{2}_{n\times 1}(0,\infty)$.

Suppose that the unitary matrix-function $S(\lambda)$ satisfies conditions 1) and 2) of Theorem 1. Take the matrix-function $\Gamma(t)$ from representation (3) and form the matrix integral equations

$$(I - \boldsymbol{\Gamma}_r\boldsymbol{\Gamma}_r^{*})\Phi_r(t) = \Gamma(r + t) \qquad (0 \leq t < \infty) \tag{12}$$

and

$$(I - \boldsymbol{\Gamma}_r^{*}\boldsymbol{\Gamma}_r)\Psi_r(t) = \Gamma^{*}(r + t) \qquad (0 \leq t < \infty) \tag{13}$$

for $r \geq 0$.

By virtue of Theorem 2, one can assert that this equations have unique solutions Φ_r and Ψ_r in the class $L^{1}_{n\times n}(0,\infty)$. Now the potential V of the equation (1) having an S-matrix equal to the given function $S(\lambda)$ is expressed in terms of these solutions via the formula

$$V(r) = \begin{pmatrix} A(r) & B(r) \\ B(r) & -A(r) \end{pmatrix} \qquad (0 \leq r < \infty), \tag{14}$$

where

$$A(r) = \frac{\Phi_r(r) - \Psi_r(r)}{2i} \quad \text{and} \quad B(r) = \frac{\Phi_r(r) + \Psi_r(r)}{2} .$$

4. A locally summable, n-th order, matrix-function $H(t) = H^*(-t)$ $(-\infty < t < \infty)$, will be called a *Hermitian accelerant* (cf. [7,8]) if the vector integral equation

$$\phi(t) + \int_0^r H(t - s)\phi(s)ds = 0 \qquad (0 \leq t \leq r)$$

has the unique solution $\phi(t) \equiv 0$, for any $r \geq 0$.

If H is an accelerant, then the resolvent kernel $\Gamma_r(t,s)$ exists for all $r > 0$ and is determined from the equation

$$\Gamma_r(t,s) + \int_0^r H(t - u)\Gamma_r(u,s)ds = H(t - s) \tag{15}$$

$(0 \leq t,s \leq r)$ as a continuous matrix-function of s taking values in $L^1_{n\times n}(0,r)$. Moreover, one can define meaningfull matrix-functions

$$A(r) = -2 \operatorname{Im} \Gamma_{2r}(0,2r) , \quad B(r) = 2 \operatorname{Re} \Gamma_{2r}(0,2r), \tag{16}$$

which turn out to be locally summable. Then one can write down the canonical differential equation (1) with the potential $V(r)$ defined by formula (14).

The following proposition holds [6,7,8].

Every canonical equation (1) *with a normalized potential is generated via the above procedure by one and only one accelerant.*

This result can be completed by the following theorem.

THEOREM 3. *In order that a locally summable matrix-function* $H(t) = H^*(-t)$ $(-\infty < t < \infty)$ *be the accelerant of some canonical differential equation* (1) *with summable, normalized potential, it is necessary and sufficient that it satisfies the* *following two conditions:*

1) $H(t) \in L^1_{n\times n}(-\infty,\infty)$,

and

2) $\det(I_n + \int_{-\infty}^{\infty} H(t)e^{i\lambda t}dt) > 0 \quad (-\infty < \lambda < \infty)$.

The following theorem explains how the accelerant generating equation (1) and the S-matrix of this equation are related.

THEOREM 4. *Let* $S(\lambda) = S_-(\lambda)S_+(\lambda)$ *be the factorization of the S-matrix of a canonical differential equation* (1) *with summable potential* V. *Then one has the equality*

$$S_+(\lambda)S_+^*(\lambda) = I_n + \int_{-\infty}^{\infty} H(t)e^{i\lambda t}dt = S_-^{-1}(\lambda)S_-^{*-1}(\lambda) \tag{17}$$

$(-\infty < \lambda < \infty)$, *where* $H(t)$ *is the accelerant of the given equation* (1).

That is to say, the second procedure for recovering the canonical differential equation (1) can be described as follows.

First factor the given matrix-function $S(\lambda)$ by means of formulas (9) and (10), i.e., determine $S_-(\lambda) = G_-(\lambda)$ and $S_+(\lambda) = G_+^{-1}(\lambda)$. Then use (17) to find the Hermitian accelerant $H(t)$. Finally, the sought for potential V is obtained by means of formulas (15), (16), and (14).

Let us note that the middle term in (17) is nothing else but $d\Sigma(\lambda)/d\lambda$, where $\Sigma(\lambda)$ is the spectral matrix-function of the canonical differential equation (1) (the spectral matrix-function of equation (1) is disscused in [8]).

REFERENCES

1. Adamyan (Adamjan), V. M.: *Canonical differential operators in Hilbert space*, Dokl. Akad. Nauk SSSR 178, No. 1 (1968), 9-12; English transl., Soviet Math. Dokl. 9 (1968), 1-5.

2. Agranovich, Z. C. and Marchenko, V. A.: *The Inverse Problem of Scattering Theory*, Khar'kov Gosud. Univ., 1960; English transl., Gordon and Breach, New York, 1963.

3. Gasymov, M.G.: *The inverse scattering problem for a system of Dirac equations of order* 2n, Dokl. Akad. Nauk SSSR 169, No. 5 (1966), 1037-1040; English transl., Soviet Physics Dokl. 11 (1967), 676-678.

4. Gasymov, M. G. and Levitan, B. M.: *Determination of Dirac's system from the scattering phase*, Dokl. Akad. Nauk SSSR 167, No. 6 (1966), 1219-1222; English transl., Soviet Math. Dokl. 7 (1966), 543-547.

5. Gohberg, I. C. and Krein, M. G.: *Systems of integral equations on a half line with kernel depending on the difference of the arguments*, Uspekhi Mat. Nauk 13, No. 2 (1958), 3-72; English transl., Amer. Math. Soc. Transl. (2) 14 (1960), 217-287.

6. Krein, M. G.: *Continuous analogs of propositions on polynomials orthogonal on the unit circle*, Dokl. Akad. Nauk SSSR 105, No. 4 (1955), 637-640. (Russian)

7. Krein, M. G.: *A contribution to the theory of accelerants and S-matrices of canonical systems*, Dokl. Akad. Nauk SSSR 111, No. 6 (1956), 1167-1170. (Russian)

8. Melik-Adamyan, F. E.: *A contribution to the theory of matrix accelerants and spectral functions of canonical differential systems*, Dokl. Akad. Nauk ArmSSR, Mat., 45, No. 4 (1967), 145-151. (Russian)